*Cyanidioschyzon merolae*

Tsuneyoshi Kuroiwa • Shinya Miyagishima
Sachihiro Matsunaga • Naoki Sato
Hisayoshi Nozaki • Kan Tanaka
Osami Misumi
Editors

# *Cyanidioschyzon merolae*

## A New Model Eukaryote for Cell and Organelle Biology

Springer

*Editors*
Tsuneyoshi Kuroiwa
Faculty of Science
Department of Chemical
and Biological Sciences
Japan Women's University
Tokyo, Japan

Sachihiro Matsunaga
Department of Applied Biological
Science, Faculty of Science
and Technology
Tokyo University of Science
Chiba, Japan

Hisayoshi Nozaki
Department of Biological Sciences
Graduate School of Science
University of Tokyo
Tokyo, Japan

Osami Misumi
Graduate School of Science and
Technology for Innovation
Faculty of Science, Department
of Biological Science and Chemistry
Yamaguchi University
Yamaguchi, Japan

Shinya Miyagishima
Department of Cell Genetics
National Institute of Genetics
Shizuoka, Japan

Naoki Sato
Department of Life Sciences
Graduate School of Arts and Sciences
The University of Tokyo
Tokyo, Japan

Kan Tanaka
Laboratory for Chemistry and Life Science
Institute of Innovative Research
Tokyo Institute of Technology
Kanagawa, Japan

ISBN 978-981-10-6100-4 ISBN 978-981-10-6101-1 (eBook)
https://doi.org/10.1007/978-981-10-6101-1

Library of Congress Control Number: 2017959728



Printed on acid-free paper

This Springer imprint is published by Springer Nature
The registered company is Springer Nature Singapore Pte Ltd.
The registered company address is: 152 Beach Road, #21-01/04 Gateway East, Singapore 189721, Singapore

# Preface

It is estimated that more than 8.7 million species of eukaryote inhabit the Earth, ranging from picoplankton, 0.7 μm in diameter, to whales, 30 m in length. These species live on land and in oceans, including extreme environments of ultrahigh or cold temperatures. Among them, plants fix and reduce carbon dioxide, liberate oxygen into the atmosphere, and generate organic compounds through photosynthesis in chloroplasts (plastids). Thus, plants sustain the base of the global ecosystem and support much of life, which includes fungi and animals, including humans.

Even though the external forms of eukaryotes, from eukaryotic picoplankton and amoebae to higher plants and animals, appear remarkably different, their constituent cells have common structures and functions. When a global environmental change of global freezing and a growing level of oxygen occurred approximately two billion years ago, an aerobic proteobacteria formed a symbiotic relationship with an anaerobic archaea host and, over time, evolved into mitochondria. Through the symbiotic process, a primitive eukaryotic cell was born. Subsequently, a photosynthetic cyanobacterium formed a symbiotic relationship with a primitive eukaryotic cell containing a mitochondrion and was transformed into a plastid, resulting in the birth of algae and plants. It is assumed that the primitive eukaryotic cell divided repeatedly and evolved into the variety of multicellular, differentiated organisms that are adapted to today's environment.

Thus, to understand eukaryotes, from the smallest eukaryotic algae of less than 1 μm to higher animals and plants, it is essential to reveal the crucial minimal structures, mechanisms, and metabolic systems shared by cells and organelles for their maintenance and division/proliferation. In general, a eukaryotic cell contains not only a double membrane-bound cell nucleus that holds its genome (all genes that characterize the organism) but also double membrane-bound mitochondria, four types of single membrane-bound organelles (endoplasmic reticulum, Golgi body, lysosomes, and peroxisomes), and the cytoskeleton, which comprises tubulin-based structures (including microtubules, centrosome, and spindle) and actin microfilaments. In addition, algae and plant cells contain double membrane-bound plastids. For eukaryotic cells to function, these double membrane-bound, single membrane-bound, and membrane-free organelles are essential, and overall cell function is established only through their interaction. Therefore, during cell proliferation, or the cell cycle, the cell nucleus and the double and the single membrane-bound organelles also divide according to their own division cycle under precise spatiotemporal control. In eukaryotes, such as the primitive algae, the endoplasmic reticulum and the Golgi body are inherited with the cell nucleus, while lysosomes and peroxisomes adhere tightly to mitochondria and are distributed and inherited by the daughter cells.

Ancestral eukaryotic cells had a basic cell nucleus, but organelles such as mitochondria, the Golgi apparatus, lysosomes, and peroxisomes evolved enhanced functions that resulted in the emergence of amoeba, plants, fungi, and animals. The cells in almost eukaryotic organisms contain many organelles (mitochondria and peroxisomes can reach 1000 to 3000 in some animal cells) that exhibit complicated irregular shapes; therefore, it is relatively difficult to define the division sites of the organelles and to examine the behavior of divided organelles that are inherited by daughter cells. To overcome this problem, it was necessary to identify single-celled organisms that possess a very small number of intracellular organelles per cell. In addition, to simultaneously investigate the proliferation and metabolic system of a cell/organelle from the viewpoint of morphology, molecular biology, and biochemistry, it was essential to develop a synchronized system for cell division. In 1953, Hiroshi Tamiya discovered that cell division of the green algae *Chlorella* is synchronized with light and darkness. We, therefore, searched for more primitive algae with the same feature. There are three species of Cyanidiophyceae, *Galdieria sulphuraria*, *Cyanidium caldarium*, and *Cyanidioschyzon merolae*, which live in hot springs all over the world. Among the three, the *C. merolae* cells are the smallest, its organelle shapes are simple, the cell wall is weak, and it can be highly synchronized by a light/dark cycle. Thus, *C. merolae* was chosen.

During the last two decades, numerous electron and fluorescence microscopy studies, combined with synchronous cultures, have revealed the spatial organization and morphological features of these organelles in each cell-cycle phase of *C. merolae*. These explorations have revealed that the timing and manner of organelle behaviors are strictly determined by the progression of the cell division cycle. In addition, the simplicity of the cell structure assists attempts to directly address the biological processes in each organelle. Genome information for *C. merolae* has provided insights into the fundamental principles and evolution of eukaryotic cells. Currently, many types of genetic techniques, as well as genome-wide and system-wide multi-omics analyses, can be carried out. Thus, the simple composition of *C. merolae* provides considerable advantages for examining the underlying molecular mechanisms related to cell proliferation, organelle biogenesis, various related metabolic pathways, and evolutionary history.

Features of *C. merolae* are listed below.

1. Cell culture is easy and has a negligible risk of contamination because it is performed at a high temperature and in acidic medium.

2. The diameter of the cell is 1.5–2.5 μm. Cells proliferate by binary division, and one cell cycle is one life cycle.
3. The cell is composed of a very small number of membranous organelles: one cell nucleus, one mitochondrion, one plastid, one peroxisome, a simple-shaped ER, one Golgi body with two cisternae, and four to five lysosomes (vacuoles). Most forms are simple and nearly spherical, and the subcellular localization is ordered. ER almost behaves as a nuclear membrane, with part of it protruding from the cell nuclear membrane into the cytoplasm. The division order of these organelles is determined.
4. It is possible to highly synchronize the division of the cell and its internal organelles with a light/dark cycle.
5. Cell walls are weak, and cell organelles and intracellular substances in each phase of the cell and organelle division cycle can be isolated.
6. It is the only eukaryotic organism whose genome is 100% decoded, and the number of genes in the cell nucleus is small, at 4775. There are few introns. It is considered one of most primitive eukaryotes from comparative genome analyses. It also contains many of the genetic systems present in fungi and animals.
7. The cell metabolism system, with 52 pathways, is the smallest among autonomous eukaryotes.
8. Transcriptome data at each cell-cycle stage has been established.
9. It is amenable to many types of genetic techniques, as well as genome-wide and system-wide multi-omics analyses. Furthermore, various gene disruption techniques are established.
10. Because it is an organism living at high temperature, it is considered that individual proteins and protein complexes are stable, and thus protein interactions in time and space can be analyzed by omics techniques.

The purpose of this book is to introduce basic knowledge and relevant experimental approaches to elucidate common phenomena of eukaryotes, such as the dynamics and proliferation of organelles essential for cell function, using the primitive eukaryote, *C. merolae*, as a model organism. This information will contribute to the future development of diverse fields, such as genomics, structural biology, and synthetic biology, and also applied research fields, including engineering.

Tokyo, Japan — Tsuneyoshi Kuroiwa
Shizuoka, Japan — Shinya Miyagishima
Chiba, Japan — Sachihiro Matsunaga

# Contents

# Part I
# Birth, Evolution and Structure of Primary Eukaryote

**Hisayoshi Nozaki**

Plastids in eukaryotic cells originate from endosymbiosis of an ancestral cyanobacterium. The single origin of almost all of the plastids is generally accepted by worldwide biological researchers. However, the origins and phylogeny of the three major groups of Archaeplastida (with primary plastids), red algae, green plants/algae and glaucophytes, have still been debated. Origin of the red algae has also been debated, ranging from 2 BYA to 1 BYA. Evolutionary significance of complete lack of flagella in the red algae has been still unclarified.

*Cyanidioschyzon merolae* belongs to the Cyanidilales of the red algae. Almost all of the cyanidialean algae are living in acid hot springs, which may represent ancient earth environments. Thus, *C. merolae* has been very attractive to evolutionary biologists. Current knowledge of biodiversity and ecology of the Cyanidiales has been nicely summarized in Chap. 1. Readers will soon understand evolutionary and ecological characteristics and significance of *C. merolae* and other members of the Cyanidiales. On the other hand, *C. merolae* has the simplest eukaryotic cell composed of one cell nucleus, one mitochondrion, one plastid, a simple ER, one Golgi body with two cisternae, a few lysosomes and one peroxisome. Based on the modern biological methods and complete genome data of *C. merolae*, various important and fundamental biological problems in a eukaryotic cell have been recently resolved. Thus, *C. merolae* is now one of the most hopeful model organisms in cell biology. Chapter 2 summarizes basic cell biological attributes of *C. merolae* and introduces recent studies that have clarified long-standing biological questions in cell-cycle regulation, organelle biogenesis, and various types of metabolic pathways, by using *C. merolae*.

# Chapter 1
# Cyanidiales: Evolution and Habitats

**Shinya Miyagishima, Jong Lin Wei, Hisayoshi Nozaki, and Shunsuke Hirooka**

**Abstract** *Cyanidioschyzon merolae* is a unicellular alga without a cell wall that inhabits sulfuric hot springs. It is a member of the red algae Cyanidiales that diverged from other red algal lineages approximately 1.3–1.4 billion years ago. Cyanidiales are the only eukaryotes and phototrophic organisms found in sulfuric hot springs, but some rare species have also been found in nonthermal and/or neutrophilic environments. They are classified into three genera, namely, *Galdieria*, *Cyanidium*, and *Cyanidioschyzon*. *Galdieria* can grow heterotrophically and mixotrophically in contrast to the (almost) obligate photoautotrophs *Cyanidium* and *Cyanidioschyzon*. Cells of *Galdieria* and *Cyanidium* are surrounded by a rigid cell wall and proliferate by forming autospores in the mother cell wall. In contrast, *Cyanidioschyzon* lacks a cell wall and proliferates by binary fission. Recent phylogenetic studies have classified Cyanidiales into four distinct lineages: *Galdieria* lineage, acidophilic *Cyanidium* lineage, neutrophilic *Cyanidium* lineage, and the lineage that comprises *G. maxima*, *Galdieria*-like algae, and *Cyanidioschyzon*.

---

S. Miyagishima (✉)
Department of Cell Genetics, National Institute of Genetics, 1111 Yata, Mishima, Shizuoka 411-8540, Japan
e-mail: smiyagis@nig.ac.jp

J.L. Wei
Department of Cell Genetics, National Institute of Genetics, 1111 Yata, Mishima, Shizuoka 411-8540, Japan

Department of Genetics, Graduate University for Advanced Studies, Shizuoka 411-8540, Japan

H. Nozaki
Department of Biological Sciences, Graduate School of Science, University of Tokyo, Hongo 7-3-1, Bunkyo-ku, Tokyo 113-0033, Japan

S. Hirooka
Department of Cell Genetics, National Institute of Genetics, 1111 Yata, Mishima, Shizuoka 411-8540, Japan

Core Research for Evolutional Science and Technology Program, Japan Science and Technology Agency, Saitama 332-0012, Japan



T. Kuroiwa et al. (eds.), *Cyanidioschyzon merolae*,
https://doi.org/10.1007/978-981-10-6101-1_1

Recent studies using high-throughput sequencing technologies have started to reveal that both gene losses and horizontal gene transfer from environmental prokaryotes have contributed to the emergence and diversification of Cyanidiales and their adaptation to respective habitats.

**Keywords** Acidophile • Cyanidiale • Cyanidiophyceae • *Cyanidioschyzon* • *Galdieria* • *Cyanidium*

## 1.1 Introduction

*Cyanidioschyzon merolae* (De Luca et al. 1978) is a unicellular alga without a cell wall that inhabits sulfuric hot springs (Fig. 1.1). Its cell exhibits a blue-green color, is 2–3 μm in diameter, and contains a mitochondrion, a chloroplast, a peroxisome, and some lysosomes. The nuclear (Matsuzaki et al. 2004; Nozaki et al. 2007), mitochondrial (Ohta et al. 1998), and chloroplastic (Ohta et al. 2003) genomes are completely sequenced. *C. merolae* possesses a simple nuclear genome (16.5 Mbp; 4775 protein-coding genes) with low genetic redundancy (Matsuzaki et al. 2004; Nozaki et al. 2007) (see Chap. 5). Therefore, this alga is suitable for various "omics" analyses. Methods for genetic manipulation, such as the transient expression of proteins from plasmids (Ohnuma et al. 2008) and gene targeting by

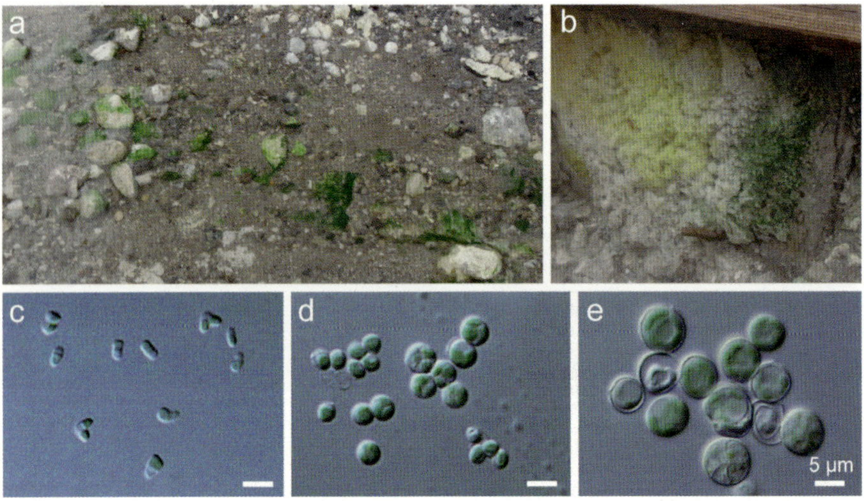


**Fig. 1.1** Cyanidialean unicellular red algae. (**a**, **b**) Images of Cyanidiales in their natural environments. Algae in water pool of an acidic hot spring are shown in (**a**), and those on the rock wall around the hot spring are shown in (**b**) (both photos were taken at Sukayu in Japan with aid of Dr. Sumiya). (**c**–**e**) Images of *Cyanidioschyzon merolae* 10D (**c**), *Cyanidium caldarium* RK-1 (**d**), and *Galdieria sulphuraria* 074 (**e**) by interference contrast (bars = 5 μm)

homologous recombination (Minoda et al. 2004), have been established (see Chap. 7). Thus, *C. merolae* is a promising model organism for studies on cell biology, metabolism, ecology, and evolution.

*C. merolae* is a member of Cyanidiales, an order of red algae. Cyanidiales are the only eukaryotes and phototrophic organisms that dominate in volcanic and thermal areas under an extremely low pH (0.05–5) and a relatively high temperature (35 °C–56 °C) (Fig. 1.1). Cyanidiales branched off from the common ancestor of other red algal lineages approximately 1.3–1.4 billion years ago (Yoon et al. 2004, 2006b). Since then, they have diverged and adapted to various isolated acidic hot springs. Certain rare species have also colonized nonthermal and/or neutrophilic environments (Reeb and Bhattacharya 2010; Castenholz and McDermott 2010). Many researchers have studied the diversity, evolution, and ecology of Cyanidiales worldwide. Recently, by virtue of the development of high-throughput sequencing technologies, it has become feasible to understand how genomic changes in Cyanidiales have resulted in the diversification of respective cyanidialean species and adaptation of these species to their natural habitats. In this chapter, we summarize the classification, physiology, ecology, and evolution of Cyanidiales.

## 1.2 Classification, Phylogenetic Relationship, and Evolution of Cyanidiales

### *1.2.1 History of Classification of Cyanidiales*

The order Cyanidiales or the class Cyanidiophyceae comprises three genera: *Galdieria*, *Cyanidium*, and *Cyanidioschyzon* (Figs. 1.1 and 1.3). As described later, *G. maxima* is phylogenetically distinct from other *Galdieria* spp., which form a monophyletic group in phylogenetic trees. The history of identification and classification of Cyanidiales is summarized in detail elsewhere (Reeb and Bhattacharya 2010; Pinto 2007). In brief, the members of algae known today as Cyanidiales were described in the nineteenth century (Meneghini 1839). However, this group had been assigned many different names and classified into various algal lineages. Later, the algae were named *Cy. caldarium* (Geitler 1933) and were finally affiliated to the red algae (Rhodophyta) (Hirose 1950).

However, it was found that several different unicellular red algae coexist in sulfuric hot springs (De Luca and Taddei 1970), although all cyanidialean strains were named *Cyanidium* (*Cy.*) *caldarium* up to that point. *Cyanidoschyzon* (*C.*) *merolae* (De Luca et al. 1978) was described as an alga without cell wall that reproduces by binary fission in contrast to other Cyanidiales that reproduce by forming four or more autospores in the mother cell wall (Fig. 1.1). *G. sulphuraria* (Merola et al. 1981) was differentiated from the obligate photoautotroph *Cy. caldarium* based on its ability to grow heterotrophically.

Based on the number of autospores produced and the shape of chloroplast, three new species of *Galdieria* were defined, namely, *G. partita*, *G. daedala*, and *G. maxima* (Gross 1999; Sentsova 1994). Finally, *G. phlegrea* was described based on molecular phylogenetic studies; it is found exclusively in endolithic habitats (Pinto 2007).

### 1.2.2 Current Understanding of Taxonomic Position of Cyanidiales in Red Algae and Eukaryotes

Cyanidiales belong to the red algae (Rhodophyta). The red algae form a monophyletic group of eukaryotes. As in case of cyanobacteria and glaucophytes (Glaucophyta), they do not possess chlorophyll *b* and *c* but contain allophycocyanin, phycocyanin, and phycoerythrin in the form of phycobilisomes on thylakoids. However, in contrast to many other eukaryotic algal groups, the red algae lack flagella and centrioles at all stages of their life (Graham and Wilcox 2000).

The red algae belong to Archaeplastida, chloroplast in which was established more than one billion years ago when a cyanobacterial endosymbiont became integrated into a previously nonphotosynthetic eukaryote (primary endosymbiosis) (Elias and Archibald 2009) (Fig. 1.2). The ancient eukaryotic algae that resulted from this primary endosymbiotic event evolved into Glaucophyta (glaucophyte algae), Rhodophyta (red algae), and Viridiplantae (green algae, charophyte algae, and land plants). After these primitive green and red algae were established, chloroplasts then spread into other eukaryote lineages through secondary endosymbiotic events, wherein red or green algae became integrated into previously nonphotosynthetic eukaryotes (Elias and Archibald 2009) (Fig. 1.2).

Cyanidiales form a monophyletic group that is phylogenetically quite distinct from other red algal lineages. It has been estimated that Cyanidiales branched off quite early in eukaryotic evolution (i.e., approximately 1.3–1.4 billion years ago) based on the molecular clock inference and fossil evidence for the first multicellular red algae (Yoon et al. 2004, 2006b). According to these results, Yoon et al. (2006b) placed Cyanidiales in their own subphylum Cyanidiophytina, class Cyanidiophyceae. In contrast to many other red algae, Cyanidiales do not contain phycoerythrin (Lin et al. 1990). Furthermore, sexual reproduction has not been reported in Cyanidiales.

### 1.2.3 Classification of Cyanidialean Members: Morphology, Physiology, and Phylogenetic Relationship

The three genera in Cyanidiales, namely, *Cyanidioschyzon*, *Cyanidium*, and *Galdieria*, are defined by differences in physiology, cell morphology, mode of cell proliferation, and presence or absence of a cell wall (Merola et al. 1981; Albertano et al. 2000) (Fig. 1.1).

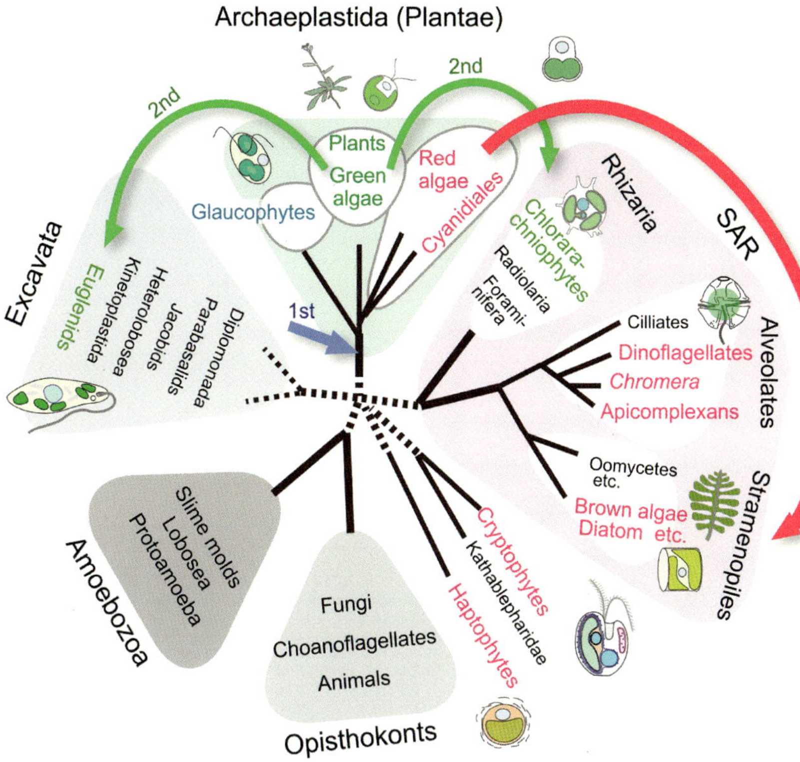


**Fig. 1.2** The position of red algae and Cyanidiales in the eukaryotic tree. The *broken lines* denote uncertainty of branch positions in the tree. Red algae and groups possessing chloroplasts (or nonphotosynthetic plastids) of red algal origin are shown in *red*. Viridiplantae (green algae and land plants) and groups possessing chloroplasts of green algal origin are shown in *green*. *Arrows* indicate the primary endosymbiotic event of a cyanobacterium (1st) and secondary endosymbiotic events (2nd). The position and number of secondary endosymbiotic events in the tree are still uncertain at present (this tree is according to Adl et al. (2012))

*C. merolae* does not possess a cell wall, exhibits an oval-to-club-like shape, and reproduces through binary fission (De Luca et al. 1978), whereas *Cy. caldarium* and *Galdieria* spp. possess cell walls, are spherical, and reproduce by forming autospores inside the mother cell wall (Merola et al. 1981). *Galdieria* is larger and forms more endospores than *Cyanidium* (Merola et al. 1981). Other differences in the cellular structure are that *Galdieria* spp. possess vacuoles (Merola et al. 1981), whereas *Cy. caldarium* and *C. merolae* possess small lysosomes (Yagisawa et al. 2007). Previous literature has described that *C. merolae* lacks Golgi apparatus. However, it was later shown that this species does contain Golgi apparatus, the structure of which is simpler than that of the apparatus observed in other Cyanidiales (see Chap. 15) (Okuwaki et al. 1996; Yagisawa et al. 2013).

The most notable and distinct physiological feature of *Galdieria* spp. is that they can grow heterotrophically and mixotrophically on more than 50 different carbon sources in contrast to the obligate photoautotrophs *C. merolae* and *Cy. caldarium* (Rigano et al. 1976, 1977; Gross and Schnarrenberger 1995; Gross 1999; Oesterhelt et al. 1999). However, a recent study has shown that *C. merolae* grows in the dark in the presence of 200 mM glycerol, although the growth is very slow compared to that in photoautotrophic conditions (Moriyama et al. 2015).

Phylogenic analyses of strains isolated worldwide as well as environmental DNA obtained from Yellowstone National Park (YNP) (USA), Italy, and Iceland have classified the Cyanidiales into four distinct lineages (Ciniglia et al. 2004, 2014; Yoon et al. 2004, 2006a; Skorupa et al. 2013) (Fig. 1.3). The first is the *Galdieria* lineage (*Galdieria* spp. excluding *G. maxima*) that is further divided into two sublineages: *Galdieria* A clade (including *G. sulphuraria* found worldwide and *G. daedala* and *G. partita* isolated in Russia) and *Galdieria* B clade (including only

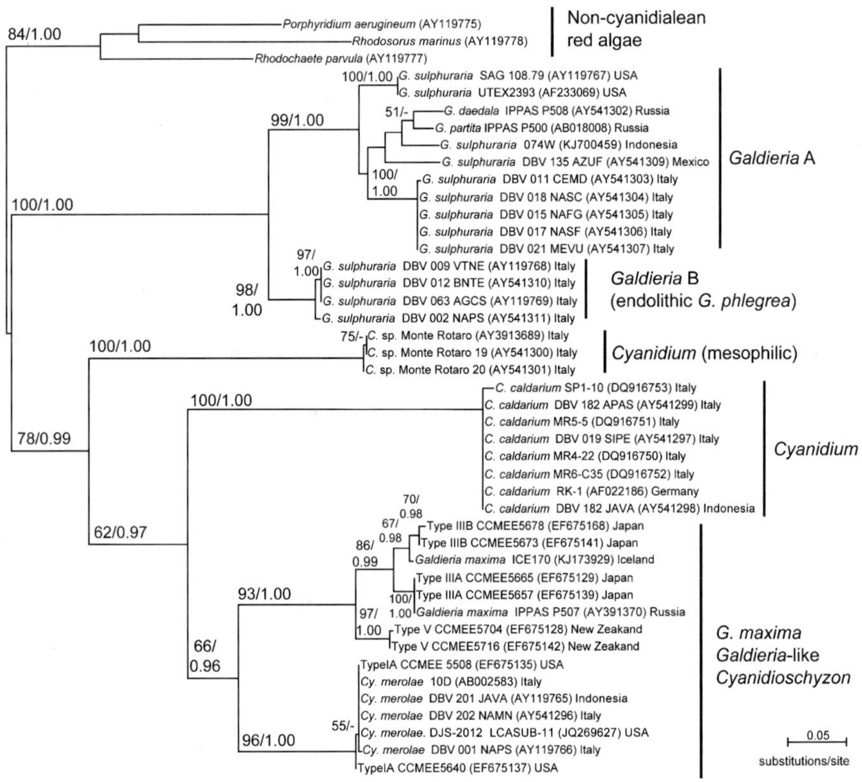


**Fig. 1.3** Evolution and divergence of Cyanidiales. A phylogenetic tree based on the *rbcL* sequence is shown. The tree was constructed by a maximum-likelihood method (RaxML 7.2.8; Stamatakis 2006). Maximum likelihood bootstrap values >50% by RaxML and Bayesian posterior probabilities >0.95 by the Bayesian analysis (MrBayes 3.2.6; Ronquist and Huelsenbeck 2003) are shown above the branches. The branch length reflects the evolutionary distances indicated by the scale bar. The accession numbers of the respective nucleotide sequences and the countries in which respective strains were isolated are indicated besides the species names

the endolithic alga *G. phlegrea* isolated in Italy). The second lineage comprises *Cy. caldarium* isolated from acidophilic sites worldwide. The third lineage comprises the mesophilic *Cyanidium* spp. that inhabit nonacidic and nonthermal sites, such as caves. Phylogenetic studies have shown that cave *Cyanidium* spp. in Chile and Italy are included in this group. Other cave *Cyanidium* spp. have not been examined through phylogenetic analyses. Finally, the fourth lineage comprises *G. maxima* and *C. merolae*. In addition, this clade contains algae with *Galdieria*-like morphology (cell-walled and producing autospores; species name not given, but defined as Type IA in a study by Toplin et al. 2008), found in YNP, and they are most closely related to *C. merolae*. Although it has not been tested whether these *Galdieria*-like algae can grow heterotrophically, as seen in *Galdieria* spp., their sequences of *rbcL* and *18S* loci are nearly identical to *C. merolae* (Skorupa et al. 2013).

Based on the above results, it has become clear that the classification of Cyanidiales into three genera (based on cellular structure, mode of proliferation, and capability of heterotrophic growth) does not correspond to their classification based on the molecular phylogeny. At this point, *Galdieria* spp. are polyphyletic, whereas *Galdieria* spp., except for *G. maxima* and the *Galdieria*-like algae in YNP (*Galdieria* A and B), are monophyletic (Fig. 1.3). In addition, the relationship between acidophilic and mesophilic *Cyanidium* spp. are unclear at present.

### *1.2.4 The Evolution of Each Member in Cyanidiales*

The above molecular phylogenetic studies have started to reveal how each member of Cyanidiales evolved. The topologies of phylogenetic trees (Ciniglia et al. 2004, 2014; Yoon et al. 2004, 2006a; Skorupa et al. 2013) (Fig. 1.3) suggest that the common ancestor of Cyanidiales was a thermoacidophilic cell-walled alga that proliferated by forming autospores in the mother cell. The first to diverge was the *Galdieria* clade (*Galdieria* A and B) (Fig. 1.3). After the *Cyanidium* clade diverged, the mesophilic *Cyanidium* clade was derived from the acidophilic *Cyanidium*. Following the divergence of the *Cyanidium* clade, the common ancestor of *G. maxima*, *Galdieria*-like algae in YNP (type IA in Toplin et al. 2008), and *C. merolae* diverged. Thus, it is suggested that *C. merolae* probably lost the cell wall and began to proliferate through binary fission, although their cell-walled ancestors proliferated through autospore formation.

A recent study suggests that the loss of the cell wall also contributed to the acquisition of crawling activity in *C. merolae*. In nature, Cyanidiales do not float in acidic water but stick to the bottom of water or to steamed rocks (Fig. 1.1). It has long been believed that Cyanidiales are immotile, but it was recently shown in the laboratory that *Cy. caldarium* delta (isolated in Japan) and *C. merolae* cells move on bottom of water toward light (Ohnuma et al. 2011). The experiment showed that only a small population of *Cy. caldarium* diffused toward light, and all the *C. merolae* cells moved (0.27 mm/h; approximately 100 times the cell diameter/h) at a much faster rate than *Cy. caldarium* (Ohnuma et al. 2011). Although the

*C. merolae* cell lacks a cell wall, it possesses a flexible extracellular matrix of unknown composition, which likely touches the surface of materials and contributes to the cellular crawling activity (Ohnuma et al. 2011). However, the precise mechanism of the crawling is unknown at present.

By virtue of the recent development of high-throughput sequencing technologies, studies have started to reveal how the genome evolution in respective cyanidialean strains has contributed to the development of their lifestyle and adaptation to their respective thermoacidophilic environments. Comparative genome analyses of *C. merolae* 10D, *G. sulphuraria* 074, *G. phlegrea* DBV 009, and other red algal lineages showed that of the 6801 orthologous genes present in the most recent common ancestor of red algae (Rhodophyta), 1448 were lost in the common ancestor of Cyanidiales (Qiu et al. 2013). Following this, *C. merolae* underwent 1312 additional gene losses, resulting in a poor gene lineage (*C. merolae* 10D has 4775 nuclear protein-coding genes, with only 27 introns) (Qiu et al. 2013). Thus, the simple genome structure of *C. merolae* is probably established by reductive evolution.

Horizontal gene transfer (HGT) from environmental prokaryotes and the expansion of gene families (increasing the copy number of a gene and subsequent diversification of gene function) for both genes derived from their eukaryotic ancestor and acquired though HGT have probably played important roles in the adaptation of Cyanidiales to their respective habitats (Schonknecht et al. 2013; Qiu et al. 2013). At least 5% of protein-coding genes of *G. sulphuraria* 074 were probably acquired through HGT. At the very least, these proteins are most likely involved in tolerance of acidic hot environments, heavy metal detoxification, and glycerol uptake and metabolism (Schonknecht et al. 2013). Overall, 5.2% of *G. sulphuraria* genes encode membrane transport proteins (mostly metabolite transporters); this value is more than that observed in most of the other eukaryotes. These metabolite transporters are likely to be essential for the ability of *G. sulphuraria* to grow heterotrophically (Schonknecht et al. 2013).

The study also showed that urease was lost in the common ancestor of Cyanidiales. However, *G. phlegrea* regained all components of urease through HGT after the divergence of *Galdieria* A and *Galdieria* B (*G. phlegrea*) (Qiu et al. 2013). Urease hydrolyzes urea into carbon dioxide and ammonia. This reaction provides an alternative source of nitrogen in environments where nitrogen is limited. In contrast to *G. sulphuraria* and *C. merolae*, *G. phlegrea* grows in less extreme habitats (i.e., moderate pH and temperature) (Ciniglia et al. 2004; Pinto 2007). There is likely to be a correlation between the adaptation to endolithic environments and the acquisition of urease, but the details are currently unknown.

## 1.3 Habitats and Physiology of Cyanidiales

Cyanidiales (except for certain rare species from nonthermal or neutrophilic environments, as described below) grow optimally at 40–45 °C and pH 2–3 but can be found in acidic environments with temperatures up to 56 °C. The lower limit in aquatic habitats is around 35 °C, but these algae are found in soils at 10 °C. Cyanidiales can also grow at pH 0.05–5 (Doemel and Brock 1971).

With a wide distribution, Cyanidiales have been found in acidic volcanic hot pools and springs in all continents except Africa and Antarctica (probably due to a shortage of survey). Most thermoacidophiles are nonphotosynthetic prokaryotes (archaea or bacteria). It is well known that cyanobacteria and other phototrophic prokaryotes are absent in volcanic waters below about pH 4 (Ward and Castenholz 2000). Thus, Cyanidiales are the only phototrophic organisms and eukaryotes found in acidic hot water. However, in cooler regions, the Cyanidiales habitat begins to overlap with that of other eukaryotic algae, such as green algae, euglenids, and diatoms (Gross 2000; Ferris et al. 2005). Cyanidiales have also been found to thrive on rocks and muddy soils around hot pools and along streams and in fissures in rock walls (interlithic) or under thin layers of rock (endolithic) that are exposed to acidic sulfur fumes (Fig. 1.1). To date, there have been no reports of Cyanidiales from basic freshwater or seawater habitats (Gross 1999).

The extremely low pH of these waters is because of the dissolution and oxidation of sulfur that is exposed to water and oxygen and produces sulfuric acid (Novis and Harding 2007). The low pH facilitates metal solubility in water; therefore, acidic waters tend to have high concentrations of heavy metals. Thus, Cyanidiales possess abilities to cope with toxic heavy metals in addition to high temperature and low pH, all of which are lethal to most eukaryotes.

In addition to acidic hot springs, certain rare species of Cyanidiales have been found in acidic water at lower temperature and from neutrophilic environments. *Cy. caldarium* was also found at nonthermal sites, such as in drainage from sulfur mines in Italy (Pinto and Taddei 1978). *Galdieria* spp. were found together with green algae and cyanobacteria in a sulfur mine, a peat bog, and a diatomite shield in the Czech Republic (the optimal temperature for growth was 35 °C in the laboratory) (Gross et al. 2002). In addition, Cyanidiales have been reported in nonthermal freshwater contaminated by acidic mine drainage (Moreira et al. 1994; Gross and Gross 2001; Aguilera et al. 2007; Baker et al. 2004).

*Cyanidium* spp. have been found in nonthermal and nonacidic caves in Chile (Schwabe 1936), France (Leclerc et al. 1983), Israel (Friedmann 1964), and Italy (Skuja 1970; Ciniglia et al. 2004). Such caves possess low temperatures (4–10 °C), high humidity, and low light intensity and are neutrophilic (pH 5–7). *Cyanidium* spp. from caves exhibit a cellular morphology that is almost indistinguishable from thermoacidophilic *Cyanidium* spp. (Hoffman 1994).

All thermoacidophilic environments are inhabited by a mixed population of Cyanidiales, but different species dominate at different sites (Ciniglia et al. 2004; Yoon et al. 2006a; Pinto 2007). *C. merolae* has been found only in the most humid environments (in pools and springs) but less frequently than others. *Galdieria* spp. and *Cyanidium* spp. are able to grow in relatively dry habitats. *Galdieria* B is found exclusively in endolithic environments.

Consistent with the distribution patterns of respective species in their natural habitats, laboratory experiments have shown that *G. sulphuraria* was found to be more resistant to water loss than *Cy. caldarium* (Pinto 2007). *G. sulphuraria* has the ability to withstand higher toxic concentrations of heavy metals (Albertano and Pinto 1986; Nagasaka et al. 2004) and higher concentrations of salt (Gross et al. 2002) than *C. merolae*. The endolithic *Galdieria* B has a maximum growth rate at a lower temperature (~25 °C). In addition, they are shade adapted and exhibit a lower maximum photosynthetic rate than *Galdieria* A. (Pinto 2007).

## 1.4 Geographical Distribution of Cyanidiales

Currently, it is unknown how extant Cyanidiales have scattered throughout the world. Thermoacidophilic environments are usually very small and geographically isolated from each other. Cyanidiales exhibit relatively low tolerance to desiccation. In addition, they are asexual and do not form resting spores throughout their life cycles. Thus, long-distance migration seems very difficult for Cyanidiales. It is expected that they are exposed to rapid divergence and the evolution of distinct strains with different features in separated hot springs. This seems to be the case for *Galdieria* spp. based on molecular phylogenetic studies. These studies have shown a clear grouping according to a geographical pattern worldwide (Ciniglia et al. 2004; Yoon et al. 2006a). However, phylogenetic studies based on *rbcL* sequence have shown that almost identical or closely related strains of *Cyanidium* spp. and *C. merolae* can be found in distantly located hot springs, regardless of their geographic origin (Ciniglia et al. 2004; Yoon et al. 2006a). Thus, Cyanidiales have probably migrated between separated acidic hot springs. Several dispersal mechanisms have been assumed (Ciniglia et al. 2014; Toplin et al. 2008), but the detail is currently unknown.

**Acknowledgments** We thank Dr. Sumiya (Keio University) for helping us to take photographs of Cyanidiales in their natural habitats. Our study was partly supported by Japan Society for the Promotion of Science Grant-in-Aid for Scientific Research 25251039 (to S.M.) and by the Core Research for Evolutional Science and Technology Program of the Japan Science and Technology Agency (S.M.).

## References

Adl SM, Simpson AG, Lane CE, Lukes J et al (2012) The revised classification of eukaryotes. J Eukaryot Microbiol 59:429–493

Aguilera A, Souza-Egipsy V et al (2007) Development and structure of eukaryotic biofilms in an extreme acidic environment, rio tinto (SW, Spain). Microb Ecol 53:294–305

Albertano P, Pinto G (1986) The action of heavy metals on the growth of the acidophilic algae. Boll Soc Nat Napoli 45:319–328

Albertano P, Ciniglia C et al (2000) The taxonomic position of *Cyanidium*, *Cyanidioschyzon* and *Galdieria*: an update. Hydrobiologia 433:137–143

Baker BJ, Lutz MA et al (2004) Metabolically active eukaryotic communities in extremely acidic mine drainage. Appl Environ Microbiol 70:6264–6271

Castenholz RW, McDermott TR (2010) The cyanidiales: ecology, biodiversity, and biogeography. In: Seckbach J, Chapman DJ (eds) Red algae in the genomic age. Springer, Drodrecht, pp 359–371
Ciniglia C, Yoon HS et al (2004) Hidden biodiversity of the extremophilic Cyanidiales red algae. Mol Ecol 13:1827–1838
Ciniglia C, Yang EC et al (2014) Cyanidiophyceae in Iceland: plastid *rbcL* gene elucidates origin and dispersal of extremophilic *Galdieria sulphuraria* and *G. maxima* (Galdieriaceae, Rhodophyta). Phycologia 53:542–551
De Luca P, Taddei R (1970) Due alghe delle fumarole acide dei Campi Flegrei (Napoli): *Cyanidium caldarium*? Delpinoa 10/11:79–89
De Luca P, Taddei R et al (1978) "*Cyanidioschyzon merolae*": a new alga of thermal acidic environments. Webbia 33:37–44
Doemel WN, Brock TD (1971) The physiological ecology of *Cyunidium caldarium*. J Gen Microbiol 67:17–32
Elias M, Archibald JM (2009) Sizing up the genomic footprint of endosymbiosis. BioEssays 31:1273–1279
Ferris MJ, Sheehan KB et al (2005) Algal species and light microenvironment in a low-pH, geothermal microbial mat community. Appl Environ Microbiol 71:7164–7171
Friedmann I (1964) Progress in the biological exploration of caves and subterranean waters in Israel. Int J Speleol 1:29–33
Geitler L (1933) Diagnoses neuer Blaualgen von den Sunda-Insela. Arch Hydrobiol Suppl 12:622–634
Graham LD, Wilcox LW (2000) Algae. Prentice Hall, Upper Saddle River
Gross W (1999) Revision of comparative traits for the acido- and thermophilic red algae *Cyanidium* and Galdieria. In: Seckbach J (ed) Evolutionary pathways and enigmatic algae: *Cyanidium caldarium* (Rhodophyta) and related cells. Springer, Dordrecht, pp 437–446
Gross W (2000) Ecophysiology of algae living in highly acidic environments. Hydrobiologia 33:31–37
Gross W, Gross S (2001) Physiological characterization of the red alga *Galdieria sulphuraria* isolated from a mining area. Nova Hedwigia Beih 123:523–530
Gross W, Schnarrenberger C (1995) Heterotrophic growth of 2 strains of the acido-thermophilic red alga *Galdieria sulphuraria*. Plant Cell Physiol 36:633–638
Gross W, Oesterhelt C et al (2002) Characterization of a non-thermophilic strain of the red algal genus *Galdieria* isolated from Soos (Czech Republic). Eur J Phycol 37:477–482
Hirose H (1950) Studies of thermal alga, *Cyanidium caldarium*. Bot Mag Tokyo 63:745–746
Hoffman L (1994) Cyanidium-like algae from caves. In: Seckbach J (ed) Evolutionary pathways and enigmatic algae: *Cyanidium caldarium* (Rhodophyta) and related cells. Kluwer, Dordrecht, pp 175–182
Leclerc JC, Coute A et al (1983) Le climat annuel de deux grottes et d'une église du Poitou, ou vivent des colonies pures d'algues sciaphiles. Algol 4:1–19
Lin S, Offner GD et al (1990) Studies on *Cyanidium caldarium* phycobiliprotein pigment mutants. Plant Physiol 93:772–777
Matsuzaki M, Misumi O et al (2004) Genome sequence of the ultrasmall unicellular red alga *Cyanidioschyzon merolae* 10D. Nature 428:653–657
Meneghini G (1839) Nuova specie di alga descritta dal Sig. Dott. Giuseppe Meneghini di Padova. Nuovo Giorn Lett Sci 39:67–68
Merola A, Castaldo R et al (1981) Revision of *Cyanidium caldarium*: three species of acidophilic algae. Gio Bot Ital 115:189–195
Minoda A, Sakagami R et al (2004) Improvement of culture conditions and evidence for nuclear transformation by homologous recombination in a red alga, *Cyanidioschyzon merolae* 10D. Plant Cell Physiol 45:667–671

Moreira D, López-Archilla A et al (1994) Characterization of two new thermoacidophilic microalgae: genome organization and comparison with *Galdieria sulphuraria*. FEMS Lett 122:109–114

Moriyama T, Mori N et al (2015) Activation of oxidative carbon metabolism by nutritional enrichment by photosynthesis and exogenous organic compounds in the red alga *Cyanidioschyzon merolae*: evidence for heterotrophic growth. Springerplus 4:559

Nagasaka S, Nishizawa NK et al (2004) Metal metabolism in the red alga *Cyanidium caldarium* and its relationship to metal tolerance. Biometals 17:177–181

Novis P, Harding JS (2007) Extreme acidophiles: freshwater algae associated with acid mine drainage. Algae and cyanobacteria in extreme environments. In: Seckbach J (ed) Algae and cyanobacteria in extreme environments. Springer, Dordrecht, pp 443–463

Nozaki H, Takano H et al (2007) A 100%-complete sequence reveals unusually simple genomic features in the hot-spring red alga *Cyanidioschyzon merolae*. BMC Biol 5:28

Oesterhelt C, Schnarrenberger C et al (1999) Characterization of a sugar/polyol uptake system in the red alga *Galdieria sulphuraria*. Eur J Phycol 34:271–277

Ohnuma M, Yokoyama T et al (2008) Polyethylene glycol (PEG)-mediated transient gene expression in a red alga, *Cyanidioschyzon merolae* 10D. Plant Cell Physiol 49:117–120

Ohnuma M, Misumi O et al (2011) Phototaxis in the unicellular red algae *Cyanidioschyzon merolae* and *Cyanidium caldarium*. Cytologia 76:295–300

Ohta N, Sato N et al (1998) Structure and organization of the mitochondrial genome of the unicellular red alga *Cyanidioschyzon merolae* deduced from the complete nucleotide sequence. Nucleic Acids Res 26:5190–5198

Ohta N, Matsuzaki M et al (2003) Complete sequence and analysis of the plastid genome of the unicellular red alga *Cyanidioschyzon merolae*. DNA Res 10:67–77

Okuwaki T, Takahashi H et al (1996) Ultrastructures of the Golgi body and cell surface in Cyanidioschyzon merolae. Cytologia 61:69–74

Pinto G (2007) Cyanidiophyceae: looking back–looking forward. In: Seckbach J (ed) Algae and cyanobacteria in extreme environments. Springer, Dordrecht, pp 387–397

Pinto G, Taddei R (1978) Le alghe delle acque e dei suoli acidi italiani. Delpinoa 18(19):77–106

Qiu H, Price DC et al (2013) Adaptation through horizontal gene transfer in the cryptoendolithic red alga *Galdieria phlegrea*. Curr Biol 23:R865–R866

Reeb V, Bhattacharya D (2010) The thermo-acidophilic Cyanidiophyceae (Cyanidiales). In: Seckbach J, Chapman DJ (eds) Red algae in the genomic age. Springer, Dordrecht, pp 411–426

Rigano C, Fuggi A et al (1976) Studies on utilization of 2-ketoglutarate, glutamate and other amino acids by the unicellular alga *Cyanidium caldarium*. Arch Microbiol 107:133–138

Rigano C, Aliotta G et al (1977) Heterotrophic growth patterns in the unicellular alga *Cyanidium caldarium*. A possible role for threonine dehydrase. Arch Microbiol 113:191–196

Ronquist F, Huelsenbeck JP (2003) MRBAYES 3: Bayesian phylogenetic inference under mixed models. Bioinformatics 19:1572–1574

Schonknecht G, Chen WH et al (2013) Gene transfer from bacteria and archaea facilitated evolution of an extremophilic eukaryote. Science 339:1207–1210

Schwabe GH (1936) Über einige Blaualgen aus dem mittleren und sudlichen Chile. Verh des Deutsch Wiss Ver Santiago de Chile 3:113–174

Sentsova OY (1994) The study of Cyanidiophyceae in Russia. Algae of genus *Galdieria*: diversity, characterization and occurrence in mixed populations with *Cyanidium caldarium*. In: Seckbach J (ed) Evolutionary pathways and enigmatic algae: *Cyanidium caldarium* (Rhodophyta) and related cells. Springer, Dordrecht, pp 167–174

Skorupa DJ, Reeb V et al (2013) Cyanidiales diversity in Yellowstone national park. Lett Appl Microbiol 57:459–466

Skuja H (1970) Alghe cavernicole nelle zone illuminate delle Grotte di Castellana (Murge di Bari). Le Grotte d'Italia 4:193–202

Stamatakis A (2006) RAxML-VI-HPC: maximum likelihood-based phylogenetic analyses with thousands of taxa and mixed models. Bioinformatics 22:2688–2690

Toplin JA, Norris TB et al (2008) Biogeographic and phylogenetic diversity of thermoacidophilic cyanidiales in Yellowstone National Park, Japan, and New Zealand. Appl Environ Microbiol 74:2822–2833

Ward DM, Castenholz RW (2000) Cyanobacteria in geothermal habitats. In: Whitton BA, Potts M (eds) Ecology of cyanobacteria: their diversity in time and space. Kluwer, Dordrecht, pp 37–59

Yagisawa F, Nishida K et al (2007) Identification and mitotic partitioning strategies of vacuoles in the unicellular red alga *Cyanidioschyzon merolae*. Planta 226:1017–1029

Yagisawa F, Fujiwara T et al (2013) Golgi inheritance in the primitive red alga, *Cyanidioschyzon merolae*. Protoplasma 250:943–948

Yoon HS, Hackett JD et al (2004) A molecular timeline for the origin of photosynthetic eukaryotes. Mol Biol Evol 21:809–818

Yoon HS, Ciniglia C et al (2006a) Establishment of endolithic populations of extremophilic Cyanidiales (Rhodophyta). BMC Evol Biol 6:78

Yoon HS, Müller KM et al (2006b) Defining the major lineages of red algae (Rhodophyta). J Phycol 42:482–492

# Chapter 2
# Cellular Structure of *Cyanidioschyzon merolae*: A Minimum Set of Organelles

**Yuuta Imoto and Yamato Yoshida**

**Abstract** The cell of the unicellular red alga *Cyanidioschyzon merolae* is composed of a very small number of membranous organelles: one cell nucleus, one mitochondrion, one chloroplast (plastid), a simple-shaped ER, one Golgi body with two cisternae, a few vacuoles (lysosomes), and one peroxisome. During the last two decades, numerous electron and fluorescence microscopic studies, combined with synchronous culture, have demonstrated spatial organization and morphological information of these organelles in each cell cycle phase of *C. merolae* cells. These explorations have revealed that the timing and manner of the organelle behaviors are strictly determined by the progression of the cell division cycle. In addition, the simplicity of the cell structure assists researchers to directly address the biological processes in each organelle. Concurrently, given the completely sequenced genome and various genetic technologies, the simple composition of *C. merolae* cells provides considerable opportunities to clarify long-standing biological questions, including cell cycle regulation, organelle biogenesis, and various types of metabolic pathways.

**Keywords** *Cyanidioschyzon merolae* • Double-membrane-bounded organelle • Single-membrane-bounded organelle • Cell nucleus • Mitochondrion • Chloroplast • Endoplasmic reticulum • Golgi body • Lysosome • Peroxisome

## 2.1 Introduction

*Cyanidioschyzon merolae* cells were first unearthed as a model organism for cell biology by Professor Tsuneyoshi Kuroiwa. The structure of *C. merolae* is basically an oval-shaped cell, with size of approximately 2.0 μm in the major axis and 1.0 μm

Y. Imoto (✉)
Division of Organelle Homeostasis, Medical Institute of Bioregulation, Kyushu University, 3-1-1 Maidashi, Higashiku, Fukuoka 812-0053, Japan
e-mail: yimoto1@jhmi.edu

Y. Yoshida
Laboratory for Single Cell Gene Dynamics, Quantitative Biology Center, RIKEN, Furuedai, Suita, Osaka 565-0874, Japan
e-mail: yamato.yoshida@riken.jp



T. Kuroiwa et al. (eds.), *Cyanidioschyzon merolae*,
https://doi.org/10.1007/978-981-10-6101-1_2

in the minor axis. As they do not contain a flagellar apparatus, the cells appear to have no swimming ability. Interestingly, each cell possesses only one nucleus, one mitochondrion, and one chloroplast, which are aligned in that order in the cell (Fig. 2.1 A and B). Other membranous organelles have been identified in very small numbers in a tiny cytosolic region, comprising one peroxisome, a simple-shaped ER, one Golgi body with two cisternae, and a few vacuoles (Kuroiwa et al. 1994; Yagisawa et al. 2007, 2013). In its life cycle, the *C. merolae* cell multiplies by binary fission, which occurs through chloroplast division, mitochondrial division, nuclear division, and cytokinesis in turn (Miyagishima et al. 1999a). Many of the other membranous organelles are also doubled and inherited to the daughter cells during these periods. The cellular structure transforms into characteristic shapes according to the progression of the cell cycle (Fig. 2.1 C). The transforming cells can be easily discriminated by optical microscopy, thereby allowing identification of individual cells in each phase of the cell cycle. Meanwhile, the spatial distribution of the organelles can also be presumed in these cells. Many endeavors during the last two decades have developed and reinforced experimental approaches for *C. merolae*. Establishment of synchronous cultivation has accelerated studies related to cell proliferation mechanisms (Suzuki et al. 1994; Miyagishima et al. 2014; see Chaps. 3 and 12). In addition, the absence of a rigid cell wall has made it possible to develop many kinds of organelle isolation techniques (Miyagishima et al. 2001a; Yoshida et al. 2006, 2009; Yagisawa et al. 2009; Imoto et al. 2013; see Chap. 4). Genome information for *C. merolae* has provided insights into the fundamental principles and evolution of eukaryotic cells (Ohta et al. 1998, 2003; Matsuzaki et al. 2004; Nozaki et al. 2007; see Chaps. 1 and 5). Currently, many types of genetic techniques (Minoda et al. 2004; Ohnuma et al. 2008; Imamura et al. 2009; Sumiya et al. 2014; Fujiwara et al. 2015; see Chap. 7), as well as genome-wide and system-wide multi-omics analyses (Fujiwara et al. 2009; Yoshida et al. 2011; Kanesaki et al. 2012; see Chap. 6), can be carried out. Thus, *C. merolae* cells provide considerable advantages for examining the underlying molecular mechanisms related to cell proliferation and biogenesis. Furthermore, studies utilizing *C. merolae* cells are developing not only in basic research fields but also in applied research fields including engineering (Misumi et al, 2008; Hirooka et al. 2009; Sakajiri et al. 2010; Sumiya et al. 2015; see Chaps. 20 and 21). In this chapter, we introduce basic knowledge related to structural features and relevant aspects for each organelle in the *C. merolae* cell.

## 2.2 Cell Nucleus

The *C. merolae* cell possesses one spherical nucleus (approximately 0.5 μm in diameter) that includes a single nucleolus. Fluorescence imaging of nuclear DNA stained with 4′,6-diamidino-2-phenylindole (DAPI) has demonstrated dynamic morphological changes of the nucleus during the cell cycle (Suzuki et al. 1994; Maruyama et al. 2007). Combined with the microtubule distribution,

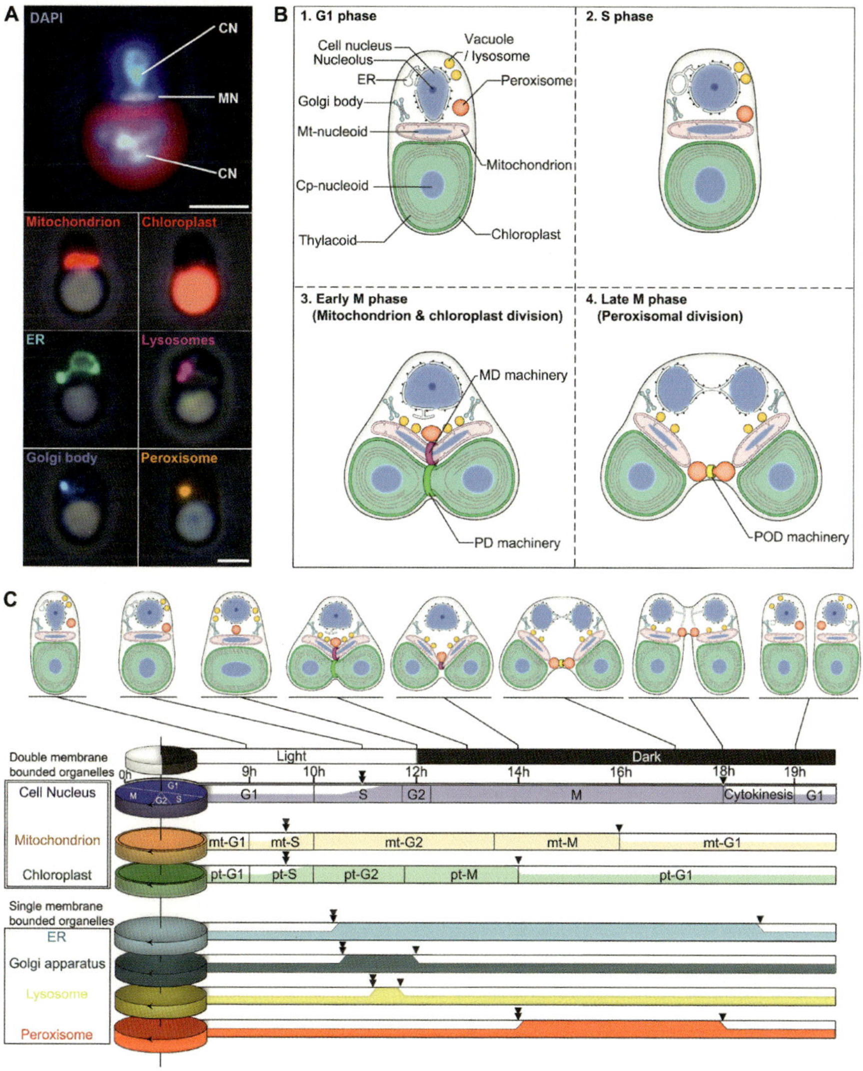


**Fig. 2.1** (**a**) Phase contrast DAPI staining images of C. merolae shows localization of DNA of cell nucleus (CN), mitochondrial (MN) and chloroplast (plastid) (CN). Phase contrast immunofluorescence images show mitochondrion (anti-POR antibody), chloroplast (autofluorescence), ER (anticalnexin antibody), Golgi body (anti-Sed5 antibody), lysosomes (anti-V-ATPase antibody) and peroxisome (anti-catalase antibody). (**b**) Schematic images representing the structures of the organelles in *C. merolae* cells during G1 phase, S phase, and M phase. *Mt-nucleoid* mitochondrial nucleoid, *Cp-nucleoid* chloroplast nucleoid, *MD machinery* mitochondrion-dividing machinery, *PD machinery* plastid-dividing machinery, *POD machinery* peroxisome-dividing machinery. (**c**) Summary of the division cycles and the temporal relationships of the three double-membranebounded organelles and the four single-membrane-bounded organelles in *C. merolae*. Mitochondrion and chloroplast divide using the MD and PD machineries, respectively. Peroxisome and lysosome are inherited using the mitochondrion as a carrier. Double arrowheads indicate the time point of the organelle replication and single arrowheads indicate the time point of the organelle division. Time shows hours after the second cell cycle of synchronization

the dynamics of the cell nucleus during the cell cycle can be represented as follows. In G1 phase, the cell nucleus takes a lightbulb shape, and its tip seems to shove the mitochondrial membrane. The structure of the cell nucleus then transforms into a spherical shape during S phase and G2 phase. At the opening of M phase, the cell nucleus forms a hemispherical shape from prophase to prometaphase. Despite possessing two major condensin complexes, chromosome condensation is not observed during mitosis in this alga (Fujiwara et al. 2013; see Chap. 11). Moving on to the next process, the cell nucleus shrinks into a slightly smaller spherical shape and the nuclear DNA becomes centered in metaphase. Mitosis is executed without breakdown of the nuclear membrane and is driven by the mitotic spindle (Maruyama et al. 2007; Nishida et al. 2005; Fujiwara et al. 2013; see Chaps. 14 and 16). The genomic profile of *C. merolae* and further details about the cell cycle are described in Chaps. 5 and 11, respectively.

## 2.3 Mitochondrion

A single mitochondrion exists in the cell of *C. merolae* and is a highly dynamic organelle (Kuroiwa et al. 1994). The mitochondrial shape is dependent on the phase of the cell cycle (Miyagishima et al. 1999b, 2001a). Throughout interphase, the mitochondrion exhibits a discoidal shape (approximately 0.5 μm in diameter and 0.2 μm in thickness). The spatial distribution of the mitochondrion is as follows. One side of the mitochondrial surface area has a dent into which the cell nucleus fits, while the other side of the mitochondrial surface physically interacts with the chloroplast surface. The simplicity of the mitochondrial shape changes significantly in M phase. The discoidal mitochondrion then elongates toward both of the two spindle poles, and the elongated mitochondrion bends at the middle in a direction away from the cell nucleus, consequently forming a V shape. The two tips of the mitochondrion become tethered to the two spindle poles (Nishida et al. 2005; see Chap. 16), and the bending point refers to the mitochondrial division site where the mitochondrion is pinched off by a ring-shaped supramolecular structure for mitochondrial division, called the mitochondrion-dividing (MD) machinery (Kuroiwa et al. 1994, 1998, 2008; Takahara et al. 2001; Nishida et al. 2003; Yoshida et al. 2009, 2013; see Chap. 14). The inner structure of the mitochondrion is not yet fully understood, but it is considered that the new-generation electron microscopy may be able to elucidate the details of the internal structure including cristae formation and distribution (see Chap. 10). The mitochondrial genome DNA assembles a nucleoid structure and accumulates at the central space of the matrix. The observed transitions in mitochondrial DNA density and spatial distribution throughout the cell cycle have raised further studies on the mitochondrial division cycle and relationship with the cell cycle (Suzuki et al. 1994; Imoto et al. 2010; see Chap. 11).

## 2.4 Chloroplast

*C. merolae* contains one chloroplast, which forms a spherical shape (approximately 1.0–1.2 μm in diameter) and accounts for more than half of the intracellular space (Kuroiwa et al. 1994). The chloroplast shape is highly dynamic during the cell cycle (Miyagishima et al. 1999b, 2001a). The distinctive chloroplast outlines in individual cells are used for practical reference to identify the cell cycle phase of the cells, because the chloroplast can easily be observed as green by white light microscopy or red by autofluorescence without any chemical staining. When the chloroplast has a capsule-like or acorn shape, the cell is assumed to be in G1 phase. Subsequently, the chloroplast assumes a spherical shape in S phase and transforms into a trapezoid shape in G2 phase. Concurrently with the start of M phase, chloroplast division occurs at the midsection symmetric to the chloroplast shape. Analogous to the MD machinery for the mitochondrion, a supramolecular complex for the chloroplast is assembled at the division site and termed the plastid-dividing (PD) machinery (Mita et al. 1986; Kuroiwa et al. 1994; Miyagishima et al. 2001b, 2003; Yoshida et al. 2006, 2010, 2012; see Chap. 14). The narrowed midsection of the dumbbell-shaped chloroplast is finally pinched off by the PD machinery, and the resultant chloroplasts are segregated into the daughter cells before cytokinesis. The internal structure of the chloroplast has been well visualized in ultrastructural studies using electron microscopy and coherentX-ray diffractionimaging (Kuroiwa et al. 1993, 1998, 2008; Takayama et al. 2015; see Chaps. 9 and 10). These studies showed that approximately five thylakoid membranes containing many phycobilisomes are concentrically stacked beneath the inner envelope membrane and that chloroplast nucleoids exist in the central space of the stroma. The photosynthetic system and further details of the phycobilisome structure are described in Chap. 17. Meanwhile, a pioneering study utilizing purified chloroplast nucleoids from *C. merolae* has characterized a major protein component of the chloroplast nucleoid, histone-like protein of chloroplast (HC), and a responsible gene, *hupA* (CMV162C), encoded in the chloroplast genome DNA (Kobayashi et al. 2002). Combined with the knowledge of the mitochondrial division cycle, the dynamic distribution of the chloroplast DNA has led to the discovery of the chloroplast division cycle. Currently, a research field for the interdependent regulation network between the cell division cycle, the mitochondrial division cycle, and the chloroplast division cycle is attracting considerable attention (Kobayashi et al. 2009, 2011; Miyagishima et al. 2014; Sumiya et al. 2016; see Chaps. 11, 12 and 13).

## 2.5 Single-Membrane-Bounded Organelles (ER, Golgi Body, Vacuoles/Lysosomes, Peroxisome)

*C. merolae* contains a simple-shaped ER, one Golgi body, a few vacuoles/lysosomes, and one peroxisome (Kuroiwa et al. 1994; Miyagishima et al. 1998; Yagisawa et al. 2013). The shape and distribution of these organelles exhibit dynamic morphological changes during the cell cycle. The structure of the ER is divided into two parts: a nuclear ER and peripheral ERs. The nuclear ER surrounds the cell nucleus throughout each cell cycle period. During mitosis, the nuclear ER becomes dumbbell-shaped and is divided at the equatorial region accompanied by cell nucleus division. The peripheral ERs protrude from different points on the nuclear ER surface. The peripheral ERs typically elongate to cytosol and to the mitochondrion or stretch toward the cell division plane. The nuclear ER and peripheral ERs are tightly attached to the spindle poles and microtubules, respectively. Thus, the morphological dynamics and inheritance of the ER are dependent on the microtubule system (Yagisawa et al. 2012; see Chap. 15). The Golgi body in *C. merolae* is composed of two or three cisternae (Yagisawa et al. 2013). The Golgi body is doubled from S phase to G2 phase and inherited to the daughter cells without fragmentation. The Golgi body is arranged near the spindle pole and the edge of the nuclear ER. Therefore, the dynamics of the Golgi body are regulated by microtubules and accompanied by segregation of the cell nucleus, similar to the case for the nuclear ER (see Chap. 15).

Unlike the ER and Golgi body, the dynamics of the vacuoles/lysosomes and peroxisome are independent of the cell nucleus but dependent on physical interaction with the mitochondrion. *C. merolae* contains two to six spherical vacuoles/lysosomes (approximately 0.5 μm in diameter) (Yagisawa et al. 2007). During G1 phase, the vacuoles/lysosomes are distributed within the cytosol and increase in number before the onset of mitosis. These organelles immediately relocate to the mitochondrial surface at the start of M phase. The vacuoles/lysosomes are released into the daughter cells from the mitochondrial surface after cell division. The apparatus and protein involved in the vacuole/lysosome–mitochondrion interaction were identified by post-genomic analysis, as described in Chap. 15 (Yagisawa et al. 2009; Fujiwara et al. 2010a, b). Similar to the vacuoles/lysosomes, the peroxisome interacts with the mitochondrion, but undergoes more dynamic morphological changes (Miyagishima et al. 1998, 1999c). One spherical peroxisome (approximately 0.3 μm in diameter) is present during G1 phase and then relocates to the equator plane of the mitochondrion during S phase to G2 phase. At the onset of M phase, the peroxisome doubles its volume and immediately elongates along the equator plane of the mitochondrion. After the sequence of morphological changes (in order: branched structure, H-shaped structure, dumbbell-shaped structure), the peroxisome is symmetrically divided into two. Analogous to the MD and PD machineries, a supramolecular complex for peroxisomal division is assembled at the division site and termed the POD machinery (Imoto et al. 2012, 2013; see Chap. 15). After the peroxisomal division, the daughter peroxisomes become spherical, as in G1 phase, and physically connect to the edge of the daughter mitochondria. The daughter mitochondria are anchored to the spindle poles for segregation into the daughter cells. Thus, the mitochondrion–vacuole/lysosome and mitochondrion–peroxisome interactions are associated to the inheritance of these organelles and regulated by the cytoskeleton (see Chaps. 15 and 16).

## 2.6 Conclusions

Optical and electron microscopy observations of the *C. merolae* cell have revealed the simple structures of double- and single-membrane-bound organelles. Moreover, the dynamics and distribution of the organelles in each cell cycle period can be evaluated by light/dark-stimulated synchronization of cell and organelle proliferation. These studies have shown that the organelles physically interact with one another, including the cell nucleus–mitochondrion–chloroplast interaction during G1 phase and the mitochondrion–peripheral ER–vacuole/lysosome or peroxisome interaction during mitosis. Recent studies have been uncovering the importance of these organelle–organelle interactions relevant to cell cycle regulation, organelle biogenesis, and various types of metabolic pathways. For example, signaling among the cell nucleus, mitochondrion, and chloroplast is involved in the regulation of their division cycles (see Chaps. 11, 12 and 13). Physical attachments between the mitochondrion and single-membrane-bound organelles are essential for coordinated inheritance of these organelle complexes (see Chaps. 15 and 16). Furthermore, the metabolic pathways found in the mitochondrion and chloroplast are linked to ER function in the synthesis of lipid droplets (Imamura et al. 2010, 2015; Kuroiwa et al. 2012; Moriyama et al. 2014; Mori et al. 2016; see Chaps. 18 and 19). Currently, *C. merolae* is the only model single-cell eukaryote with simple organelle structures whose proliferation can be synchronized, with a completely sequenced genome and established genetic manipulation techniques. Hence, further studies of *C. merolae* can offer elucidation of fundamental phenomena in organelle biology.

## References

Fujiwara T, Misumi O, Tashiro K, Yoshida Y, Nishida K, Yagisawa F, Imamura S, Yoshida M, Mori T, Tanaka K, Kuroiwa H, Kuroiwa T (2009) Periodic gene expression patterns during the highly synchronized cell nucleus and organelle division cycles in the unicellular red alga *Cyanidioschyzon merolae*. DNA Res 16:59–72

Fujiwara T, Kuroiwa H, Yagisawa F, Ohnuma M, Yoshida Y, Yoshida M, Nishida K, Misumi O, Watanabe S, Tanaka K et al (2010a) The coiled-coil protein VIG1 is essential for tethering vacuoles to mitochondria during vacuole inheritance of *Cyanidioschyzon merolae*. Plant Cell 22:772–781

Fujiwara T, Yagisawa F, Ohnuma M, Yoshida Y, Yoshida M, Nishida K, Misumi O, Kuroiwa H, Kuroiwa T (2010b) The vacuole binding to mitochondria by VIG1 contributes an equal inheritance of the vacuoles in *Cyanidioschyzon merolae*. Cytologia 75:189–194

Fujiwara T, Tanaka K, Kuroiwa T, Hirano T (2013) Spatiotemporal dynamics of condensins I and II: evolutionary insights from the primitive red alga *Cyanidioschyzon merolae*. Mol Biol Cell 24:2515–2527

Fujiwara T, Kanesaki Y, Hirooka S, Era A, Sumiya N, Yoshikawa H, Tanaka K, Miyagishima SY (2015) A nitrogen source-dependent inducible and repressible gene expression system in the red alga *Cyanidioschyzon merolae*. Front Plant Sci 6:657

Hirooka S, Misumi O, Yoshida M, Mori T, Nishida K, Yagisawa F, Yoshida Y, Fujiwara T, Kuroiwa H, Kuroiwa T (2009) Expression of the *Cyanidioschyzon merolae* stromal ascorbate peroxidase in *Arabidopsis thaliana* enhances thermotolerance. Plant Cell Rep 28:1881–1893

Imamura S, Kanesaki Y, Ohnuma M, Inouye T, Sekine Y, Fujiwara T, Kuroiwa T, Tanaka K (2009) R2R3-type MYB transcription factor, CmMYB1, is a central nitrogen assimilation regulator in *Cyanidioschyzon merolae*. Proc Nat Acad Sci 106:12548–12553

Imamura S, Terashita M, Ohnuma M, Maruyama S, Minoda A, Weber AP, Inouye T, Sekine Y, Fujita Y, Omata T et al (2010) Nitrate assimilatory genes and their transcriptional regulation in a unicellular red alga *Cyanidioschyzon merolae*: genetic evidence for nitrite reduction by a sulfite reductase-like enzyme. Plant Cell Phys 51:707–717

Imamura S, Kawase Y, Kobayashi I, Sone T, Era A, Miyagishima SY, Shimojima M, Ohta H, Tanaka K (2015) Target of rapamycin (TOR) plays a critical role in triacylglycerol accumulation in microalgae. Plant Mol Biol 89:309–318

Imoto Y, Fujiwara T, Yoshida Y, Kuroiwa H, Maruyama S, Kuroiwa T (2010) Division of cell nuclei, mitochondria, plastids, and microbodies mediated by mitotic spindle poles in the primitive red alga *Cyanidioschyzon merolae*. Protoplasma 241:63–74

Imoto Y, Yoshida Y, Yagisawa F, Kuroiwa H, Kuroiwa T (2011) The cell cycle, including the mitotic cycle and organelle division cycles, as revealed by cytological observations. J Elect Micro 60(suppl 1):S117–S136

Imoto Y, Kuroiwa H, Ohnuma M, Kawano S, Kuroiwa T (2012) Identification of peroxisome-dividing ring in *Cyanidioschyzon merolae* based on organelle partner hypothesis. Cytologia 77:515–522

Imoto Y, Kuroiwa H, Yoshida Y, Ohnuma M, Fujiwara T, Yoshida M, Nishida K, Yagisawa F, Hirooka S, Miyagishima S et al (2013) Single-membrane–bounded peroxisome division revealed by isolation of dynamin-based machinery. Proc Nat Acad Sci 110:9583–9588

Kanesaki Y, Imamura S, Minoda A, Tanaka K (2012) External light conditions and internal cell cycle phases coordinate accumulation of chloroplast and mitochondrial transcripts in the red alga *Cyanidioschyzon merolae*. DNA Res 19:289–303

Kobayashi T, Takahara M, Miyagishima SY, Kuroiwa H, Sasaki N, Ohta N, Matsuzaki M, Kuroiwa T (2002) Detection and localization of a chloroplast-encoded HU-like protein that organizes chloroplast nucleoids. Plant Cell 14:1579–1589

Kobayashi Y, Kanesaki Y, Tanaka A, Kuroiwa H, Kuroiwa T, Tanaka K (2009) Tetrapyrrole signal as a cell-cycle coordinator from organelle to nuclear DNA replication in plant cells. Proc Nat Acad Sci 106:803–807

Kobayashi Y, Imamura S, Hanaoka M, Tanaka K (2011) A tetrapyrrole-regulated ubiquitin ligase controls algal nuclear DNA replication. Nat Cell Biol 13:483–487

Kuroiwa T (1998) The primitive red algae *Cyanidium caldarium* and *Cyanidioschyzon merolae* as model system for investigating the dividing apparatus of mitochondria and plastids. BioEssays 20:344–354

Kuroiwa T, Suzuki K, Kuroiwa H (1993) Mitochondrial division by an electron-dense ring in *Cyanidioschyzon merolae*. Protoplasma 175:173–177

Kuroiwa T, Kawazu T, Takahashi H, Suzuki K, Ohta N, Kuroiwa H (1994) Comparison of ultrastructures between the ultra-small eukaryote *Cyanidioschyzon merolae* and *Cyanidium caldarium*. 341. Cytologia 59:149–158

Kuroiwa T, Misumi O, Nishida K, Yagisawa F, Yoshida Y, Fujiwara T, Kuroiwa H (2008) Vesicle, mitochondrial, and plastid division machineries with emphasis on dynamin and electron-dense rings. Int Rev Cell Mol Biol 271:97–152

Kuroiwa T, Ohnuma M, Imoto Y, Misumi O, Fujiwara T, Miyagishima SY, Sumiya N, Kuroiwa H (2012) Lipid droplets of bacteria, algae and fungi and a relationship between their contents and genome sizes as revealed by BODIPY and DAPI staining. Cytologia 77:289–299

Maruyama S, Kuroiwa H, Miyagishima SY, Tanaka K, Kuroiwa T (2007) Centromere dynamics in the primitive red alga *Cyanidioschyzon merolae*. The Plant J 49:1122–1129

Matsuzaki M, Misumi O, Shin-i T, Maruyama S, Takahara M, Miyagishima SY, Mori T, Nishida K, Yagisawa F, Nishida K et al (2004) Genome sequence of the ultrasmall unicellular red alga *Cyanidioschyzon merolae* 10D. Nature 428:653–657

Minoda A, Sakagami R, Yagisawa F, Kuroiwa T, Tanaka K (2004) Improvement of culture conditions and evidence for nuclear transformation by homologous recombination in a red alga, *Cyanidioschyzon merolae* 10D. Plant Cell Phys 45:667–671

Misumi O, Sakajiri T, Hirooka S, Kuroiwa H, Kuroiwa T (2008) Cytological studies of metal ion tolerance in the red algae *Cyanidioschyzon merolae*. Cytologia 73:437–443

Mita T, Kanbe T, Tanaka K, Kuroiwa T (1986) A ring structure around the dividing plane of the *Cyanidium caldarium* chloroplast. Protoplasma 130:211–213

Miyagishima M, Itoh R, Toda K, Takahashi H, Kuroiwa H, Kuroiwa T (1998) Visualization of the microbody division in *Cyanidioschyzon merolae* with the fluorochrome brilliant sulfoflavin. Protoplasma 201:115–119

Miyagishima SY, Itoh R, Aita S, Kuroiwa H, Kuroiwa T (1999a) Isolation of dividing chloroplasts with intact plastid-dividing rings from a synchronous culture of the unicellular red alga *Cyanidioschyzon merolae*. Planta 209:371–375

Miyagishima SY, Itoh R, Toda K, Kuroiwa H, Kuroiwa T (1999b) Real-time analyses of chloroplast and mitochondrial division and differences in the behavior of their dividing rings during contraction. Planta 207:343–353

Miyagishima SY, Itoh R, Toda K, Kuroiwa H, Nishimura M, Kuroiwa T (1999c) Microbody proliferation and segregation cycle in the single-microbody alga *Cyanidioschyzon merolae*. Planta 208:326–336

Miyagishima SY, Kuroiwa H, Kuroiwa T (2001a) The timing and manner of disassembly of the apparatuses for chloroplast and mitochondrial division in the red alga *Cyanidioschyzon merolae*. Planta 212:517–528

Miyagishima SY, Takahara M, Kuroiwa T (2001b) Novel filaments 5 nm in diameter constitute the cytosolic ring of the plastid division apparatus. Plant Cell 13:707–721

Miyagishima SY, Nishida K, Mori T, Matsuzaki M, Higashiyama T, Kuroiwa H, Kuroiwa T (2003) A plant-specific dynamin-related protein forms a ring at the chloroplast division site. Plant Cell 15:655–665

Miyagishima SY, Fujiwara T, Sumiya N, Hirooka S, Nakano A, Kabeya Y, Nakamura M (2014) Translation-independent circadian control of the cell cycle in a unicellular photosynthetic eukaryote. Nat Commun 5

Mori N, Moriyama T, Toyoshima M, Sato N (2016) Construction of global acyl lipid metabolic map by comparative genomics and subcellular localization analysis in the red alga *Cyanidioschyzon merolae*. Front Plant Sci 7

Moriyama T, Sakurai K, Sekine K, Sato N (2014) Subcellular distribution of central carbohydrate metabolism pathways in the red alga *Cyanidioschyzon merolae*. Planta 240:585–598

Nishida K, Takahara M, Miyagishima S-y, Kuroiwa H, Matsuzaki M, Kuroiwa T (2003) Dynamic recruitment of dynamin for final mitochondrial severance in a primitive red alga. Proc Nat Acad Sci 100(4):2146–2151

Nishida K, Yagisawa F, Kuroiwa H, Nagata T, Kuroiwa T (2005) Cell cycle-regulated, microtubule-independent organelle division in *Cyanidioschyzon merolae*. Mol Biol Cell 16:2493–2502

Nozaki H, Takano H, Misumi O, Terasawa K, Matsuzaki M, Maruyama S, Nishida K, Yagisawa F, Yoshida Y, Fujiwara T et al (2007) A 100%-complete sequence reveals unusually simple genomic features in the hot-spring red alga *Cyanidioschyzon merolae*. BMC Biol 5(1)

Ohnuma M, Yokoyama T, Inouye T, Sekine Y, Tanaka K (2008) Polyethylene glycol (PEG)-mediated transient gene expression in a red alga, *Cyanidioschyzon merolae* 10D. Plant Cell Phys 49:117–120

Ohta N, Sato N, Kuroiwa T (1998) Structure and organization of the mitochondrial genome of the unicellular red alga *Cyanidioschyzon merolae* deduced from the complete nucleotide sequence. Nuc Acids Res 26:5190–5198

Ohta N, Matsuzaki M, Misumi O, Miyagishima SY, Nozaki H, Tanaka K, Shin-I T, Kohara Y, Kuroiwa T (2003) Complete sequence and analysis of the plastid genome of the unicellular red alga *Cyanidioschyzon merolae*. DNA Res, 10: 67–77

Sakajiri T, Asano K, Hirooka S, Ohnuma M, Misumi O, Yoshida M, Fujiwara T, Doi H, Kuroiwa T (2010) The overexpression of *Cyanidioschyzon merolae* S-adenosylmethionine Synthetase enhances salt tolerance in transgenic *Arabidopsis thaliana*. Cytologia 75:341–352

Sumiya N, Fujiwara T, Kobayashi Y, Misumi O, Miyagishima SY (2014) Development of a heat-shock inducible gene expression system in the red alga *Cyanidioschyzon merolae*. PLoS One 9: e111261

Sumiya N, Kawase Y, Hayakawa J, Matsuda M, Nakamura M, Era A, Tanaka K, Kondo A, Hasunuma T et al (2015) Expression of cyanobacterial acyl-ACP reductase elevates the triacylglycerol level in the red alga *Cyanidioschyzon merolae*. Plant Cell Phys 56:1962–1980

Sumiya N, Fujiwara T, Era A, Miyagishima SY (2016) Chloroplast division checkpoint in eukaryotic algae. Proc Nat Acad Sci 113:E7629–E7638

Suzuki K, Ehara T, Osafune T, Kuroiwa H, Kawano S, Kuroiwa T (1994) Behavior of mitochondria, chloroplasts and their nuclei during the mitotic cycle in the ultramicroalga *Cyanidioschyzon merolae*. Eur J Cell Biol 63:280–288

Takahara M, Kuroiwa H, Miyagishima SY, Mori T, Kuroiwa T (2001) Localization of the mitochondrial FtsZ protein in a dividing mitochondrion. Cytologia 66:421–425

Takayama Y, Inui Y, Sekiguchi Y, Kobayashi A, Oroguchi T, Yamamoto M, Matsunaga S, Nakasako M (2015) Coherent X-ray diffraction imaging of chloroplasts from *Cyanidioschyzon merolae* by using X-ray free electron laser. Plant Cell Phys 56:1272–1286

Yagisawa F, Nishida K, Kuroiwa H, Nagata T, Kuroiwa T (2007) Identification and mitotic partitioning strategies of vacuoles in the unicellular red alga *Cyanidioschyzon merolae*. Planta 226:1017–1029

Yagisawa F, Nishida K, Yoshida M, Ohnuma M, Shimada T, Fujiwara T, Yoshida Y, Misumi O, Kuroiwa H, Kuroiwa T (2009) Identification of novel proteins in isolated polyphosphate vacuoles in the primitive red alga *Cyanidioschyzon merolae*. The Plant J 60:882–893

Yagisawa F, Fujiwara T, Kuroiwa H, Nishida K, Imoto Y, Kuroiwa T (2012) Mitotic inheritance of endoplasmic reticulum in the primitive red alga *Cyanidioschyzon merolae*. Protoplasma 249:1129–1135

Yagisawa F, Fujiwara T, Ohnuma M, Kuroiwa H, Nishida K, Imoto Y, Yoshida Y, Kuroiwa T (2013) Golgi inheritance in the primitive red alga, *Cyanidioschyzon merolae*. Protoplasma 250:943–948

Yoshida Y, Kuroiwa H, Misumi O, Nishida K, Yagisawa F, Fujiwara T, Nanamiya H, Kawamura F, Kuroiwa T (2006) Isolated chloroplast division machinery can actively constrict after stretching. Science 313:1435–1438

Yoshida Y, Kuroiwa H, Hirooka S, Fujiwara T, Ohnuma M, Yoshida M, Misumi O, Kawano S, Kuroiwa T (2009) The bacterial ZapA-like protein ZED is required for mitochondrial division. Curr Biol 19:1491–1497

Yoshida Y, Kuroiwa H, Misumi O, Yoshida M, Ohnuma M, Fujiwara T, Yagisawa F, Hirooka S, Imoto Y, Matsuhita K, Kawano S, Kuroiwa T (2010) Chloroplasts divide by contraction of a bundle of nanofilaments consisting of polyglucan. Science 329:949–953

Yoshida M, Yoshida Y, Fujiwara T, Misumi O, Kuroiwa H, Kuroiwa T (2011) Proteomic comparison between interphase and metaphase of isolated chloroplasts of *Cyanidioschyzon merolae* (*Cyanidiophyceae*, Rhodophyta). Phycol Res 59:1–15

Yoshida Y, Miyagishima SY, Kuroiwa H, Kuroiwa T (2012) The plastid-dividing machinery: formation, constriction and fission. Curr Opi Pant Biol 15:714–721

Yoshida Y, Fujiwara T, Imoto Y, Yoshida M, Ohnuma M, Hirooka S, Misumi O, Kuroiwa H, Kuroiwa T (2013) The kinesin-like protein TOP promotes aurora localisation and induces mitochondrial, chloroplast and nuclear division. J Cell Sci 126:2392–2400

# Part II
# Synchronous Culture of *Cyanidioschyzon merolae* and Isolation of Organelles

**Shin-ya Miyagishima**

Eukaryotic cells possess various single- and double-membraned organelles that are replicated and inherited by daughter cells during cell proliferation. However, there is very limited information on the proliferation of membranous organelles during cell cycle progression. The unicellular red alga *C. merolae* is a very useful experimental platform for studying organelle proliferation because of the following reasons: (1) The cellular architecture is extremely simple, possessing only one chloroplast, one mitochondrion, one peroxisome, and a few lysosomes. Each organelle divides during a specific period of the cell cycle. (2) The cell cycle and organelle division cycle can be highly synchronized by cultivating the cells under a 12-h light/12-h dark cycle. (3) The cells lack a rigid cell wall, facilitating the isolation of intact dividing organelles. (4) The genome and proteome compositions are simple, facilitating proteomic approaches and *in vitro* studies on the mechanisms of organelle division and inheritance.

The abovementioned characteristics of *C. merolae*, availability of its complete genome sequence information, and development of procedures for organelle isolation and genetic manipulations have clarified the mechanisms of organelle division and inheritance in studies using *C. merolae*. For readers who will start studies using *C. merolae*, Chap. 3 introduces how researchers have cultivated *C. merolae* and how the cells are synchronized under a light and dark cycle. This chapter also introduces procedures for conventional experiments involving nucleic acids and proteins. Chapter 4 summarizes procedures for the isolation of (dividing) organelles and their division machineries in synchronized cells. The procedures are applicable to the chloroplast, mitochondrion, lysosome, peroxisome, and respective division machineries (rings).

# Chapter 3
# Procedures for Cultivation, Observation, and Conventional Experiments in *Cyanidioschyzon merolae*

**Shinya Miyagishima and Jong Lin Wei**

**Abstract** *Cyanidioschyzon merolae* 10D was originally isolated from a mixture of hot spring water sampled in Naples, Italy. Currently, this strain is available in the Microbial Culture Collection at the National Institute of Environmental Studies in Japan. The strain has been cultured in 2× Allen's medium or its derivatives. The optimal growth conditions for this strain are as follows: pH 2.5, 42 °C, and ~100 μmol photons $m^{-2}$ $s^{-1}$, allowing the cell density to reach ~5 × $10^8$ cells $mL^{-1}$. *C. merolae* can also grow slowly at >20 °C. We generally store stock cultures at room temperature or at 30 °C under a low light condition (~20 μmol photons $m^{-2}$ $s^{-1}$) or as frozen stock in liquid nitrogen. Cell cycle progression can be synchronized by subjecting the culture to a 12-h light/12-h dark cycle. In addition, cells can be arrested at the S or M phases by adding relevant inhibitors. The shapes of cells and chloroplasts are clearly observed by phase-contrast or differential interference contrast microscopy. Because *C. merolae* lacks a cell wall, cellular contents (e.g., DNA, RNA, and proteins) are easily extracted.

**Keywords** Cell cycle • Cell division • Cyanidiales • *Cyanidioschyzon merolae* • Red algae • Synchronous culture

---

S. Miyagishima (✉)
Department of Cell Genetics, National Institute of Genetics, 1111 Yata, Mishima, Shizuoka 411-8540, Japan
e-mail: smiyagis@nig.ac.jp

J.L. Wei
Department of Cell Genetics, National Institute of Genetics, 1111 Yata, Mishima, Shizuoka 411-8540, Japan

Department of Genetics, Graduate University for Advanced Studies, Shizuoka 411-8540, Japan



T. Kuroiwa et al. (eds.), *Cyanidioschyzon merolae*,
https://doi.org/10.1007/978-981-10-6101-1_3

## 3.1 Introduction

Unicellular algae are potentially useful for basic biological studies in several different disciplines and also for interdisciplinary studies for the following reasons:

1. The chemical formulas of many kinds of media for algal growth are completely defined.
2. A unicellular algal culture often provides a homogeneous population in terms of cell types and surrounding environment (including pH, temperature, light strength, and inorganic nutrient condition). This contrasts with land plants in which many different types of cells and tissues differentiate and each cell is exposed to a different environment, even in the same organism.
3. Some unicellular algal species exhibit relatively simple cellular and genomic architecture among eukaryotes.
4. Many unicellular algae are relatively easy to cultivate at a low cost and at a shorter generation time than required for land plants. These simple characteristics of unicellular algae potentially offer ideal experimental platforms; however, experimental techniques, particularly those for genetic modification of eukaryotic algae, are unfortunately extremely limited. However, the recent rapid development of genetic modification tools in *C. merolae* shows promise for overcoming this issue.

The methods for several "omics" analyses (Chap. 6), genetic modification (Chap. 7), fluorescence and electron microscopy (Chaps. 8, 9 and 10), and "in silico" analyses in *C. merolae* are described in other chapters. In the present chapter, we briefly summarize procedures for cultivation and synchronization of *C. merolae* in the laboratory and other conventional experimental procedures with relevant references.

## 3.2 Cultivation of *C. merolae*

### *3.2.1 Stock Culture*

*C. merolae* 10D was originally isolated by Dr. Toda belonging to Prof. Kuroiwa's group at the University of Tokyo from a mixture of hot spring water samples collected in Naples, Italy, that was prepared by Prof. Pinto (Toda et al. 1995). Currently, this strain is available from the Microbial Culture Collection at the National Institute of Environmental Studies in Japan (NIES-3377; http://mcc.nies.go.jp).

Many laboratories culture *C. merolae* in 2× Allen's medium (Allen 1959) or its derivatives (e.g., MA medium) (Minoda et al. 2004). Although the optimal temperature for this strain is 42 °C, *C. merolae* can also grow at >20 °C at a reduced rate. Thus, we generally store stock cultures at room temperature or at ~30 °C in

Erlenmeyer flasks or plastic culture flasks which are gently agitated with illumination (~20 μmol photons $m^{-2}$ $s^{-1}$) (Fig. 3.1). Under these conditions, the cell density reaches ~5 × $10^8$ cells $mL^{-1}$, and the culture can be kept for 2 months until the next subculture.

**Fig. 3.1** Stock and asynchronous cultures of *Cyanidioschyzon merolae*. (**a**) Stock cultures of the wild type and transformants in 50 mL plastic culture flasks. Cultures are stocked at room temperature with illumination (~20 μmol photons $m^{-2}$ $s^{-1}$) on a rotator (~50 rpm). (**b**) Stock cultures of the wild type and transformants in 500 mL Erlenmeyer flasks. Cultures are stocked in a growth chamber at 30 °C with illumination (~20 μmol photons $m^{-2}$ $s^{-1}$) on a rotator (~50 rpm). (**c**) Asynchronous batch culture of *C. merolae* in an incubator at 40 °C. Cultures in 100 mL test tubes (70 mL medium) are aerated with an air pump. The air from the pump is passed through distilled water before being passed into the culture to prevent desiccation of the culture. The test tubes are illuminated by fluorescent lamps (100 μmol photons $m^{-2}$ $s^{-1}$). (**d**) Asynchronous batch culture of *C. merolae* in a 24-well plate. The plate is agitated using a shaker (~50 rpm) in a growth chamber at 40 °C with illumination (100 μmol photons $m^{-2}$ $s^{-1}$)

Alternatively, *C. merolae* 10D or its transformants can be stored as frozen stock in liquid nitrogen (Ohnuma et al. 2011); however, long-term storage (e.g., ~1 year) using this method has not yet been tested. In addition, storage in liquid nitrogen is relatively costly. Thus, cultivation under a lower temperature at a slow growth rate is much easier for the storage of many strains.

Cultivation of *C. merolae* is also feasible on a starch bed on a gellan gum plate containing 2× Allen's medium or its derivative. The cultivation of cells on gellan gum plates has been applied only to isolate colonies of transformants and is described elsewhere (Imamura et al. 2010) (Chap. 7).

Cell density in the culture is determined using a Neubauer-improved cell counting chamber or a Coulter counter (Z2, Beckman Coulter) equipped with a 50 μm aperture (Miyagishima et al. 2014). When determining the cell density with a cell counting chamber, the chamber is left for 5 min after the injection of the culture to allow the cells to adhere to the surface of the bottom glass before counting.

### 3.2.2 *Asynchronous Batch Culture*

The optimal conditions for *C. merolae* growth are as follows: temperature of 42 °C, pH 1–5, and light illumination of ~100 μmol photons $m^{-2}$ $s^{-1}$ (depending on the thickness of the culture bottle). We generally use 2× Allen's medium at pH 2.5; however, a similar growth is achieved in the medium at pH 4.6 (Nishida et al. 2005). Because the medium is acidic, the solubility of $CO_2$ is relatively low; therefore, aeration is important to yield higher cellular growth. We usually culture the cells in 100 mL test tubes (~3 cm thick; containing ~70 mL culture) or 700 mL flat bottles (~5 cm thick; containing ~500 mL culture), which are aerated with 1 L $min^{-1}$ or 3 L $min^{-1}$ ambient air by pumps (Fig. 3.1). These conditions are usually sufficient to achieve good algal growth (doubling time is ~24 h) (Miyagishima et al. 2014); however, the addition of ~5% $CO_2$ accelerates cellular growth under certain conditions. The tubes or bottles can be incubated in an air incubator as well as in a transparent water bath that is illuminated from the outside.

When many cultures of small volume are required, for example, to expose cells to a series of concentrations of inhibitors, we usually culture the cells in 24-well plates that are agitated with a shaker under illumination (Fig. 3.1). Although cellular growth under these conditions is somewhat slower than that under the methods described above, the 24-well plate culture is sufficient to test the effects of inhibitors or other chemicals on cellular growth.

### 3.2.3 *Synchronous Culture*

Cell cycle progression of *C. merolae* cells can be synchronized by a 12-h light/12-h dark cycle (Suzuki et al. 1994; Fujiwara et al. 2013; Miyagishima et al. 2014). Cells remain in the G1 phase during the light period and synchronously enter into the S and then M phases during the dark period (Suzuki et al. 1994; Fujiwara et al. 2013; Miyagishima et al. 2014). The details of the cell cycle of *C. merolae* are described elsewhere (Fujiwara et al. 2013) (Chap. 11), and the cell cycle progression is regulated by circadian rhythms (Miyagishima et al. 2014) (Chap. 12 ). Briefly, during light period, cells grow by photosynthesis, but G1/S transition is inhibited. During the dark period, cell growth ceases and only cells that have grown beyond a certain size threshold during the light period divide during the dark period (Miyagishima et al. 2014). Thus, to increase the mitotic index during the dark period, it is important to achieve good cell growth during the light period. The simplest method to accelerate cellular growth is to increase the rate of aeration.

The cells in a stationary culture (~$5 \times 10^8$ cells $mL^{-1}$ at 30 °C) are subcultured to $<1 \times 10^7$ cells $mL^{-1}$ ($OD_{750} = 0.15$–$0.20$) in 500 mL of 2× Allen's medium at pH 2.5 (in a 700 mL flat bottle) and then cultured in the dark for 12 h under aeration with ambient air (3 L $min^{-1}$) at 42 °C. The culture is then subjected to a 12-h light/ 12-h dark cycle (100 µmol photons $m^{-2}$ $s^{-1}$) (Fig. 3.2). The cells divide synchronously from the second dark cycle (Fig. 3.3). The temperature should be maintained at 40 °C–50 °C (we usually culture cells at 40 °C–42 °C). The timing of G1/S transition is temperature compensated by circadian rhythms; however, the durations of the S and M phases largely depend on the temperature (Miyagishima et al. 1999b, 2014). Therefore, at a lower temperature (<40 °C), the cell cycle does not complete within the 12 h dark period and no synchrony is evident (Miyagishima et al. 2014).

### 3.2.4 Procedures to Arrest Synchronized Cells in the S or M Phase

Inducible transgene expression systems are now available in *C. merolae* (Sumiya et al. 2014; Fujiwara et al. 2015) (Chap. 7). Thus, the arrest of synchronized cells at a desired point of the cell cycle by expressing a dominant negative form of a certain cell cycle regulator is likely feasible. However, to date, no example of such experiments has been published.

An alternative method to arrest cells in the S or M phases is the treatment of cells with cell cycle inhibitors. We have arrested cells in the S phase in the past by the addition of ampicillin (Terui et al. 1995; Itoh et al. 1996), 5-fluorodeoxyuridine (Miyagishima et al. 1999a; Nishida et al. 2005), or camptothecin (Nishida et al. 2005, 2007; Miyagishima et al. 2012) to the culture at 10–12 h after the onset of the light period in the synchronous culture (Fig. 3.3). To arrest cells in the M phase, MG132, a proteasome inhibitor, has been added to the synchronous culture at 4 h after the onset of the dark period (Nishida et al. 2005; Miyagishima et al. 2012) (Fig. 3.3). Because chloroplast division occurs in the S phase in eukaryotic algae (Miyagishima et al. 2012), the chloroplast in the S-phase-arrested cell continues to divide, producing abnormal cells that contain four to eight chloroplasts (Itoh et al.

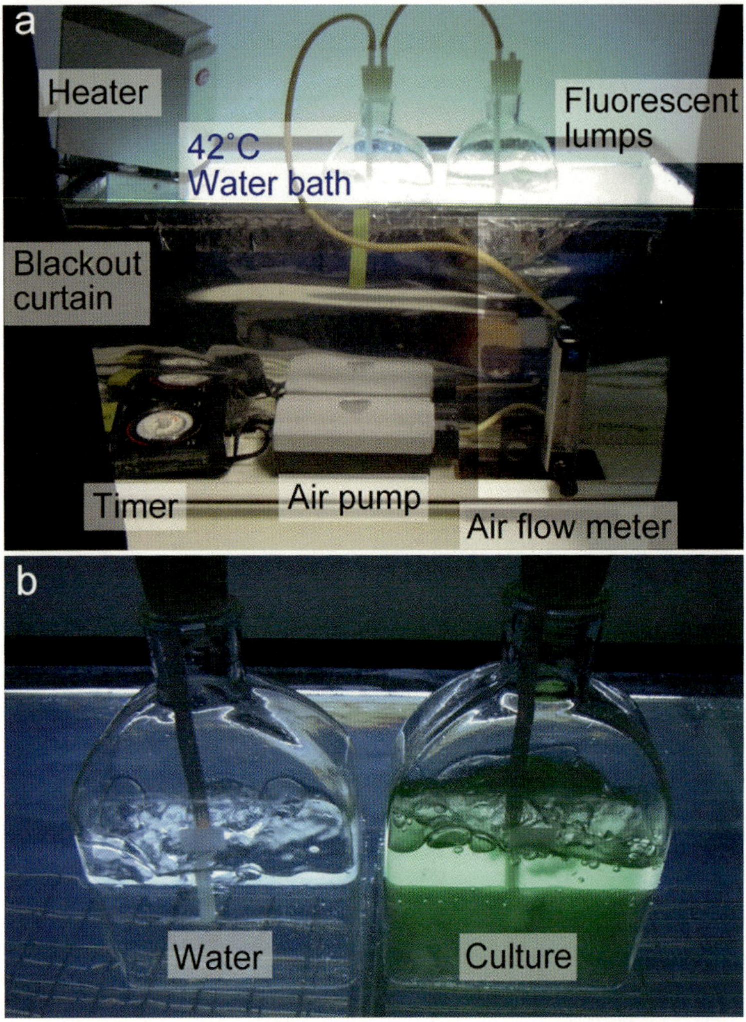


**Fig. 3.2** Synchronous culture of *Cyanidioschyzon merolae* in a water bath. A stock culture is diluted to ~$1.0 \times 10^7$ cells $mL^{-1}$ (OD750 = 0.15–0.2) in 500 mL fresh 2× Allen's medium in a 700 mL flat bottle. (**a**) The culture is aerated with ambient air (3 L air $min^{-1}$) in a water bath at 42 °C. The air from the pump is passed through distilled water before being passed into the culture to prevent desiccation of the culture. The culture is incubated in dark for >12 h and then subjected to a 12-h light/12-h dark cycle. The bottles are illuminated (100 μmol photons $m^{-2}$ $s^{-1}$) by fluorescent lamps located behind the water bath. The fluorescent lamps are regulated by a timer. (**b**) A magnified view of the bottles containing distilled water and the culture inside the water bath

1996; Miyagishima et al. 2012) (Fig. 3.3). In contrast, when cells are arrested in the M phase, the chloroplast has completed one round of division but never divides again, resulting in cells containing two chloroplasts (Miyagishima et al. 2012) (Fig. 3.3). Oryzalin, an inhibitor of microtubule polymerization, has also been

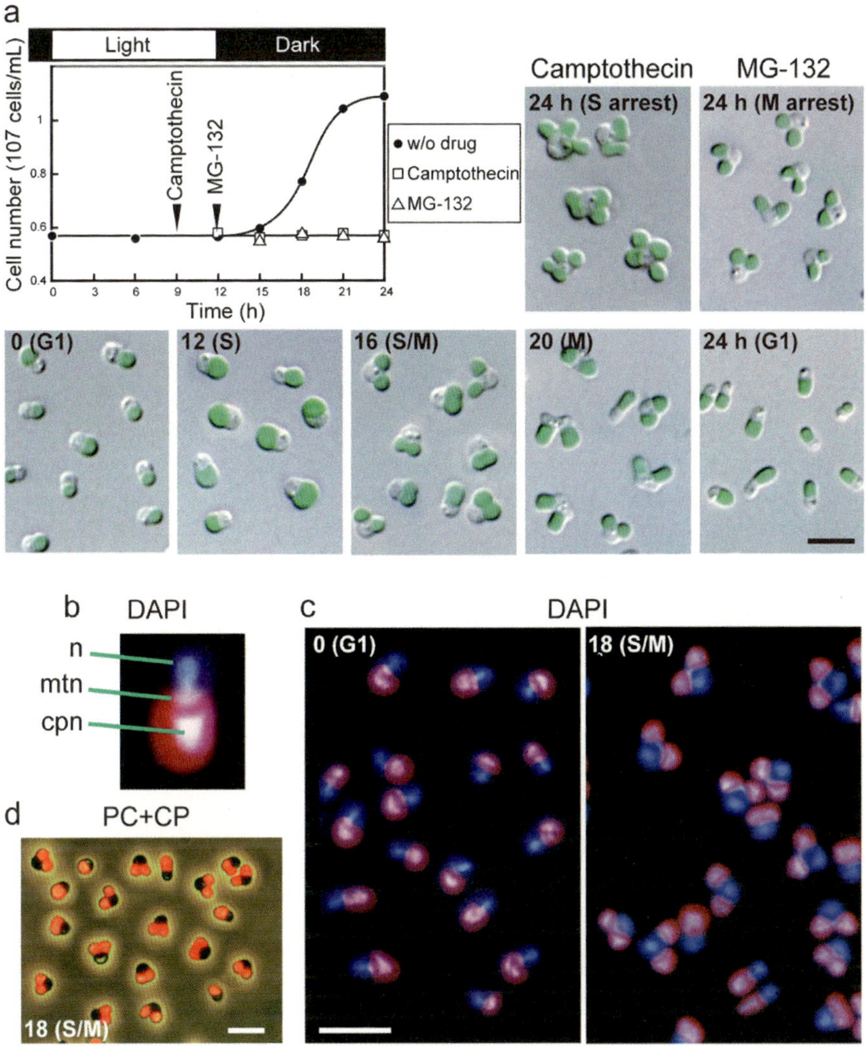


**Fig. 3.3** Observation of cell cycle progression in synchronous culture by microscopy. (**a**) Cells were entrained by a 12-h light and 12-h dark cycle; the cells in the second cycle are shown. To arrest the cells in the S phase, 1/50,000 volume of 10 mg $mL^{-1}$ camptothecin (an inhibitor of DNA topoisomerase-I) solution in dimethyl sulfoxide (DMSO) was added 8 h after the onset of the second light period. To arrest the cells in the M phase, 1/500 volume of 50 mM MG132 (an inhibitor of proteasomes) solution in DMSO was added at the onset of the second dark period. Cells were observed without fixation by differential interference contrast microscopy with a 60× objective lens. (**b**) A DAPI-stained fluorescent image of a G1 cell. n, nucleus; mtn, mitochondrial nucleoid; cpn, chloroplast nucleoid. *Blue* represents DAPI fluorescence of DNA and red represents autofluorescence of chloroplast. (**c**) DAPI-stained fluorescent images of G1 (at the onset of the light period) and S/M phase (6 h after the onset of the dark period) cells. Cells were fixed by the addition of glutaraldehyde (final concentration of 1%) to the culture. After fixation for 10 min, cells were stained with 1 μg/mL DAPI by gently squashing the fixed cells with the slide and cover glasses. (**d**) An image of phase-contrast and chloroplast autofluorescence of the S/M phase (6 h after the onset of the dark period) cells. Cells were observed without fixation with a 100× objective lens. Scale bars = 5 μm

applied to arrest synchronized cells in the M phase (Nishida et al. 2005, 2007). However, the effect does not last for a long time, and a halt of pump aeration is required when oryzalin is added to the culture (Fujiwara et al. 2013). Thus, MG132 is more suitable than oryzalin to arrest cells in the M phase.

## 3.3 Microscopic Observation

Here we briefly describe how we observe *C. merolae* cells by ordinary optical microscopy. To examine shapes of the cell and the chloroplast, we generally use phase-contrast (Miyagishima et al. 2003) or differential interference contrast microscopy (Miyagishima et al. 2012) using 40–100× objective lenses (Fig. 3.3). To clearly visualize the chloroplast, red fluorescence by green or UV excitation is useful (Terui et al. 1995; Miyagishima et al. 1999b, 2001) (Fig. 3.3). Cells are either not fixed or fixed with ~1% glutaraldehyde (added to the culture directly); however, without fixation, the cells will be crushed by the slides and cover glasses within a short period. Centrifugation at 800 × *g* is conducted when the cells have to be concentrated before observation, following which the cell pellet is resuspended with a small volume of the medium. We often add 0.01–0.001% Tween-20 to the culture before centrifugation, which prevents cells from adhering to the walls of centrifugation tubes.

## 3.4 Molecular, Genetic, and Biochemical Analyses

*C. merolae* cell lacks a rigid cell wall in contrast to its relatives, *Cyanidium* spp. and *Galdieria* spp. Therefore, there is no need to break the cells vigorously (with glass beads, ultrasonic treatment, etc.) to extract the cellular contents such as proteins, DNA, and RNA. In most cases, procedures applied for mammalian cultured cell lines are also applicable to *C. merolae*. In addition, there are no specific requirements for PCR or reverse transcription PCR because *C. merolae* nuclear genome has a moderate GC content (55%) (Matsuzaki et al. 2004; Nozaki et al. 2007). Because the *C. merolae* genome contains few introns (0.5% of the total nucleus-encoded genes) (Matsuzaki et al. 2004; Nozaki et al. 2007), genomic DNA is used to clone cDNA instead of reverse transcribing the RNA.

To extract proteins, DNA, or RNA, 0.01–0.001% of Tween-20 is often added to the culture before centrifugation to prevent cells from adhering to the walls of centrifugation tubes as above. We usually isolate DNA by ordinary phenol-chloroform extraction. For RNA extraction, TRIzol (Thermo Fisher Scientific) is often used. For ordinary PCR amplification of the genomic DNA, we usually add a small aliquot of the culture directly into the reaction mix instead of using purified DNA.

For SDS-PAGE and subsequent immunoblotting of the cellular whole proteins, the cell pellet is directly resuspended in SDS-PAGE sample buffer (50 mM Tris-HCl, pH 6.8, 6% 2-mercaptoethanol, 2% SDS, and 10% glycerol). The protein content can be determined with XL-Bradford (APRO Science) even in the presence of SDS and 2-mercaptoethanol or dithiothreitol. When the sample becomes too viscous to handle because of high concentration of DNA, samples are treated by sonication or heat (e.g., 96 °C for 10 min), which reduces the viscosity. To detect the proteins using commercially available antibodies, we have expressed proteins tagged with sfGFP (Sumiya et al. 2014), 3× FLAG (Imamura et al. 2013), or 3× HA (Ohnuma et al. 2008; Fujiwara et al. 2013; Sumiya et al. 2016) in *C. merolae* by transformation. Anti-GFP (JL-8, mouse monoclonal, Clontech), anti-FLAG (M2, mouse monoclonal, Sigma-Aldrich), and anti-HA (HA7, mouse monoclonal, Sigma-Aldrich) have yielded specific detection of the tagged proteins in *C. merolae* total proteins by immunoblotting. The anti-HA antibody has also been used for the detection of proteins by immunofluorescence microscopy (Fujiwara et al. 2013; Sumiya et al. 2016).

To extract native proteins without denaturation, the cell pellet is rapidly frozen in liquid nitrogen and resuspended in adequate buffer. This freezing and thawing step is sufficient to extract soluble proteins. Alternatively, a fresh cell pellet is resuspended in a buffer, following which sonication is applied to break the cells. We usually supplement the buffer with Complete Mini protease inhibitor mixture (Roche) to avoid protein degradation (Imamura et al. 2009; Miyagishima et al. 2014). When required, phosphatase inhibitor mixture (25 mM NaF, 2.5 mM $Na_3VO_4$, 5 mM $Na_4P_2O_7$, and 5 mM beta-glycerophosphate) is also added to the buffer (Miyagishima et al. 2014). After extraction of proteins with or without adequate detergents, the cellular lysate is ultracentrifuged to remove insoluble materials. The supernatant fraction is subjected to further analyses. Alternatively, a specific organelle can be isolated before protein extraction using methods for organelle isolation described elsewhere (Miyagishima et al. 1999a; Yagisawa et al. 2009) (Chap. 4).

**Acknowledgments** We thank Dr. Kuroiwa (Japan Women's University), Drs. Tanaka and Imamura (Tokyo Institute of Technology), Dr. Misumi (Yamaguchi University), Dr. Yoshikawa (Tokyo University of Agriculture), and their lab members and members of Miyagishima lab in National Institute of Genetics for providing information on *C. merolae* experiments. Our study was partly supported by Japan Society for the Promotion of Science Grant-in-Aid for Scientific Research 25251039 (to S.M.) and by the Core Research for Evolutional Science and Technology Program of the Japan Science and Technology Agency (to S.M.).

## References


Allen MB (1959) Studies with *Cyanidium caldarium*, an anomalously pigmented chlorophyte. Arch Mikrobiol 32:270–277

Fujiwara T, Tanaka K et al (2013) Spatiotemporal dynamics of condensins I and II: evolutionary insights from the primitive red alga *Cyanidioschyzon merolae*. Mol Biol Cell 24:2515–2527

Fujiwara T, Kanesaki Y et al (2015) A nitrogen source-dependent inducible and repressible gene expression system in the red alga *Cyanidioschyzon merolae*. Front Plant Sci 6:657

Imamura S, Kanesaki Y et al (2009) R2R3-type MYB transcription factor, CmMYB1, is a central nitrogen assimilation regulator in *Cyanidioschyzon merolae*. Proc Natl Acad Sci U S A 106:12548–12553

Imamura S, Terashita M et al (2010) Nitrate assimilatory genes and their transcriptional regulation in a unicellular red alga *Cyanidioschyzon merolae*: genetic evidence for nitrite reduction by a sulfite reductase-like enzyme. Plant Cell Physiol 51:707–717

Imamura S, Ishiwata A et al (2013) Expression of budding yeast FKBP12 confers rapamycin susceptibility to the unicellular red alga *Cyanidioschyzon merolae*. Biochem Biophys Res Commun 439:264–269

Itoh R, Takahashi H et al (1996) Aphidicolin uncouples the chloroplast division cycle from the mitotic cycle in the unicellular red alga *Cyanidioschyzon merolae*. Eur J Cell Biol 71:303–310

Matsuzaki M, Misumi O et al (2004) Genome sequence of the ultrasmall unicellular red alga *Cyanidioschyzon merolae* 10D. Nature 428:653–657

Minoda A, Sakagami R et al (2004) Improvement of culture conditions and evidence for nuclear transformation by homologous recombination in a red alga, *Cyanidioschyzon merolae* 10D. Plant Cell Physiol 45:667–671

Miyagishima SY, Itoh R et al (1999a) Isolation of dividing chloroplasts with intact plastid-dividing rings from a synchronous culture of the unicellular red alga *Cyanidioschyzon merolae*. Planta 209:371–375

Miyagishima SY, Itoh R et al (1999b) Real-time analyses of chloroplast and mitochondrial division and differences in the behavior of their dividing rings during contraction. Planta 207:343–353

Miyagishima S, Takahara M et al (2001) Plastid division is driven by a complex mechanism that involves differential transition of the bacterial and eukaryotic division rings. Plant Cell 13:2257–2268

Miyagishima SY, Nishida K et al (2003) A plant-specific dynamin-related protein forms a ring at the chloroplast division site. Plant Cell 15:655–665

Miyagishima SY, Suzuki K et al (2012) Expression of the nucleus-encoded chloroplast division genes and proteins regulated by the algal cell cycle. Mol Biol Evol 29:2957–2970

Miyagishima SY, Fujiwara T et al (2014) Translation-independent circadian control of the cell cycle in a unicellular photosynthetic eukaryote. Nat Commun 5:3807

Nishida K, Yagisawa F et al (2005) Cell cycle-regulated, microtubule-independent organelle division in *Cyanidioschyzon merolae*. Mol Biol Cell 16:2493–2502

Nishida K, Yagisawa F et al (2007) WD40 protein Mda1 is purified with Dnm1 and forms a dividing ring for mitochondria before Dnm1 in *Cyanidioschyzon merolae*. Proc Natl Acad Sci U S A 104:4736–4741

Nozaki H, Takano H et al (2007) A 100%-complete sequence reveals unusually simple genomic features in the hot-spring red alga *Cyanidioschyzon merolae*. BMC Biol 5:28

Ohnuma M, Yokoyama T et al (2008) Polyethylene glycol (PEG)-mediated transient gene expression in a red alga, *Cyanidioschyzon merolae* 10D. Plant Cell Physiol 49:117–120

Ohnuma M, Kuroiwa T et al (2011) Optimization of cryopreservation conditions for the unicellular red alga *Cyanidioschyzon merolae*. J Gen Apple Microbiol 57:137–143

Sumiya N, Fujiwara T et al (2014) Development of a heat-shock inducible gene expression system in the red alga *Cyanidioschyzon merolae*. PLoS One 9:e111261

Sumiya N, Fujiwara T et al (2016) Chloroplast division checkpoint in eukaryotic algae. Proc Natl Acad Sci U S A 113:E7629–E7638

Suzuki K, Ehara T et al (1994) Behavior of mitochondria, chloroplasts and their nuclei during the mitotic cycle in the ultramicroalga *Cyanidioschyzon merolae*. Eur J Cell Biol 63:280–288

Terui S, Suzuki K et al (1995) Synchronization of chloroplast division in the ultramicroalga *Cyanidioschyzon merolae* (Rhodophyta) by treatment with light and aphidicolin. J Phycol 31:958–961

Toda K, Takahashi H, Itoh R, Kuroiwa T (1995) DNA contents of cell nuclei in two Cyanidiophyceae: *Cyanidioschyzon merolae* and *Cyanidium caldarium* Forma A. Cytologia (Tokyo) 60:183–188

Yagisawa F, Nishida K et al (2009) Identification of novel proteins in isolated polyphosphate vacuoles in the primitive red alga *Cyanidioschyzon merolae*. Plant J 60:882–893

# Chapter 4
# Isolation of Dividing Organelles from *Cyanidioschyzon merolae* Cells

**Yamato Yoshida, Yuuta Imoto, and Fumi Yagisawa**

**Abstract** Cells of the red alga *Cyanidioschyzon merolae* contain similar types of organelles as other model eukaryotic organisms (see Chap. 2). However, *C. merolae* cells are distinct in that they contain fewer organelles, with only one nucleus, mitochondrion, chloroplast, peroxisome, Golgi, and endoplasmic reticulum (ER) per cell, and only a few lysosomes or vacuoles. *C. merolae* cells lack a rigid cell wall, and the plasma membrane is thick and elastic. These biological features facilitate the isolation of distinct types of organelles from *C. merolae* cells. Standardized protocols have been established to isolate organelles with bilayer membranes (e.g., chloroplasts and mitochondria) and single-layer membranes (e.g., peroxisomes and lysosomes). The isolated organelles can be used to address biological questions by performing in vitro experiments and biochemical analyses of purified target molecules. Together with imaging and genome information analyses, the resulting data will provide insights into key components involved in fundamental cellular mechanisms. In this chapter, we provide detailed protocols for the isolation of specific organelles from *C. merolae* cells.

**Keywords** Chloroplast • Lysosome • Mitochondria • Organelle isolation • Peroxisome

---

Y. Yoshida (✉)
Laboratory for Single Cell Gene Dynamics, Quantitative Biology Center, RIKEN, 6-2-3 Furuedai, Suita, Osaka 565-0874, Japan
e-mail: yamato.yoshida@riken.jp

Y. Imoto
Division of Organelle Homeostasis, Medical Institute of Bioregulation, Kyushu University, 3-1-1 Maidashi, Higashiku, Fukuoka 812-0053, Japan

F. Yagisawa
Center for Research Advancement and Collaboration, University of the Ryukyus, 1 Senbaru, Nishihara-cho, Nakagami-gun, Okinawa 903-0213, Japan

 
T. Kuroiwa et al. (eds.), *Cyanidioschyzon merolae*,
https://doi.org/10.1007/978-981-10-6101-1_4

## 4.1 Introduction

Photosynthetic eukaryotes contain the following organelles: nuclei, mitochondria, chloroplasts (plastids), peroxisomes, lysosomes, Golgi, endoplasmic reticulum (ER), and vacuoles. Thus, organelle studies represent important strategies to elucidate fundamental mechanisms and processes in eukaryotic cells. Organelle isolation techniques must provide undamaged samples for ultrastructural studies, in vitro biochemical analyses, genomics, and proteomics. *C. merolae* cells offer several significant advantages for organelle isolation: (1) the organelles have simple spherical shapes (see Chap. 2); (2) the cells can be disrupted by osmotic bursting or glass-bead homogenization, because they lack cell walls, thereby facilitating the isolation of intact organelles; and (3) the cells do not require frequent subcultivation to maintain cell division (usually once per 4–8 weeks), which substantially reduces labor, time, cost, and required space in rotary-shaking incubators.

In this chapter, we introduce protocols for isolating chloroplasts, mitochondria, peroxisomes, and lysosomes from *C. merolae* cells. The chloroplast isolation protocol was first established in *C. merolae* cells by Dr. Miyagishima (Miyagishima et al. 1999). The cultivated *C. merolae* cells were synchronized before isolating intact dividing chloroplasts. A series of studies using these isolated chloroplasts revealed the plastid-dividing (PD) ring ultrastructure and identified the highly conserved chloroplast division proteins FtsZ, dynamin, and PDR1 (see Chap. 14). Subsequent work developed the protocol for isolating dividing mitochondria from *C. merolae* cells (Yoshida et al. 2009). Proteomic analyses of division machinery fractions purified from dividing mitochondria and chloroplasts have revealed specific protein profiles and mechanisms associated with organellar division machinery.

Establishing protocols for isolating chloroplasts and mitochondria facilitated the development of protocols to isolate other organelles from *C. merolae*. A novel isolation method was developed for lysosomes, which were isolated for proteomics analyses (Yagisawa et al. 2009). This protocol used interphase cells, because lysosomes are bound to mitochondria during mitosis. The lysosomal fraction was free of biomarkers for other organelles except for biomarkers of the ER, which were present at low levels. Also, recent work established a protocol for isolating dividing peroxisomes. Cultured cells were synchronized and treated with the microtubule polymerization inhibitor oryzalin; then, intact dividing peroxisomes were obtained as a complex with mitochondria and chloroplasts (Imoto et al. 2013). Analyses of isolated peroxisomes identified a macroprotein complex called the peroxisome-dividing (POD) machinery (see Chap. 15).

Protocols for isolating other organelles such as nuclei and Golgi have not been established. Further research will determine if the existing organelle isolation protocols can be modified for isolating nuclei and Golgi. These studies of *C. merolae* organelles are establishing valuable tools and platforms that will elucidate fundamental knowledge on the conserved mechanisms underlying organellar biogenesis and the origin of eukaryotic cells.

## 4.2 Chloroplast Isolation

### *4.2.1 Materials*

#### 4.2.1.1 Cell Culture and Synchronization

Culture *Cyanidioschyzon merolae* strain 10D (Matsuzaki et al. 2004) cells in flasks with rotary shaking (120 rpm) under continuous light (40 W/m$^2$) at 42 °C using a constant temperature incubator shaker Bioshaker (Taitec, Aichi, Japan). Synchronize the cells as follows: subculture the cells to $<1 \times 10^7$ cells/mL in a flat flask while bubbling air into the culture medium using a tube connected to a pump. Pass the air through a sterile filter and position the opening of the long tube in the bottom of the culture. Then, incubate the cells under a 12-h light/12-h dark cycle at 42 °C using an automatic light/dark cycle CM incubator (Fujimotorika, Tokyo, Japan) (Suzuki et al. 1994) (See Chap. 3).

Prepare the following reagents in deionized water unless otherwise noted. More detailed information on cell synchronization is presented in Chap. 3:

- Allen's medium, 2×
- Glass tank covered with one-way mirrors on three sides
- Flat flask, 1 L
- Natural white fluorescent lamps (four)
- Electronic timer
- Airline tubing (4 mm inside diameter and 5 mm outside diameter)
- PFA tube (4 mm inside diameter, 6 mm outside diameter, and 40 cm length)
- Silicone plug (43 mm top diameter and 39 mm bottom diameter)
- Syringe filter unit (Millex-FG50; Merck Millipore, Massachusetts, USA) for venting filtration
- Aquarium air pump
- Immersion-type temperature controller (Thermominder; Taitec, Aichi, Japan)

#### 4.2.1.2 Isolation of Chloroplasts and Mitochondria

- Hypotonic and isotonic isolation medium (20 mM HEPES (pH 7.6), 5 mM $MgCl_2$, 5 mM KCl, 5 mM EGTA) containing 180 mM sucrose (for hypotonic isolation medium) or 300 mM sucrose (for isotonic isolation medium) and 3 mg/mL of complete EDTA-free Protease Inhibitor Cocktail (Roche); stored at 4 °C. Mix the protease inhibitor cocktail with isolation medium on the day of chloroplast isolation.
- Isotonic isolation medium containing 80% (v/v) Percoll (GE Healthcare Life Science); stored at 4 °C.
- DNaseI.
- French pressure cell press (SLM-Aminco, Rochester, NY).
- Centrifuge and fixed-angle rotor for 500 mL centrifuge bottles.

- 500 mL centrifuge bottles.
- Peristaltic pump (ATTO, Tokyo, Japan).
- Silicon tube (1 mm diameter and 1 m length).
- Glass tube (1 mm diameter and 5 cm length).
- Ultracentrifuge (Himax CP-WX; Hitachi Koki, Tokyo, Japan) and swinging-bucket rotor (P28S; Hitachi Koki, Tokyo, Japan).
- Disposable ultracentrifuge tube, 40 mL.

#### 4.2.1.3 Isolation of Division Machinery from Chloroplasts and Mitochondria

- *C. merolae* cells synchronized in early or late M phase
- Hypotonic isolation medium
- Hypotonic isolation medium containing 40% (v/v) Percoll
- Nonidet P-40 (Sigma-Aldrich)
- *n*-Octyl-β-D-glucoside (Dojindo, Kumamoto, Japan)
- Phosphate-buffered saline (PBS) pH 7.4, 1×

#### 4.2.1.4 Immunofluorescence Microscopy

- Stock solution for cell fixation [10% (w/v) paraformaldehyde, methanol, 1 mM NaOH, 10% (w/v) dimethyl sulfoxide (DMSO), in methanol]; stored at −20 °C
- Ice-cold 100% methanol; stored at −20 °C
- Permeabilization buffer [0.05% Triton X-100 (Bio-Rad) and PBS pH 7.4]
- Primary antibody
- Fluorescently labeled secondary antibody, such as Alexa Fluor 350, 488, and 555 (highly cross-adsorbed) (Molecular Probes, Thermo Fisher Scientific)
- Immunoreaction enhancer solution (Can Get Signal Immunostain Solution B; Toyobo, Osaka, Japan)
- Blocking buffer (Blocking One; Nacalai Tesque, Kyoto, Japan)

### *4.2.2 Methods*

#### 4.2.2.1 Isolation of Dividing Chloroplasts and Mitochondria from *C. merolae* Cells

1. Culture the *C. merolae* cells in 200 mL in flasks (500 mL capacity) with rotary shaking under continuous light (40 $W/m^2$) at 42 °C for 4–8 weeks.
2. Synchronize the cells as follows: subculture the cells to $<1 \times 10^7$ cells/mL with fresh medium in 700 mL in a flat flask (1 L capacity), and then incubate

the cells under a 12-h light/12-h dark cycle at 42 °C using an automatic light/dark cycle CM incubator.

3. Harvest the synchronized cells at the second dark period (early M phase) or within 3 h of starting the second dark period (late M phase) for isolating the plastid-dividing (PD) and mitochondrial-dividing (MD) machinery (see Note 1).
4. Centrifuge the synchronized cells at 750 × *g* for 7 min at room temperature, and wash the resulting pellet with 40 mL of prewarmed (42 °C) hypotonic isolation medium.
5. Centrifuge the washed cells at 750 × *g* for 5 min at room temperature, resuspend the resulting pellet in 20 mL of ice-cold hypotonic isolation medium, and then incubate the suspension for 20 min on ice.
6. Lyse the cells in a French pressure cell press at 1750 psi (see Note 2), add 100 μg/mL of DNase I, and incubate the lysate for 45 min on ice (see Note 3).
7. Layer a 10 mL aliquot of cell lysate on top of a three-step Percoll gradient prepared in 40 mL centrifuge tubes (10 mL of 80%, 10 mL of 60%, and 10 mL of 40% Percoll, respectively, dissolved in isotonic isolation medium) (see Note 4).
8. Centrifuge the Percoll gradients at 28,000 × *g* for 50 min at 4 °C in a swinging-bucket rotor (P28S; Hitachi Koki, Tokyo, Japan); intact chloroplasts resolve as a band at the 60–80% Percoll interface (see Note 5).
9. Carefully aspirate the fraction containing intact chloroplasts using a Pasteur pipette.
10. Remove the Percoll from the chloroplast fraction by mixing the fraction with 2.5 × volume isotonic isolation medium and centrifuging the fraction at 700 × *g* for 30 min at 4 °C.
11. Resuspend the isolated chloroplast pellet in isotonic isolation medium at a concentration of 1 mg total protein/mL; store at 4 °C until use (Fig. 4.1).

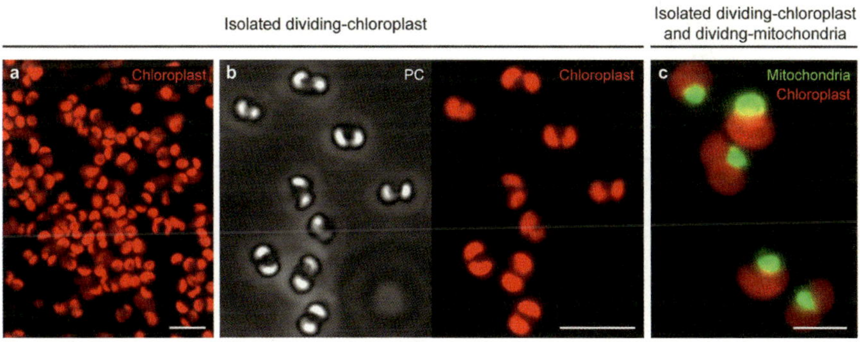


**Fig. 4.1** Isolated dividing chloroplasts and dividing mitochondria from *Cyanidioschyzon merolae* cells. (**a** and **b**) Phase contrast (PC) and fluorescence images of isolated dividing chloroplasts. Scale bars, 5 μm. (**c**) An immunofluorescence image of isolated dividing chloroplasts and dividing mitochondria. Scale bar, 2 μm. Chloroplasts emit red autofluorescence. Mitochondria were visualized by anti-mitochondrial EF-tu (*green*) (Part **c** is reproduced, with permission, from Yoshida et al. 2009)

#### 4.2.2.2 Isolation of Plastid/Mitochondrion-Dividing Machinery

1. Lyse and incubate isolated dividing chloroplasts in two volumes of hypotonic isolation medium containing 3 mg/mL of complete EDTA-free Protease Inhibitor Cocktail (Roche) and 0.2% Nonidet P-40 (NP-40); incubate the samples on ice for 30 min.
2. Centrifuge the samples at 13,200 × *g* for 20 min at 4 °C in a swinging-bucket rotor; the PD machinery with outer membrane and some thylakoid membrane with phycobilisomes are concentrated as an NP-40-insoluble fraction at the 40% Percoll interface.
3. Carefully aspirate the fraction containing the PD machinery using a Pasteur pipette.
4. Wash the PD machinery fraction once with hypotonic isolation medium, and then centrifuge this fraction at 7700 × *g* for 10 min at 4 °C.
5. Resuspend and lyse the pellet containing PD machinery in 100 mM *n*-Octyl-ß-D-glucoside (OG) in PBS on ice for 5–20 min.
6. Dissolve the chloroplast membranes in lysis buffer; the PD machinery will remain intact and this sample should be used for experiments within 30 min (Fig. 4.2).

#### 4.2.2.3 Sample Fixation and Permeabilization for Immunofluorescence Analysis

1. Prepare 1 mL of paraformaldehyde/methanol fixation solution by mixing 50 μL of stock solution with 950 μL of ice-cold methanol; gently vortex and store the solution at −20 °C until use.
2. Centrifuge cells or isolated chloroplasts at 100 × *g* for 3–5 min at room temperature and remove the supernatant.
3. Add 1 mL of ice-cold fixation solution to the pellet, immediately resuspend the pellet, and incubate for 5 min at −20 °C.
4. Centrifuge the sample at 750 × *g* for 3 min; gently aspirate supernatant using a pipette.
5. Resuspend the pellet in 1 mL of 0.05% Triton X-100 in PBS; incubate on ice for 5 min.
6. Centrifuge the sample at 750 × *g* for 3 min at 4 °C; gently aspirate the supernatant using a pipette.
7. To reduce background or non-specific staining, incubate the sample with 100 μL of blocking reagent (Blocking One; Nacalai Tesque).

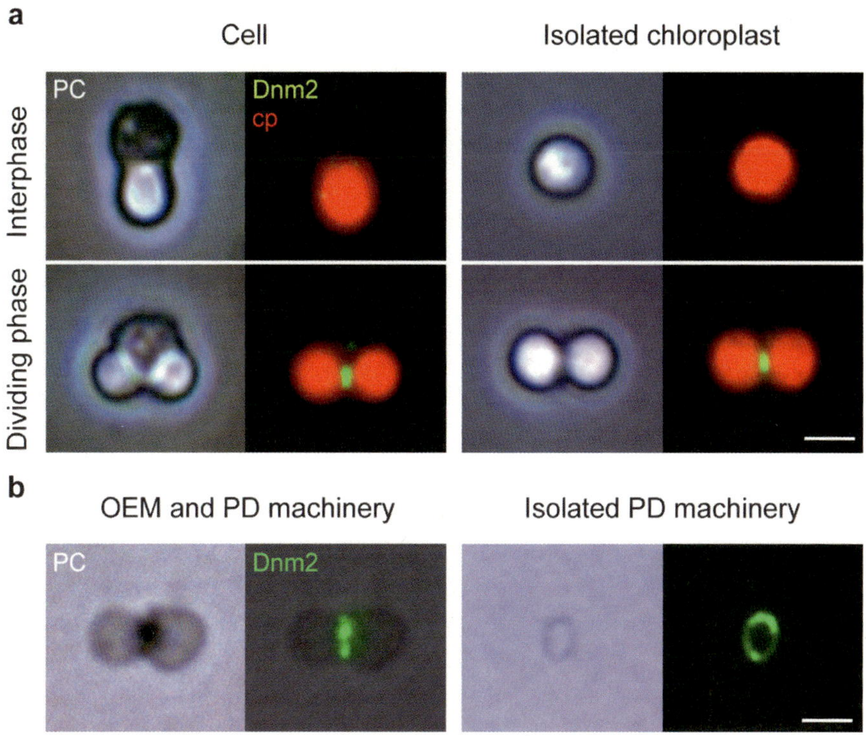


**Fig. 4.2** Isolation of chloroplasts and PD machinery from *Cyanidioschyzon merolae* cells. (**a**) Phase contrast (PC) and immunofluorescence images of the cells and the isolated chloroplasts in interphase and the M phase. Yellow fluorescence signals indicate protein localization of a chloroplast division protein Dynamin 2 (Dnm2) (see Chap. 14 for more detail). Scale bar, 1 μm. (**b**) Phase contrast (PC) and immunofluorescence images of isolated PD machinery. After the Nonidet P-40 treatment, stromal components are excluded from the chloroplasts and outer envelope membrane (OEM) and the PD machinery are retained (*left*). Additional treatment using 100 mM *n*-Octyl-β-D-glucoside (OG) dissolves OEM from the fraction, and the intact PD machinery is isolated (see Chap. 14). Scale bar, 1 μm (Images are reproduced, with permission, from Yoshida et al. 2010)

8. Centrifuge the sample at 750 × *g* for 3 min; carefully remove the supernatant; resuspend the pellet in 100 μL of diluted primary antibody in PBS and incubate at 4 °C for 1.5 h.
9. Centrifuge the sample at 750 × *g* for 3 min; gently aspirate the supernatant; resuspend the pellet in 100 μL of diluted secondary antibody in PBS and incubate at 4 °C for 1.5 h.
10. Centrifuge the sample at 750 × *g* for 3 min at 4 °C; gently aspirate the supernatant; and rinse the pellet with 100 μL of ice-cold PBS. The samples can be used directly for fluorescence microscopy or stored at 4 °C for one week (Fig. 4.2).

## 4.3 Lysosome Isolation

### 4.3.1 *Materials*

#### 4.3.1.1 Isolation of Lysosomes from Interphase *C. merolae* Cells

- *C. merolae* cells.
- Apparatus for cell synchronization (see Sect. 4.2.1).
- Complete Protease Inhibitor (Roche Diagnostics).
- Dithiothreitol (DTT).
- DNase I.
- OptiPrep (Sigma-Aldrich).
- Hypotonic isolation buffer [20 mM Tris-HCl pH 7.6, 5 mM EGTA, 5 mM KCl, 2 mM $MgCl_2$, 6% (w/v) sucrose, complete protease inhibitor (Roche Diagnostics), 2 mM DTT, and 30 μg/mL DNase I]; stored at 4 °C.
- Isotonic isolation buffer containing sucrose [20 mM Tris-HCl pH 7.6, 5 mM EGTA, 5 mM KCl, 2 mM $MgCl_2$, 10% (w/v) sucrose, complete protease inhibitor (Roche), 2 mM DTT, 30 μg/mL DNase I, and 10% (w/v) sucrose]; stored at 4 °C.
- Iodixanol 90% (w/v); completely evaporate the water from 60% (w/v) iodixanol (OptiPrep) at 70 °C; resuspend the remnant powder in distilled water; store at 4 °C.
- Iodixanol 62% and 80% (w/v); prepare these reagents by diluting the 90% (w/v) iodixanol solution; these reagents contain the same composition as isolation buffer except for the sucrose concentration, which is 116 and 65 mM sucrose in 62 and 80% iodixanol, respectively; DNase I can be omitted from these reagents.
- Corn starch 20% (w/v); wash the corn starch with 4% SDS (w/v) containing 4 mM DTT at room temperature for 15 min to remove proteins; collect starch at ~300 × *g*, extensively wash out SDS and DTT with water; replace the water with ethanol (EtOH) and store at room temperature; before use, wash the starch with water and then with hypotonic isolation buffer.
- Dounce homogenizer, 15 mL volume (Wheaton).
- Centrifuge and fixed-angle rotor for 500 mL centrifuge bottles.
- Centrifuge bottles, 500 mL.
- Centrifuge with swing rotor for 50 mL tubes.
- Centrifuge tubes, 50 mL.
- Ultracentrifuge and swinging-bucket rotor for 5 mL tubes.
- Ultracentrifuge tubes, 5 mL.

#### 4.3.1.2 Evaluate the Quality of Lysosomal Fractions by Immunoblot Analysis

- Antibodies for organellar marker proteins
- Apparatus and reagents for immunoblot analysis

#### 4.3.1.3 Microscopy of Lysosomal Fractions (Polyphosphate Staining)

- DAPI
- Cacodylate buffer (pH 7.4) containing 10% (w/v) sucrose
- Glutaraldehyde 1% (w/v)
- Fluorescence microscope with a wide band-pass filter for DAPI visualization

### *4.3.2 Methods*

#### 4.3.2.1 Isolation of Lysosomes from Interphase *C. merolae* Cells

1. Measure out 800 mL of synchronized cell culture.
2. Spin down interphase cells (3-h dark/12-h light/2-h dark cycle) at 1500 × *g*.
3. Suspend the pellet in 100 mL of Hypotonic isolation buffer warmed to 30 °C (see Note 6).
4. Incubate the sample for 1 h at 30 °C; confirm that the cells are swollen by visual inspection.
5. Pellet the cells at 1500 × *g*.
6. Suspend the pellet in 1 or 2 mL of hypotonic isolation buffer; then, transfer to a Dounce homogenizer.
7. Add hypotonic isolation buffer and prepared corn starch so that the homogenizer contains 20% (w/v) starch and cells in 10 mL of hypotonic isolation buffer.
8. Place the homogenizer and sample on ice, and wait for ~15 min until the sample and apparatus cool down.
9. Homogenize the cells with approximately 180 strokes; ensure that a tight-fitting stick is used.
10. Transfer the homogenate to a 50 mL tube; wash the homogenizer with 2 mL of hypotonic isolation buffer, and combine it with the homogenate.
11. Centrifuge the homogenate at 300 × *g* for 1 min at 4 °C (see Note 7); transfer the supernatant to another tube.
12. Resuspend the pellet in 12 mL of hypotonic isolation buffer; centrifuge again as before.
13. Combine the new supernatant with that of step 11, and then divide it into two aliquots and transfer to 50 mL centrifuge tubes.
14. Centrifuge the supernatants at 700 × *g* for 8 min at 4 °C; transfer the supernatants to new tubes.
15. Resuspend the pellet with 10 mL of hypotonic isolation buffer; centrifuge again as before.
16. Combine the new supernatant with that of step 14; divide the supernatant into two aliquots and transfer to new 50 mL tubes.
17. Layer 200 μL of 80% (w/v) iodixanol beneath the supernatant layer.
18. Centrifuge at 4700 × *g* for 15 min at 4 °C.

19. Discard the supernatant above the interphase between isolation buffer and 80% iodixanol (i.e., leave the interphase).
20. Add ~200 μL of isotonic isolation buffer and suspend the sample; adjust the volume to 600 μL with isotonic isolation buffer.
21. Prepare two 5 mL ultracentrifuge tubes as follows: layer the 600 μL sample onto a 5 mL ultracentrifuge tube containing 1-mL steps of 62 and 80% iodixanol (see Note 8).
22. Layer 1.4 mL of isotonic isolation buffer on top of the solution in the ultracentrifuge tube.
23. Centrifuge the sample at 50,000 × *g* for 1 h at 4 °C in a swing rotor.
24. Remove the lysosomes, which concentrate around the interface of the 62–80% iodixanol layers (see Note 9).

#### 4.3.2.2 Test for the Quality of Lysosomal Fractions by Immunoblots

1. Punch a hole with a needle in the bottom of the ultracentrifuge tube, and collect two to three 150 μL fractions around the interface of the 62–80% iodixanol layers.
2. Mix approximately 1 μL of each fraction with regular SDS sample buffer (see Note 10).
3. Load these samples on a polyacrylamide gel of suitable percentage; perform immunoblot analysis using antibodies against markers for lysosomes and other organelles to check the purity of lysosomal fractions (see Notes 11, 12, and 13).

#### 4.3.2.3 Microscopy of Lysosomal Fractions (Polyphosphate Staining)

1. Suspend the lysosomal fractions in three volumes of 20 mM cacodylate buffer (pH 7.4) containing 10% (w/v) sucrose.
2. Centrifuge the suspensions for 10 min at 4000 x *g* at 4 °C; suspend the pellet in 20 mM cacodylate buffer.
3. Mix the sample with an equal volume of 1% (w/v) glutaraldehyde, and allow it to stand for 1 min at room temperature.
4. Centrifuge the sample for 10 min at 4000 × *g*; suspend the pellet in distilled water.
5. Centrifuge the sample for 10 min at 4000 × *g*.
6. Suspend the pellet in distilled water containing 3 μg/mL DAPI; incubate for 30 min at 37 °C.
7. Visualize the lysosomal fractions using a fluorescence microscope with a wide band-pass DAPI filter; lysosomes retaining polyphosphate will fluoresce yellow-green (525 nm) under UV excitation.

## 4.4 Peroxisome Isolation

### *4.4.1 Materials*

#### 4.4.1.1 Isolation of Peroxisomes

- *C. merolae* cells.
- Regular set for synchronization of *C. merolae* cultures as shown in Sect. 4.2.1.2.
- Oryzalin (Dr. Ehrenstorfer standards).
- Isolation medium (20 mM HEPES pH 7.6, 5 mM $MgCl_2$, 50 mM KCl, 2 mM EGTA, 1 mM DTT) containing 180 mM sucrose (for hypotonic isolation medium) or 300 mM sucrose (for isotonic isolation medium) and 3 mg/mL of Complete EDTA-free Protease Inhibitor Cocktail (Roche); stored at 4 °C.
- Isotonic isolation medium containing Percoll [80% (v/v) Percoll (GE Healthcare Life Science), 20 mM HEPES pH 7.6, 5 mM $MgCl_2$, 50 mM KCl, 2 mM EGTA, 1 mM DTT, and 300 mM sucrose]; stored at 4 °C.
- OptiPrepTM [60% (w/v) iodixanol] (COSMO BIO).
- DNaseI.
- Triton X-100.
- French pressure cell press (SLM-Aminco, Rochester, NY).
- Centrifuge and fixed-angle rotor for 500 mL centrifuge bottles.
- Centrifuge bottles, 500 mL.
- Silicon tube (1 mm diameter, 1 m length).
- Ultracentrifuge swinging-bucket rotor and near-vertical bucket rotor.
- Disposable ultracentrifuge tubes, 40 mL.
- Capped disposable ultracentrifuge tube, 4.9 mL (OptiSeal, Beckman).

#### 4.4.1.2 Isolation of Peroxisome Dividing Machinery

- Can Get Signal Solution B (TOYOBO)
- *n*-Octyl-β-D-glucoside (OG) (Dojindo, Kumamoto, Japan)
- *n*-Dodecyl-β-D-maltoside (DDM) (Sigma-Aldrich)

#### 4.4.1.3 Immunofluorescence Microscopy

- Stock solution of fixation buffer [10% (w/v) paraformaldehyde, 1 mM NaOH, 10% (w/v) dimethyl sulfoxide (DMSO), in methanol]; stored at −20 °C
- Ice-cold 100% methanol; stored at −20 °C
- Primary antibodies
- Fluorescently labeled secondary antibodies such as Alexa Fluor 350, 488, and 555 (highly cross-adsorbed) (Molecular Probes)

- Immunoreaction enhancer solution (Can Get Signal Immunostain Solution B; Toyobo)
- Blocking buffer (Blocking One, Nacalai Tesque)

## 4.4.2 Methods

### 4.4.2.1 Isolation of Dividing Peroxisomes and Mitochondria from *C. merolae* Cells

1. Culture the *C. merolae* 10D strain in flasks with shaking under continuous light (40 W/m$^2$) at 42 °C.
2. Synchronize the cells by subculturing to $<1 \times 10^7$ cells/mL in a flat flask; then, incubate the cells under a 12-h light/12-h dark cycle at 42 °C using an automatic light/dark cycle CM incubator.
3. Add 1/2500 volume of 100 mM oryzalin stock solution to the *C. merolae* cultures at 8 h after the start of the second light period; harvest the cells 12 h later (see Note 14).
4. Collect the synchronized cells by centrifugation at 750 × *g* for 5 min, at room temperature; resuspend the cells in 50 mL of prewarmed (42 °C) hypotonic isolation medium containing 40 μM oryzalin, and then incubate the cells at room temperature for 30 min.
5. Centrifuge the treated cells at 750 × *g* for 5 min, at room temperature; resuspend the cells in 25 mL of ice-cold hypotonic isolation medium containing complete protease inhibitor, 1 mM DTT, and 30 mM KCl (see Note 15).
6. Lyse the cells in a French pressure cell press at 700 psi; add 100 μg/mL of DNase I and 2% Triton X-100; incubate the lysate for 45 min on ice.
7. Layer 12 mL of the cell lysate on top of three-step Percoll gradients prepared in 40 mL ultracentrifuge tubes [10 mL of 60%, 10 mL of 40%, and 4 mL of 20% (v/v) Percoll dissolved in isotonic isolation medium].
8. Centrifuge the lysate gradients in a swinging-bucket rotor at 17,600 × g for 60 min, at 4 °C; the organelle-membrane complex including peroxisomes, chloroplasts, and mitochondria concentrate as a band between the 40–60% Percoll layers (see Note 16).
9. Carefully aspirate the fraction containing the organelle-membrane complex using a Pasteur pipette.
10. Remove Percoll from the fraction by mixing with 2.5 volumes of isotonic isolation medium; centrifuge the samples at 700 × *g* for 30 min, at 4 °C.
11. Resuspend the organelle-membrane complex pellet in 4.5 mL of sucrose-free isolation medium containing 500 mM OG; mix by rotation for 2 h at 4 °C (see Note 17).
12. Mix 2.18 mL of the OG-treated organelle-membrane complex fraction with 2.42 mL of 50% (w/v) iodixanol; transfer this sample into an OptiSeal ultra-centrifugation 4.9 ml tube (Beckman Coulter) (see Note 18).

13. Centrifuge the sample in a near-vertical bucket rotor at 200,000 × *g* for 2 h, at 4 °C; punch a hole in the bottom of the OptiSeal tube, and collect 0.5 mL fractions.
14. Dividing peroxisomes concentrate in the fractions with iodixanol concentrations ranging from 16.0–17.0% (see Note 19).

#### 4.4.2.2 Immunolabeling

1. Wash isolated dividing peroxisomes in sucrose-free isolation medium twice.
2. Resuspend the pellet in Can Get Signal Solution B containing primary antibody, 1% BSA, 30 mM KCl, and 1 mM DTT; incubate overnight at 4 °C (see Note 20).
3. Wash the sample twice with Can Get Signal Solution B.
4. Resuspend the sample in 100 µL of Can Get Signal Solution B containing diluted secondary antibody for 1.5 h at 4 °C (for immunofluorescence microscopy).

#### 4.4.2.3 Isolation of POD Machinery

1. Wash the isolated immunolabeled dividing peroxisomes with sucrose-free isolation medium twice.
2. Lyse the isolated dividing peroxisomes in Can Get Signal Solution B containing 100 mM DDM; incubate the lysate for 20 min on ice.
3. Add equal volumes of 10 mM SDS and 20 mM urea at room temperature for 1 min.
4. Observe the sample using fluorescence microscopy, or store the sample at 4 °C for later use (within the week).

## 4.5 Notes

1. Cells in late M phase are usually used for the isolation of PD machinery (Yoshida et al. 2010). The PD machinery is physically linked with the MD machinery in early M phase cells, which enables the simultaneous isolation of dividing mitochondria containing MD machinery and dividing chloroplasts containing PD machinery. These are isolated as a complex structure from early M phase cells (Yoshida et al. 2009).
2. Organelles such as chloroplasts, mitochondria, and nuclei remain intact after cell disruption in a French pressure cell press.
3. During the 45-min incubation of the lysate and DNase I, prepare two centrifuge tubes containing the three-step Percoll gradient. Layering one Percoll solution generally requires approximately 5–7 min; therefore, allow approximately 30–42 min to prepare two tubes containing a three-step Percoll gradient.
4. The three-step Percoll gradient can be prepared using a peristaltic pump, but the lysed sample must be manually layered on top of the Percoll gradient. If the

lysate volume does not fill the ultracentrifuge tube, isotonic isolation medium can be used to adjust the sample volume in the ultracentrifuge tube. All operations should be performed in a cold room.

5. Percoll solutions generate density gradients during ultracentrifugation, and more than three layers may be observed after centrifugation. Each layer contains the following organelles (from top to bottom): cytosolic components and soluble proteins from disrupted organelles, cell membranes and nuclei, disrupted mitochondria and peroxisomes, intact chloroplasts, intact cells, and Percoll pellet.
6. Complete protease inhibitor, DTT, and DNaseI are not required at this step.
7. Use a swinging rotor for steps 11–18.
8. Prepare iodixanol step gradients while incubating cells in step 3, and leave the gradients on ice.
9. A total of 50–100 μg of protein is obtained in lysosome fractions.
10. We used ECL (Pierce) for detection. The amount of sample to load should be adjusted depending on the antibodies and detection system.
11. We used the following antibodies: vacuolar type $H^+$-ATPase large A subunit (V-ATPase A) antibody as a lysosomal marker (Yagisawa et al. 2009); plasma membrane $H^+$-ATPase (P-ATPase) antibody for cell membrane (Yagisawa et al. 2009); calnexin antibody for ER, Sed5 antibody for Golgi bodies (Yagisawa et al. 2009); EF-Tu antibody (Nishida et al. 2004) for mitochondria; HupA antibody (Kobayashi et al. 2002) for chloroplasts; anti-pumpkin catalase antibody (Ohnuma et al. 2009) for peroxisomes; and anti-human histone H3 antibody (Santa Cruz Biotechnology) for cell nuclei.
12. The V-ATPase A marker is detected in the fractions around the 62–80% iodixanol interface, which also contains low levels of calnexin marker. The V-ATPase A marker also is detected in fractions above the 62% iodixanol layer along with other organelle markers; this region of the gradient contains cell debris.
13. No protein purification is necessary for this sample. If a larger sample volume (~5 μg protein) is loaded on the gel, a trichloroacetic acid (TCA) precipitation should be performed to remove polyphosphate, which interferes with the electrophoresis.
14. Chloroplasts and mitochondria divide 12 h after the oryzalin treatment, but peroxisome division is delayed. This enables collection of cells containing dividing peroxisomes. Oryzalin readily precipitates under foaming conditions; therefore, cell cultures should be transferred to flasks and cultivated with shaking under continual dark conditions.
15. Oxidation and ionic strength affect the stability of highly oligomerized Dnm1, a major component of the POD machinery. Therefore, each buffer used in this experiment should contain DTT and should be adjusted to 50 mM KCl.
16. The organelle fraction appears as a thin green-blue layer in the gradients.
17. OG treatment disrupts the attachment between the peroxisomal membrane and mitochondria/chloroplasts. To avoid aggregation of the membrane complex, the sample should be rotated during the OG treatment.

18. Before capping the OptiSeal ultracentrifuge tube, ensure that the tube does not contain air bubbles so that it does not undergo high-pressure deformation during ultracentrifugation.
19. The isolated dividing peroxisomal fraction is invisible in the gradient; therefore, visually check fractions using a light microscope. The peroxisomal fraction generally concentrates in the 16.0–17.0% iodixanol layer, which occurs approximately 2.5–3.5 mL from the bottom of the tube. However, this location changes depending on experimental conditions. Therefore, the iodixanol concentration should be calculated by measuring the refractive index in each fraction.
20. Oxidation and ionic strength affect the stability of Dnm1 (see Note 15). Ensure that Can Get Signal Solution B contains 1 mM DTT and 30 mM KCl.

**Acknowledgments** This work was supported by a Human Frontier Science Program Long Term Fellowship (LT000356/2011-L to Y.Y.) and grants from the Japan Society for the Promotion of Science Fellowship (no. 14 J04556 to Y.I, 11044 to F.Y.).

## References

Imoto Y, Kuroiwa H, Yoshida Y et al (2013) Single-membrane-bounded peroxisome division revealed by isolation of dynamin-based machinery. Proc Natl Acad Sci U S A 110:9583–9588

Kobayashi T, Takahara M, Miyagishima SY et al (2002) Detection and localization of a chloroplast-encoded HU-like protein that organizes chloroplast nucleoids. Plant Cell 14:1579–1589

Matsuzaki M, Misumi O, Shin-I T et al (2004) Genome sequence of the ultrasmall unicellular red alga *Cyanidioschyzon merola*e 10D. Nature 428:653–657

Miyagishima SY, Itoh R, Aita S et al (1999) Isolation of dividing chloroplasts with intact plastid-dividing rings from a synchronous culture of the unicellular red alga *Cyanidioschyzon merolae*. Planta 209:371–375

Nishida K, Misumi O, Yagisawa F et al (2004) Triple Immunofluorescent labeling of FtsZ, Dynamin, and EF-Tu reveals a loose association between the inner and outer membrane mitochondrial division machinery in the red alga *Cyanidioschyzon merolae*. J Histochem Cytochem 52:843–849

Ohnuma M, Misumi O, Fujiwara T et al (2009) Transient gene suppression in a red alga, *Cyanidioschyzon merolae* 10D. Protoplasma 236:107–112

Suzuki K, Ehara T, Osafune T et al (1994) Behavior of mitochondria, chloroplasts and their nuclei during the mitotic cycle in the ultramicroalga *Cyanidioschyzon merolae*. Eur J Cell Biol 63:280–288

Yagisawa F, Nishida K, Yoshida M et al (2009) Identification of novel proteins in isolated polyphosphate vacuoles in the primitive red alga *Cyanidioschyzon merolae*. Plant J 60:882–893

Yoshida Y, Kuroiwa H, Hirooka S et al (2009) The bacterial ZapA-like protein ZED is required for mitochondrial division. Curr Biol 19:1491–1497

Yoshida Y, Kuroiwa H, Misumi O et al (2010) Chloroplasts divide by contraction of a bundle of nanofilaments consisting of polyglucan. Science 329:949–953

# Part III
# Genomic and Post-genomic Analyses of *Cyanidioschyzon merolae*

**Shin-ya MiyagishimaTsuneyoshi Kuroiwa**

Unicellular algae are potentially useful in biological studies because many unicellular algae provide relatively homogeneous populations (cell types and surrounding environment) and are relatively easy to cultivate at low cost and with a short generation time. Among unicellular algae, the red alga *C. merolae* has become a promising model organism for the following reasons: (1) The cell possesses a nucleus, one mitochondrion, one chloroplast, and a minimal set of single-membrane-bound organelles. (2) These organelles divide in a cell cycle-dependent manner, and the cell/organelle division cycle is synchronized by a 12-h light/12-h dark cycle. (3) The cell lacks a rigid cell wall; therefore, it is relatively easy to extract or isolate cellular contents. (4) The nuclear and organelle (i.e., the mitochondrial and chloroplast) genomes have been completely sequenced. The nuclear genome is extremely compact with little genetic redundancy (16.5 Mb; 4,775 protein-coding genes). Recently, procedures for genetic manipulation, such as transient and stable expression of transgenes and nuclear targeting by homologous recombination, have been developed. These features have facilitated genetic, transcriptomic, and proteomic analyses regarding a variety of cellular processes, such as organelle division and inheritance, circadian rhythms, and metabolism.

Chapter 5 summarizes the features of the mitochondrial, chloroplast, and nuclear genomes. In addition, the structure of centromeres is also described. Chapter 6 introduces examples of -omics analyses, such as transcriptome studies by oligonucleotide DNA microarrays and RNA-seq and proteomic analyses by TOF-MS/MS. Examples of transcriptomic analyses include those for changes in the transcriptome during synchronous culture and cellular adaptation to nitrogen-depleted conditions. Examples of proteomic analyses include those for the isolation of chloroplasts and the organelle division machinery (ring). Chapter 7 summarizes experimental procedures for genetic manipulation in detail for readers who will start studies using *C. merolae*.

# Chapter 5
# The 100%-Complete Nuclear and Organellar Genome Sequences of the Ultrasmall Red Algal Species *Cyanidioschyzon merolae* 10D

**Hisayoshi Nozaki, Yu Kanesaki, Motomichi Matsuzaki, and Shunsuke Hirooka**

**Abstract** We were the first group to successfully sequence the 100%-complete entire eukaryotic genome in 2007, mainly using automated Sanger sequencing of the unicellular, ultrasmall red algal species *Cyanidioschyzon merolae* 10D. This world record was principally based on the ultrasmall size of the *C. merolae* genome (ca. 16 megabase pairs) as well as on three excellent previous studies: the 100%-complete mitochondrial genome in 1998, the 100%-complete plastid genome in 2003, and the first algal cell nuclear genome in 2004. The 100%-complete nuclear sequences demonstrated that this ultrasmall red alga contains unusually simple sets of genes and genetic sequences. For example, because introns are lacking in almost all of the protein-coding nuclear genes of *C. merolae*, the 100%-complete sequence can be used to directly deduce the sequences of all *C. merolae* proteins, which will be extremely valuable in further proteomics research. Thus, this small red alga represents an ideal model organism for studying the fundamental relationships among the plastid, mitochondrial, and nuclear genomes. The 100%-complete nuclear genome sequence has greatly improved the precision and value of biological analyses of *C. merolae*.

H. Nozaki (✉)
Department of Biological Sciences, Graduate School of Science, University of Tokyo, Hongo 7-3-1, Bunkyo-ku, Tokyo 113-0033, Japan
e-mail: nozaki@bs.s.u-tokyo.ac.jp

Y. Kanesaki
Research Institute of Green Science and Technology, Shizuoka University, 836 Ohya, Suruga-ku, Shizuoka 422-8529, Japan

M. Matsuzaki
Department of Parasitology, National Institute of Infectious Diseases, Shinjuku-ku, Tokyo 162-8640, Japan

S. Hirooka
Department of Cell Genetics, National Institute of Genetics, 1111 Yata, Mishima, Shizuoka 411-8540, Japan

Core Research for Evolutional Science and Technology Program, Japan Science and Technology Agency, Saitama 332-0012, Japan

 
T. Kuroiwa et al. (eds.), *Cyanidioschyzon merolae*,
https://doi.org/10.1007/978-981-10-6101-1_5

**Keywords** Algal genome • *Cyanidioschyzon merolae* • Eukaryotic cell • Mitochondrial genome • Nuclear genome • 100%-complete genome • Plastid genome • Red alga

## 5.1 Introduction

The eukaryotic cell has three types of nuclei that contain nuclear, mitochondrial, and plastid genomes (Kuroiwa 1982). Thus, biological attributes and/or functions in each species of photosynthetic eukaryotes are fundamentally based on consortia of these three genomic information sources, except for some nucleomorph-containing secondary phototrophs such as cryptophytes and chlorarachniophytes (Curtis et al. 2012). Therefore, complete determination of the information from all of the three genomes in a single eukaryotic species was considered tantamount to revealing the complete molecular blueprint of the eukaryote. Thanks to the skillful and patient work of Japanese researchers, the first complete plastid genome was successfully determined through manual DNA sequencing based on the radiolabeling method (Ohyama et al. 1986; Shinozaki et al. 1986) and the first cyanobacterial genome through the automated Sanger DNA sequencing method (Kaneko et al. 1996).

Even in the current era of next-generation sequencing, the 100%-complete sequencing of all three eukaryotic genomes may seem as crazy an endeavor as "eating a bicycle," which was once recorded in the Guinness Book of World Records (although this disturbing record has since been prohibited from nomination to the book). However, we indeed succeeded in the first 100%-complete eukaryotic genome in 2007, using automated Sanger sequencing of the unicellular red algal species *Cyanidioschyzon merolae* 10D (Nozaki et al. 2007). This world record was principally based on the ultrasmall genome size (ca. 16 megabase pairs) of *C. merolae*, as well as on three excellent previous studies: the 100%-complete mitochondrial genome sequence (Ohta et al. 1998), the 100%-complete plastid genome sequence (Ohta et al. 2003), and the first algal cell nuclear genome sequence (Matsuzaki et al. 2004).

Although our original nuclear genomic sequence of *C. merolae* 10D revealed some unique features such as very few introns, only three copies of ribosomal DNA (rDNA), and a relatively small number of total genes (Matsuzaki et al. 2004), the exact features of certain important repeated elements such as histone gene clusters and telomeres remained uncertain. Due to the functional significance of these elements, it was extremely desirable to complete the sequence and resolve all ambiguities to summarize all of the unique genomic features of *C. merolae*. Thus, we painstakingly elucidated all of the nucleotide sequences remaining as gaps and telomere-lacking chromosomal ends, to construct the first 100%-complete nuclear genome sequence (Nozaki et al. 2007).

We believed that the establishment of the first 100%-complete eukaryotic genome would be of great widespread interest to the biological sciences as a whole and that demonstration of the simplest set of genomic features from the hot spring red alga *C. merolae* would represent an important advancement in the fields of genomics and evolutionary biology. More than 10 years have passed since we determined the 100%-complete genome sequences of the mitochondrial, plastid, and nuclear genomes in *C. merolae* (Ohta et al. 1998, 2003; Nozaki et al. 2007). Recent advancements in next-generation sequencing methods have contributed much to the establishment of the organellar and nuclear whole genomes of numerous eukaryotes (e.g., Hanschen et al. 2016; Session et al. 2016). However, no other 100%-complete eukaryotic genomes have been determined. Thus, our studies (Ohta et al. 1998, 2003; Nozaki et al. 2007) remain the first and only achievement of a 100%-complete eukaryotic genome sequence.

## 5.2 Mitochondrial Genome

The 100%-complete nucleotide sequence of the mitochondrial genome of *C. merolae* 10D is on circular DNA containing 34 or 35 genes encoding proteins (including unidentified open reading frames [ORFs]) and 28 RNA genes (3 rRNAs and 25 tRNAs) (Table 5.1, Fig. 5.1.; Ohta et al. 1998; Yang et al. 2015). The genes are encoded on both strands. This is the largest number of protein-coding genes within the current 37 sequenced red algal mitochondrial genomes (Yang et al. 2015). The G + C content of the *C. merolae* mitochondrial genome is 27.2%. The genome size is 32,211 base pairs (bp), exceeding the average size (26,866 bp) of the 37 red algal mitochondrial genomes (Yang et al. 2015). The genome size and number of protein-coding genes of another cyanidialean species, *Galdieria sulphuraria*, are 21,428 bp and 19 genes, respectively. Thus, the large size and gene richness of the *C. merolae* mitochondrial genome do not represent Cyanidiales-specific characteristics.

## 5.3 Plastid Genome

The 100%-complete nucleotide sequence of the plastid genome of *C. merolae* 10D is on circular DNA composed of 149,987 bp lacking inverted repeats (Table 5.1, Fig. 5.2). The G + C content is 37.6% (Ohta et al. 2003). This organelle genome contains 243 genes on both strands and consists of 207 protein-coding genes (including unidentified ORFs) and 36 RNA genes (a ribonuclease P RNA component, 3 rRNAs, 31 tRNAs, and tmRNA). Approximately 40% of the protein-coding genes overlap, and none of the genes in this plastid genome have introns (Ohta et al. 2003).

Recently, the plastid genomes of various red algal species have been sequenced (Janouškovec et al. 2013; Tajima et al. 2014). The *C. merolae* plastid genome is similar to other red algal plastid genomes in that it is gene rich: 223–250 unique genes are found in plastid genomes of red algae (Janouškovec et al. 2013). The

**Table 5.1** Key attributes of the 22 chromosomes constituting the 100%-complete three genomes of the ultrasmall red alga *Cyanidioschyzon merolae* 10D

| Genome/ chromosome | No. of nucleotides (bp) | Shape of chromosome | No. of protein-coding genes |
|---|---|---|---|
| Cell nucleus[a] | | | |
| 1 | 422,616 | Linear | 102 |
| 2 | 457,013 | Linear | 125 |
| 3 | 481,791 | Linear | 144 |
| 4 | 513,455 | Linear | 140 |
| 5 | 528,682 | Linear | 161 |
| 6 | 536,163 | Linear | 131 |
| 7 | 584,452 | Linear | 173 |
| 8 | 739,753 | Linear | 213 |
| 9 | 810,151 | Linear | 231 |
| 10 | 839,707 | Linear | 247 |
| 11 | 852,849 | Linear | 236 |
| 12 | 859,119 | Linear | 258 |
| 13 | 866,983 | Linear | 249 |
| 14 | 852,727 | Linear | 256 |
| 15 | 902,900 | Linear | 265 |
| 16 | 908,485 | Linear | 261 |
| 17 | 1,232,258 | Linear | 355 |
| 18 | 1,253,087 | Linear | 360 |
| 19 | 1,282,939 | Linear | 384 |
| 20 | 1,621,617 | Linear | 484 |
| Total | 16,546,747 | | 4775 |
| Unassigned | 0 | | 0 |
| Plastid[b] | 149,987 | Circular | 208 |
| Mitochondrion[c] | 32,211 | Circular | 34 |
| Total of three genomes | 16,728,945 | | 5017 |

[a]Nozaki et al. (2007)
[b]Ohta et al. (2003)
[c]Ohta et al. (1998)

striking feature of the *C. merolae* genome within the red algae is the high degree of gene compaction: it has the smallest genome size and the shortest intergenic region (Janouškovec et al. 2013). The median intergenic distance of the *C. merolae* plastid genome is extremely small (only 10 bp), whereas the intergenic distances of other red algal plastid genomes (including that of another cyanidialean species, *Cyanidium caldarium*) range from 61 to 85 bp (Janouškovec et al. 2013). Thus, the extraordinary compaction of the plastid genome of *C. merolae* has likely resulted from species-specific factors following the divergence of *Cyanidioschyzon*

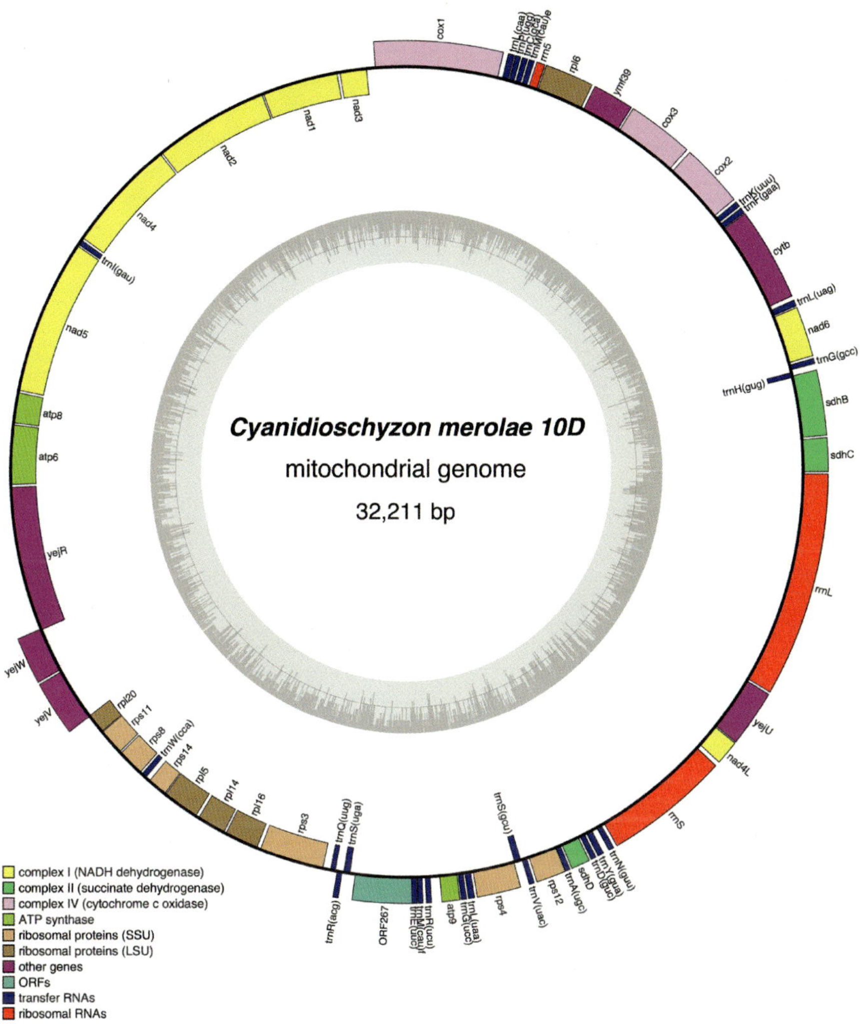


**Fig. 5.1** Genetic map of the *C. merolae* mitochondrial genome. Note the *C. merolae* mtDNA is a circular-mapping molecule. The map was drawn based on the 100%-complete sequence (NC_000887; Ohta et al. 1998) using OrganellarGenomeDRAW http://ogdraw.mpimp-golm.mpg.de/index.shtml

from other cyanidialean algae, such as *Cyanidium* (Janouškovec et al. 2013). The plastid genome of *C. merolae* encodes several genes that are rarely present in other plastid genomes (Ohta et al. 2003). Recently, 16 cyanobacterial genes were resolved to be present in the plastid genomes of only two Cyanidiales (*C. merolae* and *Cyanidium caldarium*) out of all known red algal plastid genomes (Janouškovec et al. 2013). Because plastid gene loss (endosymbiotic gene transfer) is slow in red algal plastid genomes (Nozaki et al. 2003; Janouškovec et al. 2013)

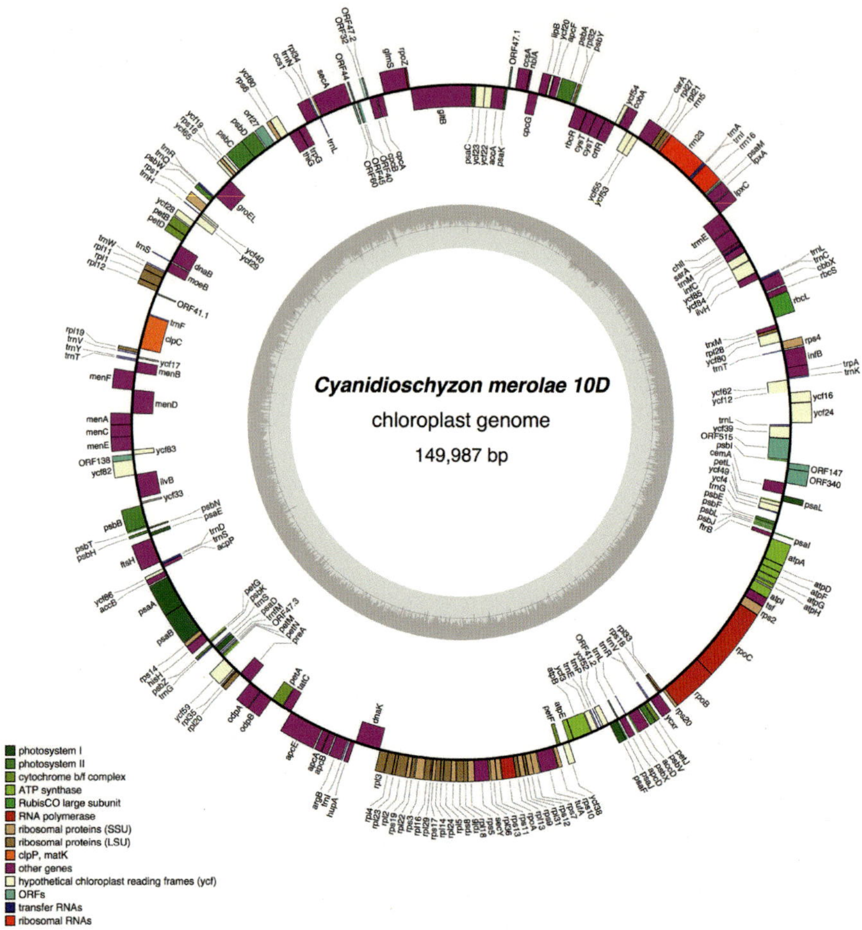


**Fig. 5.2** Genetic map of the *C. merolae* chloroplast genome. The map was drawn based on the 100%-complete sequence (NC_004799; Ohta et al. 2003) using OrganellarGenomeDRAW http://ogdraw.mpimp-golm.mpg.de/index.shtml

and these two cyanidialean species are distantly related to other red algae with published plastid genomes, the 16 genes may be ancestral cyanobacterial genes that were lost in the latter red algae.

## 5.4 Nuclear Genome

To construct a 100%-complete nuclear genome sequence of *C. merolae* 10D, unresolved gaps between contigs and undetermined chromosome ends were filled using our previously constructed *C. merolae* bacterial artificial chromosome (BAC) clones (Matsuzaki et al. 2004). Polymerase chain reactions (PCRs) of BAC clones containing these gaps were performed using specific primers complementary to sequences flanking the gaps (Nozaki et al. 2007). We used DNA walking annealing control primer technology in order to directly amplify unknown sequences adjacent to known sequences within a contig. PCR products were sequenced by using the cycle sequencing methodology, except for a single gap with extremely high G + C content on chromosome 10, which was filled by an in vitro transcription sequencing reaction (Nozaki et al. 2007). Chromosomal ends were sequenced by the inverse PCR method, polyC-tailing and the anchor primer method, and the asymmetric PCR method (Nozaki et al. 2007). To completely determine the sequences of the histone cluster area in this red algal genome, we performed *Not*I and *Apa*I subcloning of the BAC clone GESZ2-b20, which includes possible histone clusters on chromosome 14 (Matsuzaki et al. 2004). We performed Southern blot analysis with histone-related probes, restriction enzyme analysis, and end sequencing of the subclones to reveal the relative positions of the subclones on this chromosome. We completely filled six gaps between contigs/fragments in the area of histone cluster (Matsuzaki et al. 2004) using primer walking of the subclones (Nozaki et al. 2007). Our complete nuclear genome sequence of *C. merolae* consisting of 16,546,747 nucleotides covers 100% of the 20 linear chromosomes from telomere to telomere (Table 5.1, Fig. 5.3). These 20 unambiguous DNA molecules represent the simple and unique structures of eukaryotic chromosomes (Nozaki et al. 2007): a histone gene cluster of the smallest known size, all chromosomal ends with a unique telomeric repeat, and an extremely low number of transposons. Based on these genomic features and others that were discovered previously (Matsuzaki et al. 2004), *C. merolae* appears to contain the simplest nuclear genome of the nonsymbiotic species of eukaryotes. These unusually simple genomic features found in this 100%-complete genome sequence were considered very useful for further biological studies of cells of eukaryotes.

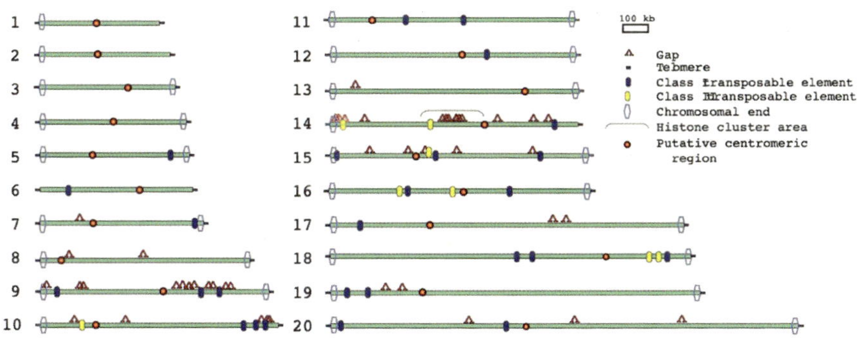


**Fig. 5.3** Bird's-eye view of the 100%-complete structures of 20 chromosomes of *C. merolae* showing regions that were filled by Nozaki et al. (2007; "gap" and "chromosomal end"), the histone cluster area, telomere repeats, putative centromeric regions, and transposable elements ("class I" and "class II") (From Nozaki et al. (2007))

## 5.5 Centromere Regions and Their Dynamics

The centromere is a conserved chromosomal site responsible for attachment of spindles and accurate segregation of chromosomes during mitosis and meiosis. Each chromosome in a eukaryotic cell has a unique centromere where the kinetochore complex assembles and captures the spindle (Cleveland et al. 2003). In addition, the centromere core contains the histone H3 variant, centromere protein A (CENP-A), which replaces canonical H3 at the centromere. The 100%-complete genome sequence of *C. merolae* clarified that each chromosome also possesses a single A + T-rich region, which is predicted to be the putative centromere region (Matsuzaki et al. 2004; Maruyama et al. 2008). By ChIP-on-chip analysis using an anti-CENP-A antibody and a whole-genome tiling array, core centromere sequences were determined at the predicted region (Kanesaki et al. 2015). The identified centromeres were of the regional type, 1–3 kb in length, and with no consensus sequences or repeat elements. The expression of the CENP-A protein rapidly increased during the S phase of the cell cycle; subsequently a drastic reconstitution into two discrete foci adjacent to the spindle poles occurred at metaphase (Maruyama et al. 2007). The dynamics of condensins I and II during the cell cycle were also well analyzed (Fujiwara et al. 2013a, b). In *C. merolae*, condensin II is enriched at centromere regions and is absent along chromosome arms during metaphase of the M phase, whereas condensin I is not enriched at centromere regions. Furthermore, *C. merolae* possesses the centromere proteins CENP-A, CENP-E, and the hypothetical CENP-C, but does not possess CENP-B or CENP-T. The characteristics of the centromere/kinetochore in lower plants and algae remain largely unknown. Thus, *C. merolae* is the most well-studied model system for clarifying the detailed mechanisms of centromere dynamics in primitive eukaryotic plant cells.

## 5.6 Conclusions

A histone gene cluster of the smallest known size, all chromosomal ends with a unique telomeric repeat, and an extremely low number of transposons in *C. merolae* (Nozaki et al. 2007), as well as other simple features of the *C. merolae* nuclear genome (Matsuzaki et al. 2004; Misumi et al. 2005), are extremely distinctive and represent the simplest set of features of genomes recognized in any nonsymbiotic eukaryote yet studied (Fig. 5.4). It is generally considered that such simple genomic features are the consequences of reductive evolution of a eukaryote of ultrasmall size (Derelle et al. 2006). However, none of these features are shared by *Ostreococcus*, a similarly ultrasmall green alga, in which histone genes are distributed among 6 or more chromosomes, 39% of the genes harbor introns, and 8166 protein genes and 417 transposable elements are dispersed across the chromosomes (Derelle et al. 2006). These characteristics suggest differences in modes of

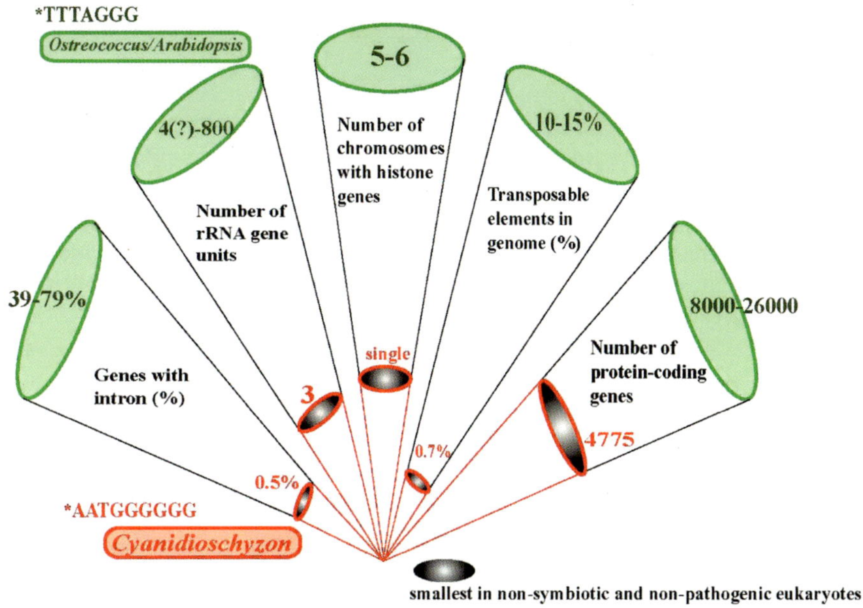


**Fig. 5.4** Comparison of the three nuclear genomes of photosynthetic eukaryotes, *Cyanidioschyzon*, *Ostreococcus* (a green alga of ultrasmall size), and *Arabidopsis* (a seed plant). Telomere repeat sequences are indicated by asterisks above the generic names (Based on Nozaki et al. (2007))

reduction of genomes between the ancestors of *Cyanidioschyzon* (red algae) and *Ostreococcus* (Chloroplastida [green plants and algae]). On the other hand, algae growing in acidic hot springs (pH 1.5, 45 °C) may be candidates for retaining ancient or ancestral attributes of plants, because throughout Earth's history, volcanic activity is thought to have provided such an extreme environment. According to Cunningham et al. (2006), *C. merolae* has perhaps the simplest chlorophyll and carotenoid assortment found in any photosynthetic eukaryote. In addition, the *C. merolae* plastid and mitochondrial genomes contain large numbers of genes, which are thought to be ancestral features, because it is generally considered that reversal of plastid gene loss is impossible (Martin et al. 2002; Nozaki et al. 2003). Thus, we hypothesized that some of the unusual or simple characteristics of the *C. merolae* genome may represent ancestral features that have been conserved in the *Cyanidioschyzon* lineage but have become modified extensively during the evolution of other lineages of plants/algae (Nozaki et al. 2007). Alternatively, the unique features of the *C. merolae* genome may reflect adaptations to their extreme environment. However, many simple features found in the *C. merolae* genome, such as the rare presence of introns and few rRNA regions in the nuclear genome, are not present in the nuclear genome sequence of another hot spring red alga, *Galdieria* (Barbier et al. 2005; Schönknecht et al. 2013). Organellar genomic features are also different between *Cyanidioschyzon* and *Galdieria*, as discussed above. Based on the

100%-complete nuclear, mitochondrial, and plastid genome sequences (Ohta et al. 1998, 2003; Matsuzaki et al. 2004; Nozaki et al. 2007), all of the major types of genetic information in eukaryotes are present in *C. merolae*. Furthermore, *C. merolae* contains unusually simple sets of genes and sequences as revealed by the 100%-complete genome. Because introns are lacking in almost all protein-coding nuclear genes of this ultrasmall red alga, the 100%-complete genome can be used to directly deduce the sequences of all red algal proteins, which will be very valuable in future research of proteomics. Thus, *C. merolae* represents an ideal model for studying the fundamental relationships among the chloroplast, mitochondrial, and nuclear genomes. The 100%-complete nuclear genome sequence (Nozaki et al. 2007) has greatly improved the precision of biological analyses of *C. merolae*, including studies of the dividing machineries of plastids (Yoshida et al. 2010) and peroxisomes (Imoto et al. 2013). The 100%-complete genome sequence also enables us to design various tools for genome-wide analyses, such as an organellar DNA microarray (Minoda et al. 2005; Kanesaki et al. 2012), nuclear DNA microarray (Fujiwara et al. 2009; Kanesaki et al. 2009; Imamura et al. 2009), and high-density whole-genome tiling array (Kanesaki et al. 2015). Furthermore, the 100%-complete genome sequence greatly contributed to the establishment of homologous recombination techniques (Minoda et al. 2004; Fujiwara et al. 2013a, b; Taki et al. 2015) and conditional gene expression systems (Sumiya et al. 2014; Fujiwara et al. 2015). The establishment of these advanced molecular biological techniques has made *C. merolae* one of the most exceptional model organisms among the eukaryotic algae.

**Acknowledgments** HN was supported by a Grant-in-Aid for Scientific Research (grant number 16H02518) from the Ministry of Education, Culture, Sports, Science and Technology (MEXT)/ Japan Society for the Promotion of Science (JSPS) KAKENHI.

## References

Barbier G, Oesterhelt C et al (2005) Comparative genomics of two closely related unicellular thermo-acidophilic red algae, *Galdieria sulphuraria* and *Cyanidioschyzon merolae*, reveals the molecular basis of the metabolic flexibility of *Galdieria sulphuraria* and significant differences in carbohydrate metabolism of both algae. Plant Physiol, 137:460–474. https://doi.org/10.1104/pp.104.051169

Cleveland DW, Mao Y et al (2003) Centromeres and kinetochores: from epigenetics to mitotic checkpoint signaling. Cell 112:407–421. https://doi.org/10.1016/S0092-8674(03)00115-6

Cunningham FX Jr, Lee H et al (2006) Carotenoid biosynthesis in the primitive red alga *Cyanidioschyzon merolae*. Eukaryot Cell 6:533–545. https://doi.org/10.1128/EC.00265-06

Curtis BA, Tanifuji G et al (2012) Algal genomes reveal evolutionary mosaicism and the fate of nucleomorphs. Nature 492:59–65. https://doi.org/10.1038/nature11681

Derelle E, Ferraz C et al (2006) Genome analysis of the smallest free-living eukaryote *Ostreococcus tauri* unveils many unique features. Proc Natl Acad Sci U S A 103:11647–11652. https://doi.org/10.1073/pnas.0604795103

Fujiwara T, Misumi O et al (2009) Periodic gene expression patterns during the highly synchronized cell nucleus and organelle division cycles in the unicellular red alga *Cyanidioschyzon merolae*. DNA Res 16:59–72. https://doi.org/10.1093/dnares/dsn032

Fujiwara T, Ohnuma M et al (2013a) Gene targeting in the red alga *Cyanidioschyzon merolae*: single- and multi-copy insertion using authentic and chimeric selection markers. PLoS One 8: e73608. https://doi.org/10.1371/journal.pone.0073608

Fujiwara T, Tanaka K et al (2013b) Spatiotemporal dynamics of condensins I and II: evolutionary insights from the primitive red alga *Cyanidioschyzon merolae*. Mol Biol Cell 24:2515–2527. https://doi.org/10.1091/mbc.E13-04-0208

Fujiwara T, Kanesaki Y et al (2015) A nitrogen source-dependent inducible and repressible gene expression system in the red alga *Cyanidioschyzon merolae*. Front Plant Sci 6:657. https://doi.org/10.3389/fpls.2015.00657

Hanschen ER, Marriage TN et al (2016) The *Gonium pectorale* genome demonstrates co-option of cell cycle regulation during the evolution of multicellularity. Nat Commun 7:11370. https://doi.org/10.1038/ncomms11370

Imamura S, Kanesaki Y et al (2009) R2R3-type MYB transcription factor, CmMYB1, is a central nitrogen assimilation regulator in *Cyanidioschyzon merolae*. Proc Natl Acad Sci U S A 106:12548–12553. https://doi.org/10.1073/pnas.0902790106

Imoto Y, Kuroiwa H et al (2013) Single-membrane-bounded peroxisome division revealed by isolation of dynamin-based machinery. Proc Natl Acad Sci U S A 110:9583–9588. https://doi.org/10.1073/pnas.1303483110

Janouškovec J, Liu S-L et al (2013) Evolution of red algal plastid genomes: ancient architectures, introns, horizontal gene transfer, and taxonomic utility of plastid markers. PLoS One 8:e59001. https://doi.org/10.1371/journal.pone.0059001

Kaneko T, Sato S et al (1996) Sequence analysis of the genome of the unicellular cyanobacterium *Synechocystis* sp. strain PCC6803. II. Sequence determination of the entire genome and assignment of potential protein-coding regions. DNA Res 3:109–136

Kanesaki Y, Kobayashi Y et al (2009) Mg-protoporphyrin IX signaling in *Cyanidioschyzon merolae*: multiple pathways may involve the retrograde signaling in plant cells. Plant Signal Behav 4:1190–1192. https://doi.org/10.4161/psb.4.12.10061

Kanesaki Y, Imamura S et al (2012) External light conditions and internal cell cycle phases coordinate accumulation of chloroplast and mitochondrial transcripts in the red alga *Cyanidioschyzon merolae*. DNA Res 19:289–303. https://doi.org/10.1093/dnares/dss013

Kanesaki Y, Imamura S et al (2015) Identification of centromere regions in chromosomes of a unicellular red alga, *Cyanidioschyzon merolae*. FEBS Lett 589:1219–1224. https://doi.org/10.1016/j.febslet.2015.04.009

Kuroiwa T (1982) Mitochondrial nuclei. Int Rev Cytol 75:1–59

Martin W, Rujan T et al (2002) Evolutionary analysis of *Arabidopsis*, cyanobacterial, and chloroplast genomes reveals plastid phylogeny and thousands of cyanobacterial genes in the nucleus. Proc Natl Acad Sci U S A 99:12246–11251. https://doi.org/10.1073/pnas.182432999

Maruyama S, Kuroiwa H et al (2007) Centromere dynamics in the primitive red alga *Cyanidioschyzon merolae*. Plant J 49:1122–1129. https://doi.org/10.1111/j.1365-313X.2006.03024.x

Maruyama S, Matsuzaki M et al (2008) Centromere structures highlighted by the 100%-complete *Cyanidioschyzon merolae* genome. Plant Signal Behav 3:140–141. https://doi.org/10.4161/psb.3.2.5066

Matsuzaki M, Misumi O et al (2004) Genome sequence of the ultrasmall unicellular red alga *Cyanidioschyzon merolae* 10D. Nature 428:653–657. https://doi.org/10.1038/nature02398

Minoda A, Sakagami R et al (2004) Improvement of culture conditions and evidence for nuclear transformation by homologous recombination in a red alga, *Cyanidioschyzon merolae* 10D. Plant Cell Physiol 45:667–671. https://doi.org/10.1093/pcp/pch087

Minoda A, Nagasawa K et al (2005) Microarray profiling of plastid gene expression in a unicellular red alga, *Cyanidioschyzon merolae*. Plant Mol Biol 59:375–385. https://doi.org/10.1007/s11103-005-0182-1

Misumi O, Matsuzaki M et al (2005) *Cyanidioschyzon merolae* genome. A tool for facilitating comparable studies on organelle biogenesis in photosynthetic eukaryotes. Plant Physiol 137:567–585. https://doi.org/10.1104/pp.104.053991

Nozaki H, Ohta N et al (2003) Phylogeny of plastids based on cladistic analysis of gene loss inferred from complete plastid genome sequences. J Mol Evol 57:377–382. https://doi.org/10.1007/s00239-003-2486-6

Nozaki H, Takano H et al (2007) A 100%-complete sequence reveals unusually simple genomic features in the hot-spring red alga *Cyanidioschyzon merola*e. BMC Biol 5:28. https://doi.org/10.1186/1741-7007-5-28

Ohta N, Sato N et al (1998) Structure and organization of the mitochondrial genome of the unicellular red alga *Cyanidioschyzon merolae* deduced from the complete nucleotide sequence. Nucleic Acids Res 26:5190–5198

Ohta N, Matsuzaki M et al (2003) Complete sequence and analysis of the plastid genome of the unicellular red alga *Cyanidioschyzon merolae*. DNA Res 10:67–77

Ohyama K, Fukuzawa H et al (1986) Chloroplast gene organization deduced from complete sequence of liverwort *Marchantia polymorph*a chloroplast DNA. Nature 322:572–574

Schönknecht G, Chen WH et al (2013) Gene transfer from bacteria and archaea facilitated evolution of an extremophilic eukaryote. Science 339:1207–1210. https://doi.org/10.1126/science.1231707

Session AM, Uno Y et al (2016) Genome evolution in the allotetraploid frog *Xenopus laevis*. Nature 538:336–343. https://doi.org/10.1038/nature19840

Shinozaki K, Ohme M et al (1986) The complete nucleotide sequence of the tobacco chloroplast genome: its gene organization and expression. EMBO J 5:2043–2049

Sumiya N, Fujiwara T et al (2014) Development of a heat-shock inducible gene expression system in the red alga *Cyanidioschyzon merolae*. PLoS One 9:e111261. https://doi.org/10.1371/journal.pone.0111261.

Tajima N, Sato S et al (2014) Analysis of the complete plastid genome of the unicellular red alga *Porphyridium purpureum*. J Plant Res 127:389–397. https://doi.org/10.1007/s10265-014-0627-1

Taki K, Sone T, Kobayashi Y, Watanabe S, Imamura S, Tanaka K (2015) Construction of a URA5.3 deletion strain of the unicellular red alga *Cyanidioschyzon merolae*: a background less host strain for transformation experiments. J Gen Appl Microbiol 61:211–214. https://doi.org/10.2323/jgam.61.211

Yang EC, Kim KM et al (2015) Highly conserved mitochondrial genomes among multicellular red algae of the Florideophyceae. Genome Biol Evol 7:2394–23406. https://doi.org/10.1093/gbe/evv147

Yoshida Y, Kuroiwa H et al (2010) Chloroplasts divide by contraction of a bundle of nanofilaments consisting of polyglucan. Science 329:949–953. https://doi.org/10.1126/science.1190791

# Chapter 6
# "Omics" Analyses of *Cyanidioschyzon merolae* Based on Complete Genome Information

**Masaki Yoshida and Takayuki Fujiwara**

**Abstract** The cells of *Cyanidioschyzon merolae* have advantageous features for omics analyses. The genomes of the cell nucleus, mitochondrion, and chloroplast have been completely sequenced in *C. merolae* and are thus ready for omics analyses utilizing complete genome information. Phase-specific microarray analysis of a synchronous culture of *C. merolae* revealed the gene expression profiles nearing the organelle division cycle and mitotic cycle. In ammonium-depleted cultures of *C. merolae*, the gene expression patterns of cells were analyzed, and the nitrate assimilation pathway was revealed. Microarray analysis for chloroplast-encoded and mitochondrion-encoded genes of *C. merolae* was also conducted. The results revealed that organelle gene expressions correlate to the light period or the progression of the cell cycle. Proteomic analyses of organelles and sub-organelles of *C. merolae* have been promoted using MALDI TOF-MS. Comparative proteomic analysis of chloroplasts between interphase and metaphase revealed the dynamic changes in protein composition. Proteins composing the dividing machineries of chloroplasts and peroxisomes were identified using isolation techniques for sub-organellar structure. Isolated vacuoles and storage glucans were also analyzed, and proteins associated with the functions of these organelles were found.

**Keywords** Gene expression • MALDI TOF-MS • Microarray • mRNA • Organelle isolation • Protein • Proteome • Transcriptome

---

M. Yoshida (✉)
Integrative Environmental Sciences, Graduate School of Life and Environmental Sciences/ Algae Biomass Energy System Development Research Center (ABES), University of Tsukuba, 401, Research Project Building, 1-1-1 Tennodai, Tsukuba, Ibaraki 305-8572, Japan
e-mail: yoshida.masaki.gb@u.tsukuba.ac.jp

T. Fujiwara
Department of Cell Genetics, National Institute of Genetics, 1111 Yata, Mishima, Shizuoka 411-8540, Japan

Department of Genetics, Graduate University for Advanced Studies (SOKENDAI), 1111 Yata, Mishima, Shizuoka 411-8540, Japan

Japan Science and Technology Agency, CREST, 4-1-8 Honcho, Kawaguchi, Saitama 332-0012, Japan



T. Kuroiwa et al. (eds.), *Cyanidioschyzon merolae*,
https://doi.org/10.1007/978-981-10-6101-1_6

## 6.1 Introduction

The genomes of the cell nucleus, mitochondrion, and chloroplast have been completely sequenced in *Cyanidioschyzon merolae* (Matsuzaki et al. 2004; Nozaki et al. 2007; Ohta et al. 1998, 2003). The extremely simple composition of the cell nuclear genome has facilitated various omics analyses such as transcriptomics and proteomics. A small total number of genes (4775 protein-encoding genes) make omics efficient and limit target genes to be analyzed in cellular processes of interest. Very few introns (27 introns in protein-encoding genes) facilitate determining coding sequences (CDSs) of proteins and identifying a functional form of proteins because alternative splicing of premature mRNA does not need to be considered. In addition, *C. merolae* is biochemically and cytologically suitable: (1) each cell can be exposed to relatively homogenous conditions in cultivation, e.g., nutrient concentrations and light intensity, because C. *merolae* is a unicellular, non-colonial organism; (2) degradable cell covering facilitates obtaining and isolating cellular contents or organelles; and (3) cell cycle synchronized culture is useful for examining temporal dynamics of gene expression, protein localization, and organelle behaviors. These features have also enhanced effectiveness of omics analyses.

The aim of transcriptome analyses and proteomic analysis is the comprehensive detection of mRNA (transcriptome) or proteins (proteome) in a specific cell state, cell phase, or intercellular organelle. Though these omics experiments differ from traditional morphological studies, detailed microscopic observations and organelle isolation techniques produced in past studies are indispensable for advancing omics experiments and interpreting their results, especially in proteomic analysis of specific cell organelles. Furthermore, controlled synchronized cultivation of *C. merolae* greatly enhances the obtaining of definitive results with high time-axis resolution of time course analysis. These conditions facilitate omics analyses using *C. merolae*.

## 6.2 Transcriptome

The transcriptome is the total mRNA in a cell, and it reveals the genes that are actively expressed at a given time. Transcriptome analyses are a powerful tool for identifying differentially expressed genes depending on cellular states and extracellular conditions. We have performed transcriptome analyses using DNA microarrays that cover 4586 nuclear protein-encoding genes (~96% of protein-encoding genes). The 50-mer oligonucleotide probes are spotted on microarray slides. The experiments using this nuclear microarray have contributed to examining gene expression during the S and M phases to produce the analytical platform for the division and inheritance of the organelles (Fujiwara et al. 2009) and identifying the central nitrogen assimilation regulator CmMYB1 (Imamura et al. 2009). In the former approach, microarray analysis of the synchronous culture of

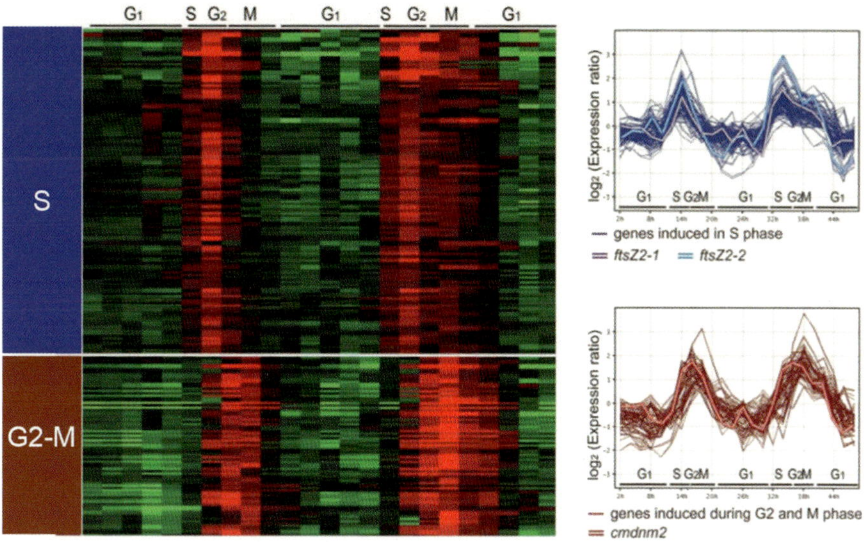


**Fig. 6.1** Heatmaps and oscillations of genes in S or G2-M groups. (Aa) The heatmap of transcription levels of genes in the S-phase group which showed specific induction in S phase. Genes correspond to the rows, and the time points form the columns. Map colors indicate induction (*red*) and suppression (*green*) of gene expression. (Ab) The oscillations of transcription levels in the S-phase group. This group contained *ftsZ2-1* and *ftsZ2-2*. (Ba) The heatmap of transcription levels of genes in the G2-M-phase group which were induced during G2 and M phase. (Bb) The oscillations of transcription levels in the G2-M-phase group. This group contained *cmdnm2* (Fujiwara et al. 2009)

cell cycle in *C. merolae* produces gene expression profiles nearing the organelle division cycle as well as the mitotic cycle. The microarray analysis of two rounds of the cell cycle has showed that 95 and 63 genes are time specifically induced during the S and G2-M phases, respectively (Fujiwara et al. 2009) (Fig. 6.1). For example, *FtsZ1*/CMS361C (mitochondrial division) and *FtsZ2*/CMS004C (chloroplast division), as well as other S-phase genes, such as proliferating cell nuclear antigen *PCNA*/CMS101C (a processivity factor for DNA polymerase δ) and *cyclin 1*/CML219C (G1/S cyclin), are induced in the S phase, and *CmDnm2*/*DRP5B*/CMN262C (chloroplast division) as well as other G2-M-phase genes, such as a condensin subunit *SMC4*/CME029C (mitotic chromosome organization and segregation) and *cyclin B*/CMI227C (M phase cyclin), are also induced in the G2-M phase. The expression of these genes is coincident with mitochondrial and chloroplast divisions that take place once a cell cycle. These microarray data are useful for identifying novel genes related to the division and inheritance of organelles. The data contributed to identifying three novel genes: *ZED*/CMQ228C (mitochondrial division, Yoshida et al. 2009), *PDR1*/CMR358C (chloroplast division, Yoshida et al. 2010), and *VIG1*/CML153C (vacuole inheritance, Fujiwara et al. 2010). Further, Imanuma et al. (2009) revealed the nitrate assimilation pathway in *C. merolae*. They analyzed the gene expression pattern of cells in ammonium-depleted culture by using the nuclear microarray to identify genes that respond to ammonium depletion and the *CmMYB1* gene that positively regulates nitrate assimilation genes.

In addition, Minoda et al. (2005) has reported a microarray for analyzing the expression of genes which are encoded in the chloroplast genome of *C. merolae*. The chloroplast microarray can analyze expression of 193 protein-encoding genes, out of a total 208 protein-encoding genes in the chloroplast genome (Nozaki et al. 2007). It has been reported that responsibility of chloroplast gene expression shifts to the light from the dark condition, and it has been suggested that the sigma factor for chloroplast RNA polymerase CmSIG2 is highly light responsive. Furthermore, Kanesaki et al. (2012) has reported a microarray for analyzing expression of genes which are encoded in the organelle (mitochondrion and chloroplast) genomes of *C. merolae*. The organelle microarrays can analyze expression of all organelle genome-encoded open reading frames. They performed microarray analysis of organelle gene expression during a 6-h light and 18-h dark cycle. They have documented organelle gene expressions which highly correlate to the light period or cell cycle progression. In addition, a strong correlation has been reported between the gene expression profiles in the mitochondrion and chloroplast, which suggests a regulatory network for co-expression of those genes during the light and dark cycle.

Currently, RNA-seq analysis has also started to be utilized. For example, Tardu et al. (2016) reported blue and red light-mediated transcriptional responses in *C. merolae* using the Illumina MiSeq sequencing platform. *C. merolae* is an ideal model for investigating the effect of specific light qualities on genome-wide transcription because this organism has no genes encoding photoreceptors other than the cryptochrome/photolyase gene family in its compact genome (Matsuzaki et al. 2004). They demonstrated that blue and red light solely regulated transcription of 683 genes (13% of total genes) and transcription of 521 genes (10%), respectively. In addition, 595 genes (12%) were commonly regulated by both blue and red light. Gene ontology (GO) term and Kyoto Encyclopedia of Genes and Genomes (KEGG) pathway enrichment analyses of differentially expressed genes revealed that specific cellular processes and pathways are upregulated depending on exposure to either blue or red light. For example, "biosynthesis of amino acids/Cme01239; KEGG pathway ID" and "aminoacyl-tRNA biosynthesis/Cme00970" are significantly upregulated by blue light, while "ribosome/Cme03010" and "carbon fixation in photosynthetic organisms/Cme00710" are upregulated by red light. In addition, transcription of three cryptochrome genes (*CmPHR2*/CML130C, *CmPHR3*/CMM076C, and *CmPHR7*/CMQ453) is regulated by blue light, and their physiological function is discussed.

## 6.3 Proteomics

Proteomic analyses of *C. merolae* have been advanced in conjunction with the development of organelle isolation techniques. Given that the proteome is the set of all expressed proteins in a whole cell or specific organelle, the target fraction should be isolated and purified in advance of proteomic analysis. As mentioned in Chap. 4, the organelle or even the sub-organellar isolation of the division machinery has become possible in *C. merolae* cells. Taking advantages of these isolation techniques, organellar and sub-organellar proteomic analyses have been advanced.

### *6.3.1 Proteomic Analysis of Chloroplast*

The chloroplast is one of the most attractive targets of analysis in *C. merolae*. The main role of chloroplasts is to conduct photosynthesis; thus, the chloroplast of *C. merolae* contains the photosynthetic pigment chlorophyll *a* and phycobiliproteins to capture the energy from light. Molecules associated with energy conversion and storage are also contained in chloroplasts. In addition, the chloroplast has its own division process, and chloroplast division-related proteins, e.g., *FtsZ* family proteins, are also incorporated especially in metaphase.

Through comparative proteomic analysis of chloroplasts between interphase and metaphase, dynamic changes in proteins during chloroplast division of *C. merolae* were revealed (Yoshida et al. 2011). From the soluble fraction of chloroplasts at both phases, a total of 355 spots (170 identical spots) were recognized in 2D polyacrylamide gel electrophoresis (Fig. 6.2). Recognized spots are quantified to estimate the amount of protein and then analyzed using matrix-assisted laser desorption/ionization time-of-flight mass spectrometry (MALDI TOF-MS). A total of 105 proteins were identified, including 18 proteins for posttranslational functions, 17 photosynthesis-related proteins, 10 carbohydrate-related proteins, 15 proteins of unknown functions, and 8 proteins predicted to be contaminates from other organelles. On the basis of spot quantity, photosynthesis-related proteins, e.g., phycobiliproteins and RuBisCO subunit proteins, were estimated to be the most dominant group (45.3% in interphase and 56.4% in metaphase). In particular, phycobiliproteins (phycocyanin alpha chain/CMV063C, beta chain/CMV064C, allophycocyanin alpha chain/CMV158C, and beta chain/CMV159C) were abundant. Phycobiliproteins compose the light-harvesting antenna complex of photosystem II called the phycobilisome, which is contained in the chloroplast of rhodophytes, glaucophytes, and cyanobacteria.

Comparison of interphase with metaphase revealed that aminomethyltransferase/CMG086C notably increased in interphase. CP24, "similar to chlorophyll a/b-binding protein/CMN235C," increased in metaphase in agreement with a previously performed microarray analysis. Both cell division protein *FtsH*/CMQ295C and plastid division protein *FtsZ*2/CMS004C increased in interphase.

The chloroplast contains its own DNA and, in *C. merolae*, contains 208 protein-coding genes (Ohta et al. 2003). As for other photosynthetic organisms, the chloroplast genome is heavily reduced after endosymbiosis and partially lost or transferred its genome during endosymbiotic the gene transfer process. As a result of the proteomic analysis of the chloroplast of *C. merolae*, chloroplast-coded

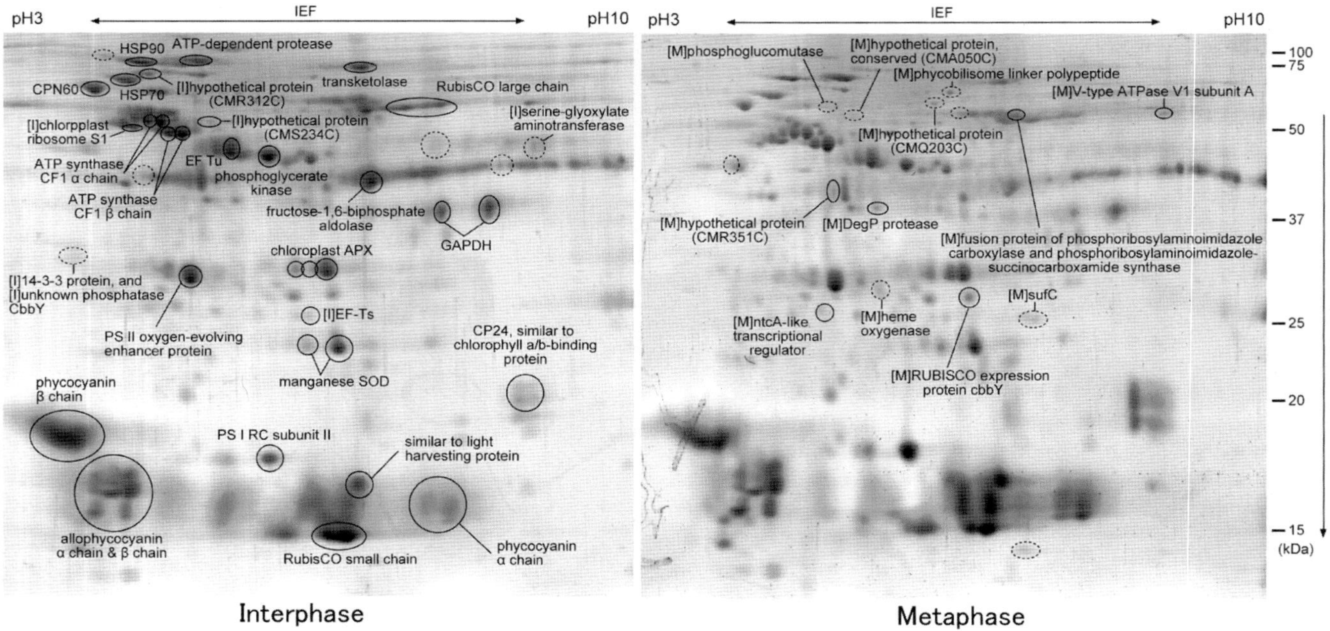


**Fig. 6.2** Overview of 2D gel-separated proteins in the soluble fraction of the chloroplast of *C. merolae* in interphase and metaphase. Proteins were separated by isoelectric focusing (horizontal axis, pH 3–10) followed by SDS-PAGE (vertical axis) and were then stained with SYPRO Ruby fluorescent dye and scanned using a Typhoon 9210. An area that was crowded with spots (pH range of 4–7) was expanded using separate gels for protein identification. *Dotted-line* circles show phase-specific spots, whereas solid-line circles show common spots. Selected names of major proteins and phase-specific proteins are shown. Phase-specific proteins are labeled with [I] (interphase-specific) or [M] (metaphase-specific). *APX* ascorbate peroxidase, *CF* coupling factor, *CPN* chaperonin, *EF* elongation factor, *RC* reaction center, *SOD* superoxide dismutase (Yoshida et al. 2011)

proteins accounted for 18.1% (19 proteins) of all detected proteins. On the basis of spot quantity, chloroplast-coded proteins accounted for 37.6% (interphase) or 43.7% (metaphase) of all detected proteins. The chloroplast of *C. merolae* substantially depends on nucleus-encoded proteins (Fig. 6.3).

The mechanisms of chloroplast division have been selected as priority phenomenon in the progress of the research of *C. merolae*, as described in previous sections. As a milestone in the research of chloroplast division of *C. merolae*, filamentous structure of plastid-dividing (PD) machinery and its related proteins were identified in 2010 (Yoshida et al. 2010). By proteomic analysis of PD machineries isolated from *C. merolae*, the glycosyltransferase protein plastid-dividing ring 1 (PDR1/CMR358C), which constructs the PD ring, was identified. Though various proteins have been found in the PD machineries by proteomic analysis with MALDI TOF-MS, PDR1 was predicted to be an essential PD ring protein because it is

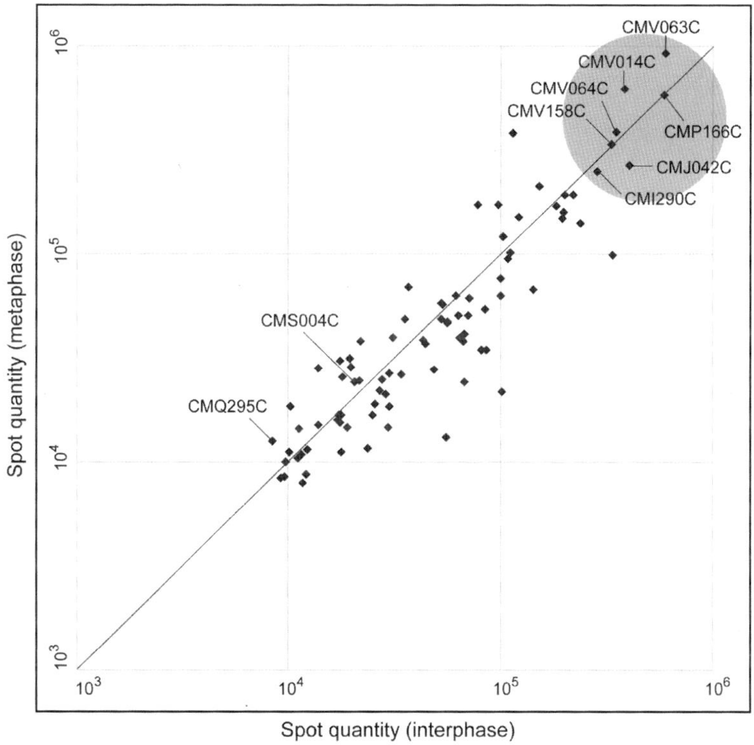


**Fig. 6.3** Scatter plot of spot quantities for interphase and metaphase. An abundant protein group (*gray circles*; CMI290C, manganese-stabilizing protein precursor; CMJ042C, GAPDH; CMP166C, phycocyanin-associated rod linker protein; CMV014C, RuBisCO small chain; CMV063C, phycocyanin (PC) alpha chain; CMV064C, phycocyanin (PC) beta chain; CMV158C and allophycocyanin (APC) alpha chain) and cell division-related proteins (CMQ295C, FtsH and CMS004C, FtsZ) are shown in particular. Note that the phase-specific proteins whose quantities fell to zero in either phase were not plotted (Yoshida et al. 2011)

part of a subgroup of glycosyltransferase proteins. It is hypothesized that PDR1 constructs polyglucan filaments of PD ring as a priming glycosyltransferase by a similar synthetic mechanism to that of glycogenin. This hypothesis was confirmed using electron microscopy and fluorescent microscopy with immunostaining technique. Moreover, after removal of proteins from isolated PD machineries, the residual filamentary fraction contained only glucose as a result of chromatography analyses. Thus, the chloroplast of *C. merolae* is divided by the PD ring, which is a bundle of PDR1-mediated polyglucan filaments, and the PDR1 protein formed a PD ring with glucose-containing carbohydrates at the chloroplast division site. Though proteomic analysis is a powerful tool to identify novel proteins in specific organelle or sub-organellar structure, the obtained information is still statistically significant "candidate." Immunochemical techniques teamed with microscopy and other visualization technique are essential to determine the in situ localization of the target protein.

### 6.3.2 Proteomic Analysis of Other Organelles

Peroxisomes are ubiquitous single-membrane-bounded organelles and fulfill essential roles in the cellular metabolism in eukaryotic cells. However, the dividing machinery involved in peroxisome division had been elusive compared with double-membrane-bounded organelles, e.g., chloroplast or mitochondrion. Imoto et al. (2013) identified the peroxisome-dividing (POD) machinery through the use of proteomic analysis of isolated dividing peroxisomes. Because the peroxisomal division completes within a very short time, it is very difficult to isolate dividing peroxisomes in non-treated cells. Since a method to collect cells containing dividing peroxisomes by oryzalin treatment had been established (Imoto et al. 2010), comparative proteomic analysis between static peroxisomes and dividing peroxisomes using *C. merolae* was enabled. Isolated peroxisomal fractions were obtained from non-treated (control) and oryzalin-treated cells and examined by MALDI TOF-MS. As a result, proteins in 10 (control) and 13 (oryzalin-treated) bands were identified. The major bands specific to oryzalin-treated cells were identified as catalase/CMI050C and Dnm1/CME019C. Catalase is one of the major peroxisomal enzymes and harnesses $H_2O_2$ to oxidize other substrates by means of the peroxidation reaction. Dnm1 is one of only two dynamins encoded in the *C. merolae* genome. These dynamins, Dnm1 and Dnm2, are both important components of the mitochondrion-dividing (MD) machinery and chloroplast-dividing machinery, respectively. This suggests the possibility that the Dnm1 molecules control peroxisomal division after mitochondrial division. Because the Dnm1 is continuously expressed throughout the cell cycle of *C. merolae*, an unknown factor that can recruit Dnm1 molecules to the division sites may play a role for the key regulator in the divisions of peroxisome and mitochondrion. Nevertheless, there were some differences between peroxisomal and mitochondrial divisions. The size of the POD machinery was smaller than those of the MD or PD machinery, and the isolated POD machinery contains a rigid filamentous ring unlike the isolated MD machinery. In addition, the width of the filamentous ring in the POD machinery remained constant during peroxisome contraction, whereas those of the MD machinery increased during contraction. Undetected proteins could still exist in POD machinery of *C. merolae*.

A method for isolating vacuoles from *C. merolae* is also established (Yagisawa et al. 2009). Vacuoles of *C. merolae* are polyphosphate-rich compartments in cytoplasm and are predicted to have functions of intracellular digestion, maintenance of ion homeostasis, and phosphate storage and supply. The protein sample of isolated vacuoles was subjected to SDS-PAGE and stained to be visible to the protein bands; a total of 58 bands were recognized and analyzed by MALDI TOF-MS. Among these bands, proteins were successfully identified from 50 of the bands, and 35 different proteins were identified. Eleven additional proteins were also detected from the gel slices without staining. The identified proteins were classified as 6 lytic enzymes, 13 transporters (including 8 subunits of V-type ATPase), 6 proteins involved in membrane fusion or vesicle trafficking, 5 non-lytic enzymes, 13 proteins of unknown function, and 3 miscellaneous proteins. The most intense band contained the endothelin-converting enzyme-1/CMP249C. In terms of the localization of identified proteins, 14 subunit proteins of V-type ATPase, vacuolar-type H + -translocating inorganic pyrophosphatase/CMO102C, sphingomyelinase/endopolyphosphatase/CMG087C, lysosomal acid phosphatase/CMT279C, zinc transporter/CMF058C, $Fe^{2+}/Mn^{2+}$ transporter/CMT466C, and Vtc1p/CMP062C were predicted to be homologues of known proteins which show vacuolar or lysosomal localizations. A further 12 proteins were homologous to proteins characteristically localized in other compartments, such as cell nucleus, endosomes, Golgi vesicles, ER, chloroplast, and mitochondrion. On the other hand, the fact that only 8 of the 12 subunits of V-type ATPase were identified indicates that the MALDI TOF-MS analysis did not cover the complete proteome of the vacuoles. An exhaustive proteomic analysis of specific organelle or subcellular compartment is of assistance to not only identification of constitutive proteins but also evaluation of the isolation and purification quality of the target compartment.

Hirabaru et al. (2010) reported the properties of the storage glucans of *C. merolae*. A protein extract of the purified glucan granules was separated by SDS-PAGE and showed three obvious bands corresponding to about 40, 47, and 90 kDa. As a result of MALDI TOF-MS analysis, the proteins of 40, 47, and 90 kDa were predicted to be phycocyanin-associated rod linker protein/CMP166C, eukaryotic translation elongation factor 1α/CMH226C, and glycogen phosphorylase/CMD184C, respectively. This result suggests that phosphorylase is a major protein among the glucan granule-bound proteins of *C. merolae*. The existence of proteins identified to be phycobiliprotein and eukaryotic translation elongation factor is possibly due to contamination by the proteins that abundantly exist in the plastids and cytosol.

### 6.3.3 Proteomics by MALDI TOF-MS

We used matrix-assisted laser desorption/ionization time-of-flight mass spectrometry (MALDI TOF-MS) device and peripherals (Fig. 6.4) for our proteomic analyses of *C. merolae*. MALDI is a relatively soft ionization technique, so it enables the analysis of large fragile biomolecules, e.g., DNA, proteins, peptides, polysaccharides, etc. The most innovative breakthrough for large fragile biomolecules ionization came in 1987 by Tanaka et al. of Shimadzu Corporation (Tanaka et al. 1988). They used the mixture of 30-nm cobalt particles and glycerol as a matrix for a 337-nm nitrogen laser ionization, and the way has been paved to analyze various biomolecules. In a conventional method, protein samples should be solubilized, purified, and separated by polyacrylamide gel electrophoresis prior to MS analysis. Proteins on the gel were stained using specific dye, e.g., Coomassie Brilliant Blue, SYPRO Ruby Protein Gel Stain, and other compatible dyes. Silver staining is also available for the preparation. Then the stained gel could be scanned using a gel scanner device suitable for the dye to recognize and quantify the protein spots. Background correction, spot outline recognition, normalizing of spot quantities, and some other image processing are applied in this process as a general step. Stained spots on the polyacrylamide gel were excised manually or automatically

**Fig. 6.4** Proteomic analysis system at Rikkyo University. MALDI TOF-MS AXIMA-TOF$^2$ (Shimadzu), high-performance gel processer Xcise (Proteome Systems/Shimadzu), and protein identification software and database MASCOT (Matrix Science)

using the automated gel processor Xcise (Fig. 6.4). The gel slices were then destained and digested with trypsin or other proteases to make peptide fragments. After the digestion, peptide fragments were extracted and desalted to be ready for MALDI ionization in a small spot.

MALDI ionization of peptide fragments is a three-step process. (1) The samples are mixed with a suitable matrix molecules, e.g., α-cyano-4-hydroxycinnamic acid (CHCA), 2,5-dihydroxybenzoic acid (DHB, DHBA), or mixture of these chemicals, and mounted on a conductive MALDI plate. (2) A pulsed laser irradiates the sample, triggering the desorption of the sample and matrix molecules. (3) The sample molecules are ionized by protonation or deprotonation and are accelerated by electric field and fly into a mass spectrometer. The type of a mass spectrometer commonly combined MALDI is the TOF (time-of-flight) mass spectrometer, mainly due to its large mass range. This characteristic has an important advantage in the analysis of proteins, peptide fragments, and other biomolecules. MALDI-TOF combined instrument usually equips the ion mirror called reflectron, doubling the ion flight path and increasing the resolution. After the flight, ionized sample molecules are caught on the detector and converted into electric signals. Microchannel plate and photomultiplier tube are generally used for the detector device of MALDI TOF-MS. Finally, m/z of the sample molecule is calculated based on the flight time and electric charge number.

The advantages of MALDI TOF-MS compared with other instruments used for proteomic analysis, e.g., LC/MS, are flexible ionization condition, laser intensity and kind of matrix molecules, and repeatability of analysis. Because MALDI ionization consumes only a small amount of sample molecules, we can obtain the mass spectra repeatedly and accumulate spectra to improve the sensitivity and quality of analysis. On the other hand, the time efficiency of analysis for large numbers of samples is inferior to the inline-automated LC/ESI TOF-MS system. However, because the number of contained proteins in organelle or sub-organelle compartment of *C. merolae* is limited (ca. $< 200$), this disadvantage is not so serious in the study of *C. merolae*. In addition, for the purpose of combining HPLC and MALDI TOF-MS, sample autoloading device for MALDI plate AccuSpot (Shimadzu) is available. This device enables "LC-MALDI" system and can establish a time-effective proteomics application using a conventional MALDI TOF-MS.

### *6.3.4 Protein Database for Proteomics*

For the downstream process of MALDI TOF-MS, peptide mass fingerprinting (PMF) and MS/MS ion search are commonly used to identify the target protein based on m/z of the peaks in spectra. PMF is the most popular analytical application of MALDI TOF-MS. As a result of MALDI TOF-MS analysis of a protein spot on the gel, a list of m/z of the peptide fragments is derived from one to several proteins, which is often called a peak list. Database searching software, i.e., Mascot (Matrix Science), performs in silico digests on proteins in the database with the same enzyme (e.g., trypsin) used in the preparation of the sample. In case of

*C. merolae*, FASTA file of whole-genome information is available from the *Cyanidioschyzon merolae* Genome Project website (http://merolae.biol.s.u-tokyo.ac.jp/). Database searching software converts genome information into protein lists by means of in silico translation, and protein database is generated. The obtained m/z is compared to generated protein database. The mass of peptide fragments is then calculated and compared to the peak list of actually measured m/z. The results are statistically analyzed, and possible candidate proteins are shown in a results table. MS/MS ion search is used to reveal partial amino acid sequence information of target peptides, using post-source decay (PSD) or collision-induced dissociation (CID), when the result of PMF is not decisive to identify target proteins.

In PMF and MS/MS ion search of *C. merolae* in our work, MASCOT software in a database built from 5014 open reading frames of the *C. merolae* genome was used. The total size of generated databases of *C. merolae* is only 2.73 megabytes, and this small-sized database can contribute to save computation power and calculation time drastically. In comparison, comprehensive protein database, e.g., SwissProt or NCBI-nr, reach gigabyte order in size, and the searching process against these databases is very computer resource consuming. Furthermore, the complete genome information of *C. merolae* provides adequate search space for PMF or MS/MS ion search and may also contribute to the repression of pseudo-positive/pseudo-negative results. As a tip of database search, we combined the peak list of known major contaminants, e.g., keratins and cytokeratins of human (keratin I, II, 2e, 67 k; cytokeratin 9, 10), trypsin autolysis fragments, and Coomassie Brilliant Blue molecules, to detect the common contaminations rapidly. These peaks are also used for a kind of internal standards to calibrate MS during the analysis.

## 6.4 Perspective

The organelle division system of *C. merolae* seems to be simplified in both transcriptome and proteome level, but its essential mechanisms should be common to not only *C. merolae* and other microalgae but also all eukaryotic organisms. *C. merolae* is easy to cultivate and able to synchronize cell cycle; thus, cell phase-specific comparative omics analysis could be a greatly useful approach to elucidate the cell cycle and cell division system of eukaryotic organisms. Furthermore, small-sized genome and its complete information of *C. merolae* enable fast and efficient downstream computing process of omics analyses. These omics approach will reinforce the structural experiment in collaboration with immunochemical visualization techniques. Revealing the localization and function of further proteins of *C. merolae* should contribute to the comprehensive understanding of the function, evolution, and dynamics of eukaryotic cell.

**Acknowledgments** Our study was partly supported by JSPS KAKENHI (no. 23770234 to M.Y.; JP15K18588 to T.F.; no. 25251039 to S.M.), by the Maekawa Houonkai Foundation (M.Y.), and by the Core Research for Evolutional Science and Technology Program of the Japan Science and Technology Agency (S.M.).

## References

Fujiwara T, Misumi O, Tashiro K et al (2009) Periodic gene expression patterns during the highly synchronized cell nucleus and organelle division cycles in the unicellular red alga *Cyanidioschyzon merolae*. DNA Res 16:59–72

Fujiwara T, Kuroiwa H, Yagisawa F et al (2010) The coiled-coil protein VIG1 is essential for tethering vacuoles to mitochondria during vacuole inheritance of *Cyanidioschyzonmerolae*. Plant Cell 22:77–781

Hirabaru C, Izumo A, Fujiwara S et al (2010) The primitive rhodophyte *Cyanidioschyzon merolae* contains a semiamylopectin-type, but not an amylose-type, α-glucan. Plant Cell Physiol 51:682–693

Imamura S, Kanesaki Y, Ohnuma M et al (2009) R2R3-type MYB transcription factor, CmMYB1, is a central nitrogen assimilation regulator in *Cyanidioschyzon merolae*. Proc Natl Acad Sci U S A 106:12548–12553

Imoto Y, Fujiwara T, Yoshida Y et al (2010) Division of cell nuclei, mitochondria, plastids, and microbodies mediated by mitotic spindle poles in the primitive red alga *Cyanidioschyzon merolae*. Protoplasma 241:63–74

Imoto Y, Kuroiwa H, Yoshida Y et al (2013) Single-membrane-bounded peroxisome division revealed by isolation of dynamin-based machinery. Proc Natl Acad Sci U S A 110:9583–9588

Kanesaki Y, Imamura S, Minoda A et al (2012) External light conditions and internal cell cycle phases coordinate accumulation of chloroplast and mitochondrial transcripts in the red alga *Cyanidioschyzon merolae*. DNA Res 19:289–303

Matsuzaki M, Misumi O, Shin T et al (2004) Genome sequence of the ultrasmall unicellular red alga *Cyanidioschyzon merolae* 10D. Nature 428:653–657

Minoda A, Nagasawa K, Hanaoka M et al (2005) Microarray profiling of plastid gene expression in a unicellular red alga, *Cyanidioschyzon merolae*. Plant Mol Biol 59:375–385

Nozaki H, Takano H, Misumi O et al (2007) A 100%-complete sequence reveals unusually simple genomic features in the hot-spring red alga *Cyanidioschyzon merolae*. BMC Biol 5:28

Ohta N, Sato N, Kuroiwa T (1998) Structure and organization of the mitochondrial genome of the unicellular red alga *Cyanidioschyzon merolae* deduced from the complete nucleotide sequence. Nucleic Acids Res 26:5190–5198

Ohta N, Matsuzaki M, Misumi O et al (2003) Complete sequence and analysis of the plastid genome of the unicellular red alga *Cyanidioschyzon merolae*. DNA Res 10:67–77

Tanaka K, Waki H, Ido Y et al (1988) Protein and polymer analyses up to m/z 100 000 by laser ionization time-of flight mass spectrometry. Rapid Commun Mass Spectrom 2:151–153

Tardu M, Dikbas UM, Baris I et al (2016) RNA-seq analysis of the transcriptional response to blue and red light in the extremophilic red alga, *Cyanidioschyzon merolae*. Funct Integr Genomics 16:657–669

Yagisawa F, Nishida K, Yoshida M et al (2009) Identification of novel proteins in isolated polyphosphate vacuoles in the primitive red alga *Cyanidioschyzon merolae*. Plant J 60:882–893

Yoshida Y, Kuroiwa H, Hirooka S et al (2009) The bacterial ZapA-like protein ZED is required for mitochondrial division. Curr Biol 19:1491–1497

Yoshida Y, Kuroiwa H, Misumi O et al (2010) Chloroplasts divide by contraction of a bundle of nanofilaments consisting of polyglucan. Science 329:949–953

Yoshida M, Yoshida Y, Fujiwara T et al (2011) Proteomic comparison between interphase and metaphase of isolated chloroplasts of *Cyanidioschyzon merolae* (Cyanidiophyceae, Rhodophyta). Phycol Res 59:1–15

# Chapter 7
# Procedures for Transformation and Their Applications in *Cyanidioschyzon merolae*

**Takayuki Fujiwara and Mio Ohnuma**

**Abstract** The red alga *Cyanidioschyzon merolae* exhibits an extremely simple cytological genomic architecture. This feature offers advantages in cytological and biochemical studies. Methods for nuclear gene targeting have been developed during the last decade. Currently, gene knockout, knock-in, and a stable expression of transgene are feasible in *C. merolae*. In addition, to examine the function of genes that are essential for cell viability, heat-shock inducible gene expression and nitrogen source-dependent inducible/repressible gene expression systems have also been developed. The combination of emerging genetic tools and the simple features of *C. merolae* facilitate various "omics" analyses across several disciplines.

**Keywords** Transformation • Homologous recombination • Gene targeting, biotechnology • Algae • Genetic engineering

## 7.1 Introduction

The unicellular red alga *Cyanidioschyzon merolae* has been developed as a promising model eukaryotic alga. A *C. merolae* cell is haploid, and its genome was fully sequenced (16.5 Mbp, 4775 protein-cording genes) (Matsuzaki et al. 2004; Nozaki et al. 2007). In *C. merolae*, nuclear gene targeting techniques have been developed based on homologous recombination (Imamura et al. 2009). The integrated

T. Fujiwara (✉)
Department of Cell Genetics, National Institute of Genetics, 1111 Yata, Mishima, Shizuoka 411-8540, Japan

Department of Genetics, Graduate University for Advanced Studies (SOKENDAI), 1111 Yata, Mishima, Shizuoka 411-8540, Japan

Japan Science and Technology Agency, CREST, 4-1-8 Honcho, Kawaguchi, Saitama 332-0012, Japan
e-mail: tkfujiwara@nig.ac.jp

M. Ohnuma
National Institute of Technology, Hiroshima College, 4272-1 Higashino, Osaki-Kamijima, Toyota, Hiroshima 725-0231, Japan
e-mail: mioohnuma@hiroshima-cmt.ac.jp



T. Kuroiwa et al. (eds.), *Cyanidioschyzon merolae*,
https://doi.org/10.1007/978-981-10-6101-1_7

transgenes are stably expressed without any silencing (Fujiwara et al. 2013b, 2015). The emerging genetic modification techniques and the extremely simple cellular and genome architecture of *C. merolae* (Chap. 2) are contributing to studies on a variety of cellular processes including nitrogen assimilation (Imamura et al. 2009, 2010), circadian regulation of cell cycle progression (Miyagishima et al. 2014), and reciprocal regulation between cell cycle progression and chloroplast division (Sumiya et al. 2016). In addition, trials for the production of biofuels have also started by metabolic engineering in *C. merolae* (Sumiya et al. 2015; Imamura et al. 2015, 2016).

In this chapter, we will introduce the procedures for genetic modifications and their applications in *C. merolae*.

## 7.2 First Success in *C. merolae* Transformation

Minoda et al. (2004) succeeded for the first time in the nuclear transformation of *C. merolae*. They isolated the spontaneous uracil-auxotrophic mutant M4 strain, which possesses one nucleotide insertional mutation in the midportion of the *URA5.3* gene (CMK046C). *URA5.3* encodes a protein composed of the orotidine-5′-phosphoribosyltransferase (OPRTase) domain and the orotidine 5′-monophosphate (OMP)-decarboxylase domain. Due to the frameshift mutation, a stop codon arises at the beginning of the second domain, which truncates the second domain of *URA5.3* in the M4 strain (Fig. 7.1a). By using the M4 strain, they demonstrated that the introduction of wild-type *URA5.3* comprised on a plasmid vector by electroporation repairs the mutated allele, possibly by a homologous recombination event (Minoda et al. 2004).

## 7.3 Transient Expression of a Transgene by Polyethylene Glycol (PEG)-Mediate DNA Introduction

Ohnuma et al. (2008) then demonstrated that DNA is easily introduced by a polyethylene glycol (PEG)-mediated method. It was also reported that efficient transformation was archived when the cells were diluted into medium immediately after mixing a plasmid DNA vector, PEG4000, and *C. merolae* cells for a very short time (~10 s). This technique enabled transient expression of an HA-tagged version of $\beta$-tubulin/CM263C from a plasmid (Ohnuma et al. 2008) and endogenous *Catalase* (CMI050C) gene suppression by expression of the antisense RNA (Ohnuma et al. 2009). The former procedure has mostly been utilized for examination of subcellular localization proteins fused with epitope tags or fluorescent

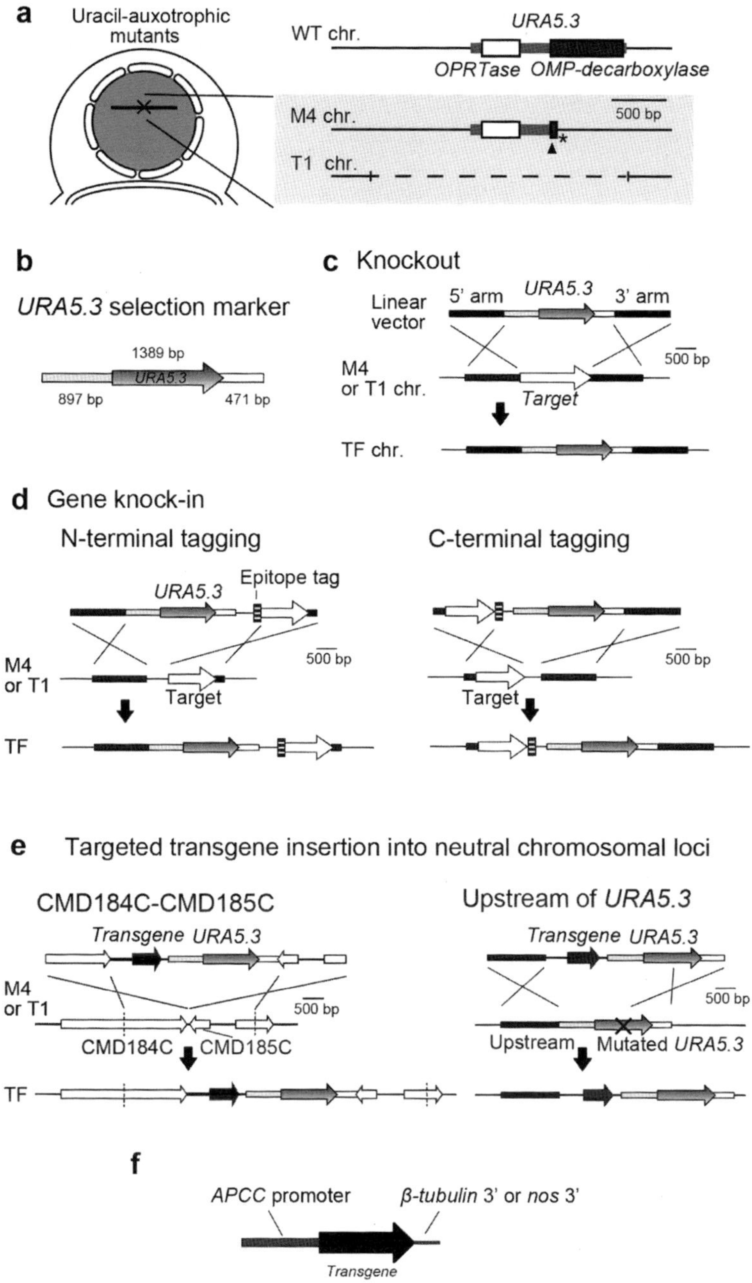


**Fig. 7.1** Schematic diagrams of the uracil-auxotrophic mutants and vector design for gene targeting. (**a**) The chromosomal structure of the uracil-auxotrophic mutants M4 and T1 along with the wild type (WT). The wild-type *URA5.3* gene consists of the *OPRTase* domain and the *OMP-decarboxylase* domain. The *URA5.3* of the M4 strain has an insertional mutation, and the second domain is truncated. The arrowhead and asterisk show the position of insertional mutation and the emerging stop codon, respectively. The flanking upstream region and the *URA5.3 orf* are deleted in the T1 strain. The dash line shows the deleted region of the T1 chromosome. (**b**) The structure of the *URA5.3* transformation selection marker. The marker is composed of the wild-type

proteins (Yagisawa et al. 2009; Moriyama et al. 2014), and the latter has contributed to studies on plastid-division machinery (Yoshida et al. 2010), peroxisomal division (Imoto et al. 2013), and regulation of chloroplast-encoding gene transcription (Fujii et al. 2013).

## 7.4 Nuclear Gene Targeting

Imamura et al. (2009) succeeded in gene knockout by homologous recombination in *C. merolae*. To knockout the *MYB1* gene (CMJ282C), they introduced the linear DNA vector, which consists of ~1.5 kb homologous arms flanked with the *URA5.3* transformation selection marker into the *MYB1* locus of the uracil-auxotrophic mutant M4 by the PEG-mediate protocol (Fig. 7.1a–c). The transformed cells were selected on a uracil-free MA2 gellan gum plate.

The gene targeting technique has been improved through several studies (Imamura et al. 2010; Fujiwara et al. 2013b). The time required to obtain transformants has been shortened to a month: ~2 weeks for colony formation and ~2 more weeks for preparation of the liquid culture of a transformed clone. Other genetic manipulations such as gene knock-in (Fujiwara et al. 2013a) and targeted transgene insertion into chromosomal neutral loci for its stable expression have been developed (Fujiwara et al. 2013b). The gene knock-in technique has been used for epitope-tagging of an endogenous gene, for example, the *CAP-D3* gene encoding a subunit of the condensin II complex (Fujiwara et al. 2013a) (Fig. 7.1d), which facilitates the detection of the protein by antibodies against a tag. Two chromosomal loci have been used to insert a transgene into the chromosomal neutral loci: the convergent intergenic region between CMD184C and CMD185C and the upstream region of *URA5.3* (Fig. 7.1e). It was reported that the level of green fluorescent protein (GFP) expressed from both loci is almost the same. For efficient expression of the transgene, the 600 bp of the 5′-flanking sequence of *APCC*/CMO250C *orf* has mostly been used as a promoter because the *APCC* gene exhibits stable and high expression in *C. merolae* (Fig. 7.1f) (Watanabe et al. 2011; Fujiwara et al. 2015).

Recently, Taki et al. (2015) newly established a host strain, uracil-auxotrophic mutant T1, in which the *URA5.3* gene was completely deleted. This strain is more useful than the M4 strain possessing a single nucleotide mutation in *URA5.3*

**Fig. 7.1** (continued) URA5.3 orf, 897-bp upstream, and 471-bp downstream flanking sequences. (**c–e**) Schemas of linear DNA vectors for gene knockout (**c**), gene knock-in (**d**), and targeted transgene insertion (**e**). (**f**) The promoter and 3′-*utr* for the efficient transgene expression. The 600-bp upstream flanking sequence of *APCC orf* is used as a promoter. The 3′-*utr* of $\beta$-tubulin or the 3′-*utr* of *nos* is utilized for putative polyadenylation signal sequence (Ohnuma et al. 2008; Watanabe et al. 2011)

because selection of transformants of T1 excludes contamination of clones with a reverse mutation in *URA5.3* (Fig. 7.1a).

## 7.5 Inducible and Repressible Transgene Expression Systems

Currently, a heat-shock inducible gene expression system (Sumiya et al. 2014) (Fig. 7.2a) and a nitrogen source-dependent inducible/repressible gene expression system (Fujiwara et al. 2015) (Fig. 7.2b) are available. The former enables a rapid and high level of gene induction (within 1 h after heat shock). It was applied to the expression of a dominant-negative form of dynamin-related protein Dnm2/DRP5B (Sumiya et al. 2016). We are currently attempting to apply the latter system for conditional knockdown of a gene.

## 7.6 Procedures for Nuclear Gene Targeting

Here, we mainly describe a procedure for gene targeting because transient transgene expression can be performed in a similar manner. The flowcharts of the both experiments are shown in Fig. 7.3.

### *7.6.1 Materials*

5-Fluoroorotic Acid (5-FOA) (Fluorochem, #003234)
Air flow meter (KOFLOC, RK1650)
Air pump (Suisaku, SSPP-2S)
Corn starch (Nacalai Starch Corn, #32127-25)
Cotton (Iwatsuki, #006-062010)
Cotton filter tube (SANSYO, Φ 21 × H 70× H 250 mm, custom-made)
Ethanol (WAKO, #057-00456)
Fluorescent lights (Technica, FL/TK12C-S)
Gellan gum (WAKO, #075-03075)
Heater for thermobath (Fine, FTB-01)
Incubator shaker (TAITEC, BioShaker BR-43FL)
LED light (TAITEC, LC-LED450W)
Petri dish (IWAKI, 9-cm diameter SH90-15)

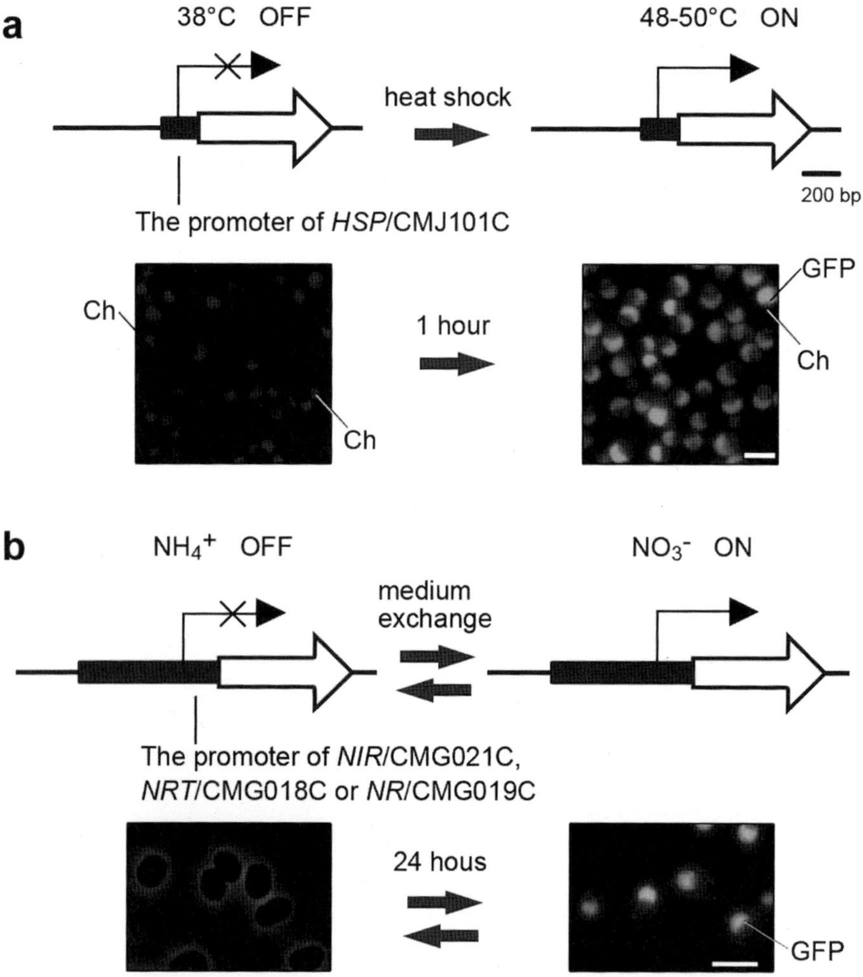


**Fig. 7.2** Inducible and repressible gene expression systems. (**a**) The heat-shock inducible gene expression system. The 200-bp flanking upstream sequence of *HSP20*/CMJ101C *orf* is utilized for a heat-shock promoter which induces transgene expression by temperature shift from 38°C to 48–50°C. The left and right images show the fluorescence of sfGFP before and 1 h after the heat-shock treatment, respectively. Scale bar, 5 μm. (**b**) The nitrogen source-dependent inducible and repressible gene expression system. The 800-bp flanking upstream sequence of *NIR* (nitrite reductase), *NRT* (nitrate transporter), or *NR* (nitrate reductase) *orf* is utilized as a promoter which is activated by exchange of the nitrogen source from ammonium to nitrate. The signal of sfGFP is detectable in the transformant within 24 h after the medium exchange. In contrast, addition of ammonium to the culture in the nitrate medium represses the promoter activity within 1 h. The left and right images show the fluorescence of sfGFP in the ammonium and the nitrate media, respectively. Scale bar, 5 μm. Reproduced from Sumiya et al. (2014)

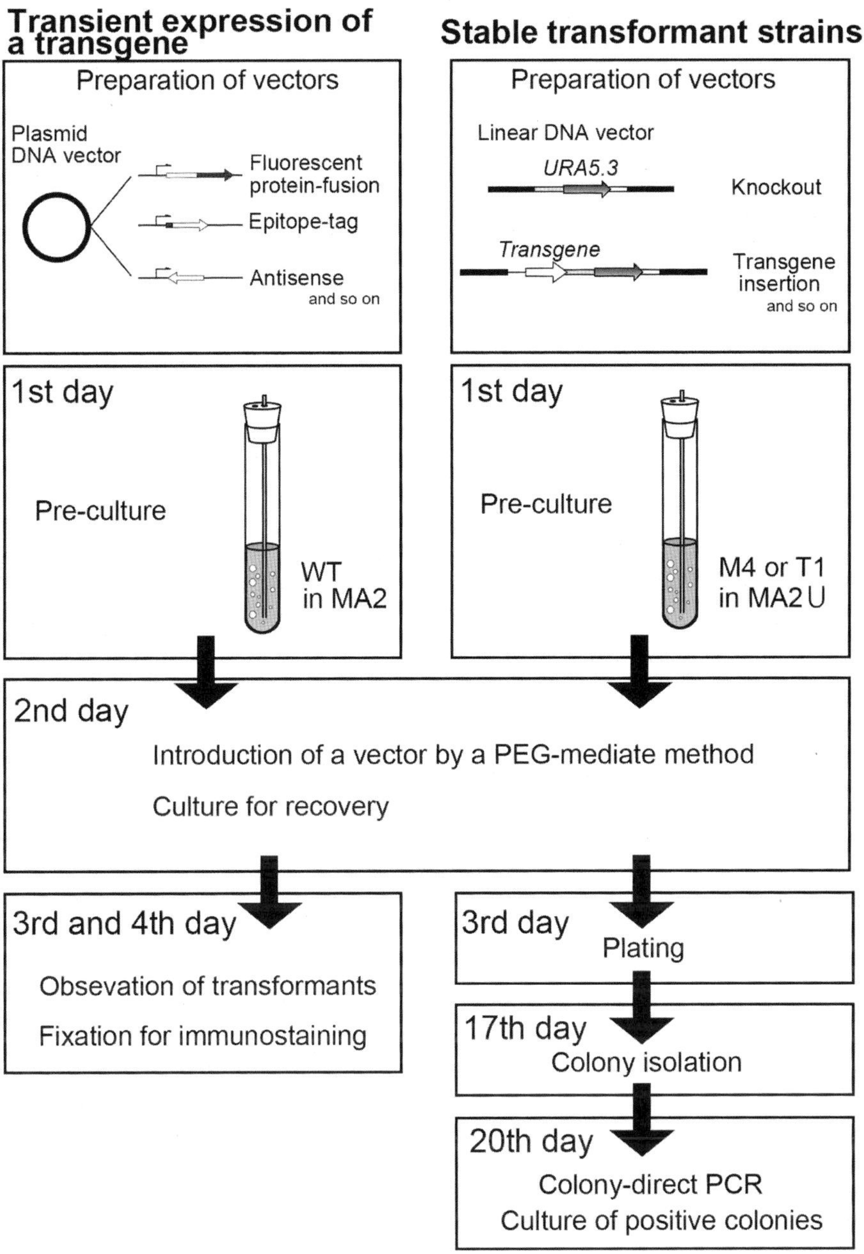


**Fig. 7.3** Flowcharts showing procedures for transient expression of a transgene and producing a stable transformant

Polyethylene glycol (PEG) 4000 (Aldrich, Poly(ethylene glycol) average Mn 4000, platelets; #81240)
Quantum flux meter (Apogee Instruments MQ-200 Quantum Flux Meter)
Silicone stopper (ASONE size no. 6 Φ 23 × Φ 20 × H 26 mm, 6-336-06)
Silicone stopper (ASONE size no. 9 Φ 30 × Φ 25 × H 30 mm, 6-336-09)
Silicone tube (ASONE 4 × 8 mm, #6-586-14)
Sterilization filter (EMD Millipore, #SLGP033RS; Corning, #431096)
Test tube 100 mL (IWAKI, TEST30NP)
Test tube stand (Sanwakaken, #6-302-19)
Tissue culture flasks with filter cap (TPP 25 cm2, # 90026)
Uracil (SIGMA-ALDRICH, #U0750)

**Medium**

MA2 medium (Ohnuma et al. 2008) (Table 7.1)
Metal solutions (Ohnuma et al. 2008) (Table 7.2)
MA2U: MA2 supplemented with uracil (0.5 mg/mL)
2x MA2 medium for gellan gum plates (Table 7.3)
10x transformation (TF) solution (Table 7.4)

## *7.6.2 Methods*

**Algal Culture**

The wild-type *C. merolae* 10D (NIES-3377) is used for transient gene expression or as nurse cells in producing stable transformants. The wild-type strain is maintained in MA2 medium (Table 7.1) at a cellular concentration of OD750 = 1.5–3.0 at

**Table 7.1** MA2 medium

| | | |
|---|---|---|
| $(NH_4)_2SO_4$ | 5.28 g | 40 mM |
| $MgSO_4 \cdot 7H_2O$ | 0.98 g | 4 mM |
| $KH_2PO_4$ | 1.09 g | 8 mM |
| $CaCl_2$ | 0.11 g | 1 mM |
| $H_2SO_4$ | ~300 μL | ~0.03% |
| [a]250x Fe | 4 mL | 320 μM |
| [a]500x Zn | 2 mL | 2.8 μM |
| [a]2500x Co | 400 μL | 0.7 μM |
| [a]2500x Mo | 400 μL | 6.4 μM |
| [a]2500x Mn | 400 μL | 36 μM |
| [a]2500x Cu | 400 μL | 1.3 μM |
| Distilled water | Up to 1 L | |

[a]The metal solutions are supplemented after autoclave (121°C, 20 min)

**Table 7.2** Metal solutions

| | | |
|---|---|---|
| 250x Fe | | |
| $Fe_3Cl_3{\cdot}6H_2O$ | 3.4 g | 79 mM |
| $Na_2$-EDTA·$2H_2O$ | 3.5 g | 106 mM |
| $H_2SO_4$ | A few drops | |
| Distilled water | Up to 500 mL | |
| 500x Zn | | |
| $ZnCl_2$ | 95 mg | 1.4 mM |
| Distilled water | Up to 500 mL | |
| 2500x Co | | |
| $CoCl_2{\cdot}6H_2O$ | 40 mg | 1.7 mM |
| Distilled water | Up to 100 mL | |
| 2500x $Na_2MoO_4$ | | |
| $Na_2MoO_4{\cdot}2H_2O$ | 390 mg | 16 mM |
| Distilled water | Up to 100 mL | |
| 2500x Mn | | |
| $MnCl_2{\cdot}4H_2O$ | 1.78 g | 90 mM |
| Distilled water | Up to 100 mL | |
| 2500x Cu | | |
| $CuCl_2$ | 43 mg | 3.2 mM |
| Distilled water | Up to 100 mL | |

Each solution is sterilized with filtration

**Table 7.3** 2x MA2 medium

| | | |
|---|---|---|
| $(NH_4)_2SO_4$ | 2.64 g | 80 mM |
| $MgSO_4{\cdot}7H_2O$ | 0.49 g | 4 mM |
| $KH_2PO_4$ | 0.54 g | 16 mM |
| $CaCl_2$ | 0.56 g | 2 mM |
| [a]$H_2SO_4$ | ~250 μL | ~1% |
| Distilled water | Up to 250 mL | |
| [b]250x Fe | 2 mL | 320 μM |
| [b]500x Zn | 1 mL | 2.8 μM |
| [b]2500x Co | 200 μL | 0.7 μM |
| [b]2500x Mo | 200 μL | 6.4 μM |
| [b]2500x Mn | 200 μL | 36 μM |
| [b]2500x Cu | 200 μL | 1.3 μM |

[a]$H_2SO_4$ is added to correct a pH shift by gellan gum
[b]The metal solutions are supplemented after autoclave (121 °C, 20 min)

**Table 7.4** 10x transformation solution

| | | |
|---|---|---|
| $(NH_4)_2SO_4$ | 5.28 g | 400 mM |
| $MgSO_4{\cdot}7H_2O$ | 0.98 g | 40 mM |
| $H_2SO_4$ | ~300 μL | ~0.3% |
| Distilled water | Up to 100 mL | |

The solution is sterilized by autoclave (121 °C, 20 min)

40–42 °C under light condition (TAITEC, LED; 30 μE/m$^2$s) in a shaker (TAITEC, BR-43FL, 130 rpm). The uracil-auxotrophic mutant *C. merole* M4 or T1 is used as a parental strain for production of stable transformants. The M4 or T1 culture is maintained in MA2U medium supplemented with 5-FOA (0.8 mg/mL). The other conditions are the same as for the wild type.

**Cotton Filter Tubes**
Cotton filter tubes are used to prevent contamination of other organisms from provided air.

1. Make two holes in a silicone stopper (size, no. 9).
2. Set a cotton filter tube into the silicone stopper.
3. Stuff cotton into the cotton filter tube.
4. Set the cotton filter tube on a 100-mL test tube.
5. Cover (4) with aluminum foil.
6. Make a hole in a silicone stopper (size, no. 6).
7. Set 4 cm of PTFE tube into the silicone stopper no. 6.
8. Prepare 30 cm of silicone tube.
9. Sterilize (5), (7), and (8) by autoclave at 121 °C for 20 min.

**MA2 Gellan Gum Plates**
MA2 medium is solidified with gellan gum as described in Imamura et al. (2010) with some modifications as follows:

1. Prepare 200 mL of "2x MA2" medium (Table 7.3).
2. Suspend 2 g of gellan gum in 200 mL of distilled water in a 500-mL bottle.
3. Autoclave (1) and (2) separately.
4. Add the metal solutions to 2x MA2 after it gets cold (below 65 °C).
5. Combine (4) with (2) at 58–63 °C and mix well by a stirrer for 2 min.
6. Pour ~20 mL of (5) into a 9-cm petri dish.
7. Dry extra moisture of a solidified medium without a cover.
8. Store the MA2 gellan gum plates in a bag with zipper at room temperature.
   Note: (4) to (7) should be performed in a clean bench.

**MA2U Gellan Gum Plates**
Add 2 g of uracil into 200 mL of "2x MA2" (the final concentration of uracil is 0.5 g/L) and then prepare gellan gum plates as described in the procedure for MA2 gellan gum plates.

**Spotting Starch on a Gellan Gum Plate**
Starch is prepared as described in Imamura et al. (2010) with some modifications. A 20% slurry of starch is prepared as follows:

**Starch Suspended in MA2 (MA2-Starch)**

1. Suspend 10 g of corn starch in 50 mL of ethanol (99.5%), centrifuge at 500 *g* for 2 min, and discard the supernatant.
2. Repeat (1).

3. Resuspend the starch pellet into 50 mL of sterilized distilled water, centrifuge at 500 *g* for 2 min, and discard the supernatant.
4. Resuspend the starch pellet into 50 mL of 10x transformation buffer, centrifuge at 500 *g* for 2 min, and discard the supernatant.
5. Resuspend the starch pellet into 50 mL of sterilized distilled water, centrifuge at 500 *g* for 2 min, and discard the supernatant.
6. Resuspend the starch pellet into 50 ml of MA2 medium. The MA2-starch suspension can be preserved at 4 °C.
7. Drop a 15-μl aliquot of the MA2-starch suspension on a MA2 gellan gum plate at a regular interval to produce 20–30 spots per dish according to the sheet (Fig. 7.4a).

**Starch Suspended in MA2U (MA2U-Starch)**

1. Transfer 1 mL of MA2-starch suspension into a 1.5-mL tube.
2. Centrifuge by a tabletop centrifuge and discard the supernatant.
3. Add MA2U up to 1 mL and suspend it.
4. Drop 5-μl aliquot of the MA2U-starch suspension on a MA2U gellan gum plate at a regular interval to produce 84 spots according to the sheet (Fig. 7.4b).

**60% PEG4000 Solution**

1. To prepare 500 μL of 60% PEG4000 solution, dissolve 300 μg of PEG4000 in 225 μL of MA2 medium at 95 °C for 10 min in a 1.5-mL tube.
2. Pour 125 μL of 60% PEG4000 solution into a 1.5-mL tube.
3. Keep it at 42 °C on a heat block until use.
   Note: The PEG4000 solution should be prepared just before use.

**Production of Stable Transformants**
Preparation of stable transformants of *C. merolae* is performed as described in Ohnuma et al. (2008) and Imamura et al. (2010) with some modifications.

**Preparation of a Linear DNA Vector**
To integrate a DNA sequence into a chromosomal locus, a linear DNA vector, which consists of the *URA 5.3* selection marker flanked with 1–1.5 kb of homologous arms is prepared (Fig. 7.1c–e). Then, amplify the vector by PCR with a high-fidelity PCR enzyme. The reaction volume of PCR is 100 μL (25 μL x 4). Purify the PCR product with a DNA purification column and elute the product with 200 μL of distilled water.

**Protocol**

– First day, pre-culture-

  1. Dilute the uracil-auxotrophic mutant M4 or T1 culture to give a cellular concentration of OD750 = 0.4 in 50 mL of MA2U medium.
  2. Incubate the culture in a 100-mL test tube with aeration (600 mL air/min) under light (100 $\mu E/m^2s$) at 42°C for 16–20 h (Figs. 7.3 and 7.5a).
     Note: Provide air through a cotton filter.

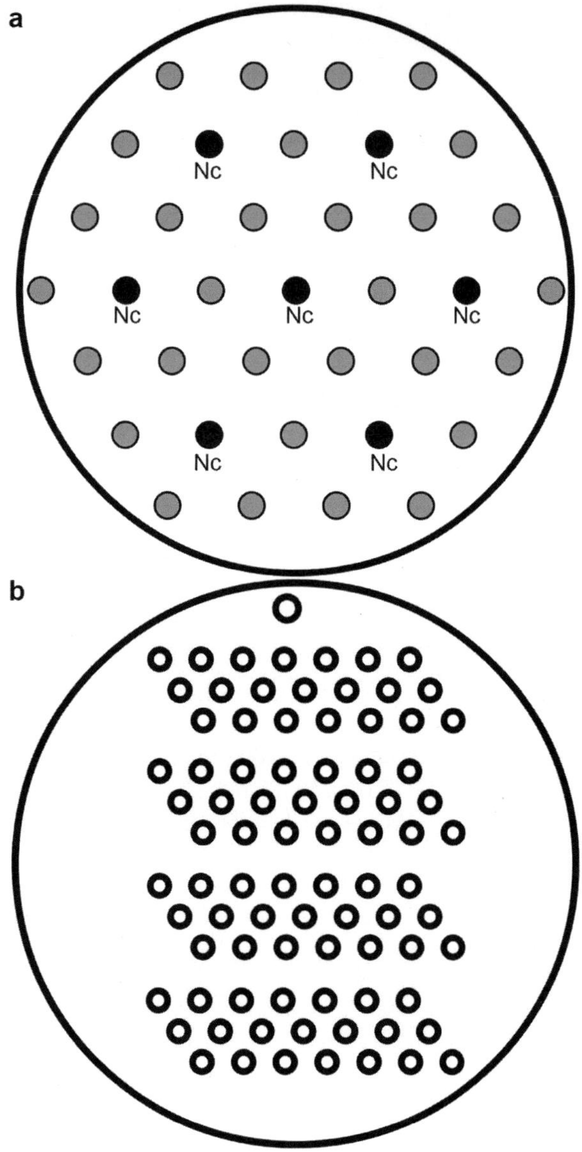


**Fig. 7.4** Templates for spotting starch. (**a**) A template for spotting MA2-starch on a MA2 gellan gum plate. Circles are marks for spotting MA2-starch. *Gray* and *black circles* indicate positions at which transformed cells and the wild-type cells as a nurse cell (Nc) are inoculated, respectively. (**b**) A template for spotting MA2U-starch on a MA2U gellan gum plate. Open circles are marks for spotting MA2U-starch

– Second day, transformation-

3. Prepare the 60% PEG4000 solution.
4. Measure the OD750 of the culture and confirm that the value of OD750 is ~0.7.
5. Harvest the cells by centrifugation (1500 *g*, for 5 min) after addition of Tween-20 (0.001%), and discard the supernatant.

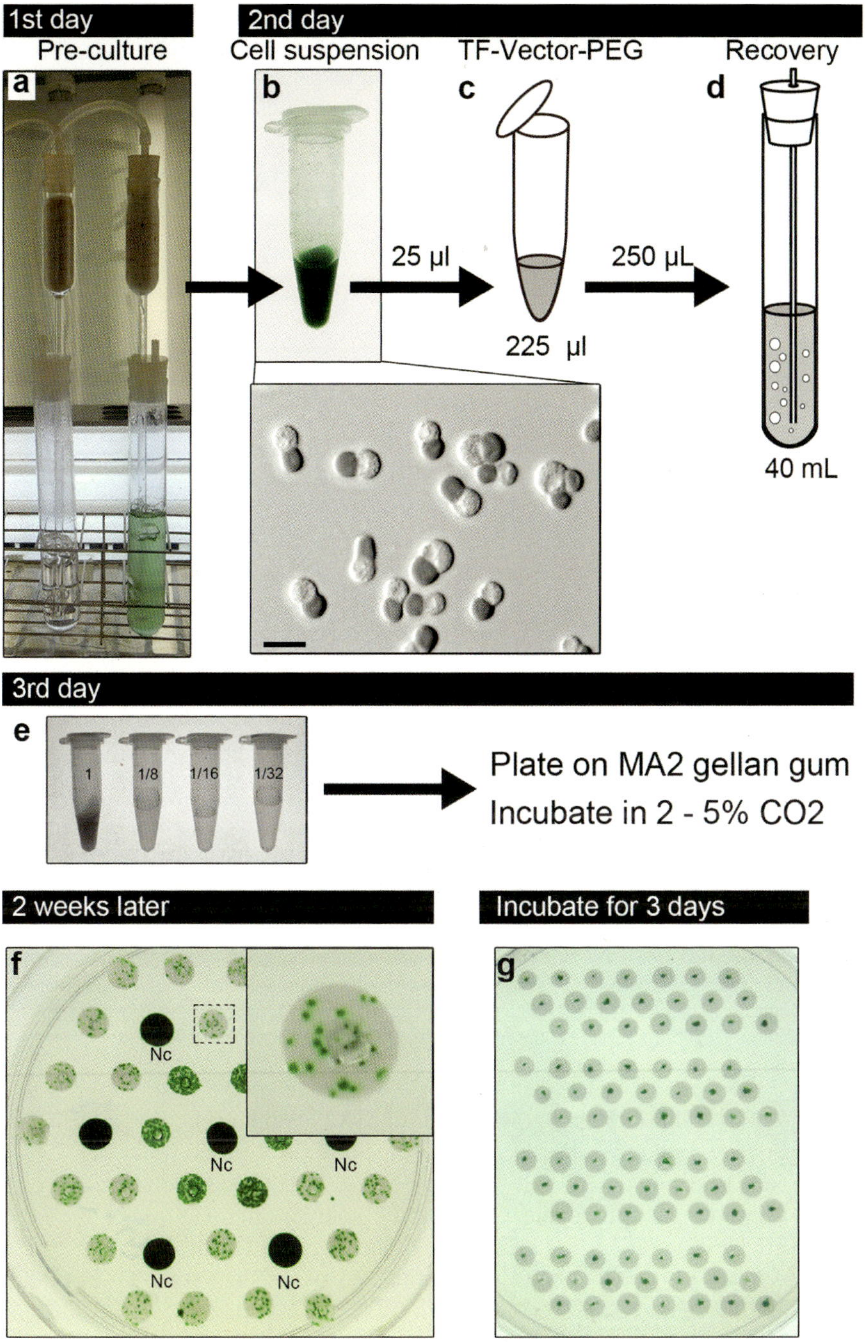


**Fig. 7.5** A procedure for producing a stable transformants. (**a**) A pre-culture for transformation. To prevent contamination of other organisms, a filter cotton tube is set onto 100-mL test tubes, and air is provided through the filter. To prevent drying the culture, air charged with moisture is provided through the test tube in which desterilized water is poured (the *left test tube*).

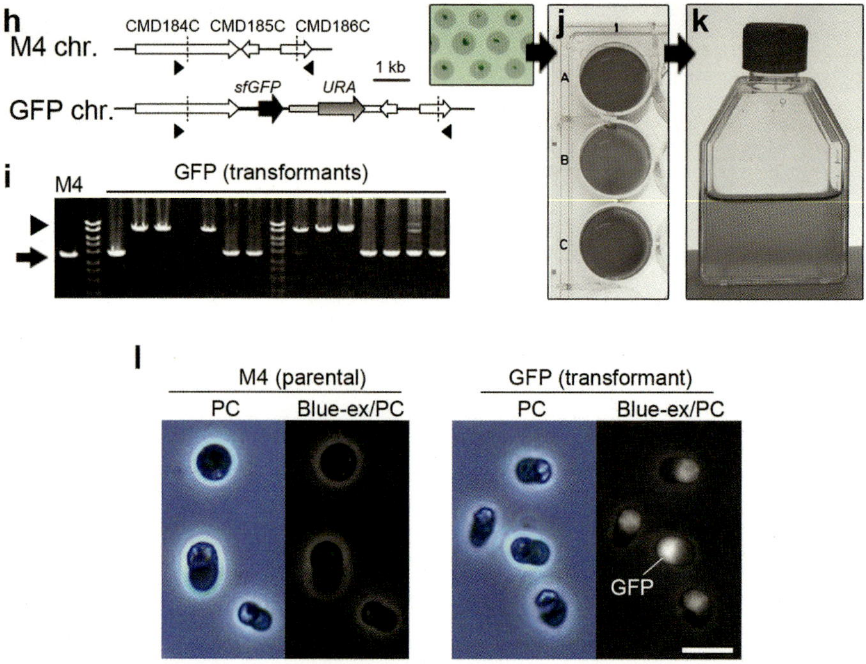


**Fig. 7.5** (continued) (**b**) Collected cells in a 1.5-mL tube. (**c**) A 25 μL of the concentrated culture is added to the TF-vector-PEG mixture. (**d**) Cells-TF-vector-PEG mixture is immediately transferred into 40 mL of MA2 medium. The culture is incubated for recovery overnight. (**e**) Transformed cells are concentrated into 400 μL of MA2 medium and then diluted to 1/8, 1/16, and 1/32. The diluted cells are inoculated onto MA2 gellan gum plate. (**f**) Colony formation of transformants on the MA2 gellan gum plate. (**g**) Isolated colonies. (**h**) Colony-direct PCR for an examination of targeted transgene insertion. Schema indicating the chromosomal structure of the M4 and, for example, the transformant in which *sfGFP* gene is inserted into the chromosomal neutral locus (intergenic region between CMD184C and CMD185C). Arrowheads indicate priming sites of primers used in the PCR analysis. (**i**) Agarose gel electrophoresis analysis of PCR product of the GFP transformed colonies. The arrowhead and arrow indicate targeted *sfGFP* insertion-positive and negative bands, respectively. (**j** and **k**) Culture of the transformants in a 24-well plate (**j**) and in a tissue culture flask (**k**). (**l**) Phase-contrast (PC) and fluorescent images of the M4 and the GFP transformants. GFP fluorescence is observed in the GFP transformant. Bars = 5 μm

6. Resuspend the cell pellet in 320 μL of MA2 medium (Fig. 7.5b).
7. Prepare 4 μg of a linear DNA vector in 90 μL of distilled water.
8. Add 10 μL of 10x TF solution to the 90 μL of the vector solution.
9. Combine 100 μL of the vector in TF solution and 125 μL of the 60% PEG4000 solution, and mix well by pipetting in a 1.5-mL tube.
10. Add 25 μL of the cell suspension to 225 μL of the vector-TF-PEG4000 mixture (Fig. 7.5c).
11. Invert the mixture 3 to 4 times vigorously within 10 s.
12. Immediately transfer the mixture into 40 mL of MA2U in a 100-mL test tube (Fig. 7.5d).

13. Incubate the culture with aeration (300 mL air/min) under light (50 μE/$m^2$s) at 42 °C for ~24 h.
    Note: Provide air through a cotton filter.

– Third day, plating for selection and colony formation-

14. Harvest the cells by centrifugation (1500 *g* for 5 min) after addition of Tween-20 (0.01%) and discard the supernatant.
15. Resuspend the cell pellet in 400 μL of MA2 medium, and transfer the culture into a 1.5-mL tube.
16. Centrifuge by a tabletop rotor and discard the supernatant.
17. Suspend the cell pellet in 400 μL of MA2 by pipetting and then by a micro tube mixer (TAITEC) at ~2000 rpm for 2 min.
18. Dilute the cell suspension to 1/8, 1/16, and 1/32 with MA2 medium (Fig. 7.5e).
19. Inoculate 8 μl of the diluted cell suspension onto starch spots on a MA2 gellan gum plate. Spot the wild-type cells at an appropriate interval as nurse cells (Fig. 7.5f).
20. Incubate the plate under light (50 μE/$m^2$s) at 42 °C with 2–5% $CO_2$ until colony becomes visible by naked eyes (for ~2 weeks).

– Two weeks later-

21. Transfer 42 or 84 colonies onto new starch spots on a MA2U gellan gum plate. The MA2U accelerates growth of colonies.
22. Incubate transformed colonies under light (50 μE/$m^2$s) at 42 °C with 2–5% $CO_2$ until colonies become larger (for 3–5 days) (Fig. 7.5g).
23. Examine the targeted insertion by colony-direct PCR with a primer set priming outside the introduced DNA (Fig. 7.5h and 7.5i).
24. Transfer respective targeted insertion-positive clones into 2 mL of MA2U in a 24-well plate and suspend the cells (Fig. 7.5j).
25. Incubate the cells under light at 42 °C with 5% $CO_2$ for 3–5 days.
26. Transfer the culture to 20 mL of MA2 into a 25-$cm^2$ tissue culture flask, and incubate cells under light (50 μE/$m^2$s) at 42 °C with 2–5% $CO_2$ until an adequate quantity of transformant cells is obtained (Fig. 7.5k).
27. Examine a phenotype. In the case of targeted *sfGFP* insertion, sfGFP fluorescence is observed in transformants (Fig. 7.5l).
    Note: If you produce a strain in which a cytotoxic gene is induced by the heat-shock inducible system, incubate the cells at 36–38 °C after transformation to prevent leaky expression of the transgene.

**Transient Transgene Expression**
A procedure for transient transgene expression is almost the same as for production of stable transformants (Fig. 7.3).

1. Culture the wild-type cells for 16–20 h.
2. Harvest the cells by centrifugation.

3. Resuspend the cells in 320 μL MA2 medium.
4. Prepare 20–50 μg of a plasmid DNA vector in 90 μL of distilled water, and mix well with 10x TF and 60% PEG solution.
5. Mix 25 μL of cell suspension with the TF-vector-PEG mixture within 10 s.
6. Immediately transfer the cell-TF-vector-PEG mixture into 40 mL of MA2 medium, and incubate the culture for recovery for 1 or 2 days.
7. Observe the transformed cells.

**Acknowledgments** Our study was partly supported by JSPS KAKENHI (no. JP15K18588 to T. F.; no. 25251039 to S.M.) and by the Core Research for Evolutional Science and Technology Program of the Japan Science and Technology Agency (S.M.). We thank S. Miyagishima in National Institute of Genetics for his careful and critical reading of our manuscript.

## References

Fujii G, Imamura S et al (2013) Nuclear-encoded chloroplast RNA polymerase sigma factor SIG2 activates chloroplast-encoded phycobilisome genes in a red alga, *Cyanidioschyzon merolae*. FEBS Lett 587:3354–3359

Fujiwara T, Tanaka K et al (2013a) Spatiotemporal dynamics of condensins I and II: evolutionary insights from the primitive red alga *Cyanidioschyzon merolae*. Mol Biol Cell 16:2515–2527

Fujiwara T, Ohnuma M et al (2013b) Gene targeting in the red alga *Cyanidioschyzon merolae*: single- and multi-copy insertion using authentic and chimeric selection markers. PLoS One 8(9):e73608

Fujiwara T, Kanesaki Y et al (2015) A nitrogen source-dependent inducible and repressible gene expression system in the red alga *Cyanidioschyzon merolae*. Front Plant Sci 6:657

Imamura S, Kanesaki Y et al (2009) R2R3-type MYB transcription factor, CmMYB1, is a central nitrogen assimilation regulator in *Cyanidioschyzon merolae*. Proc Natl Acad Sci U S A 106:12548–12553

Imamura S, Terashita M et al (2010) Nitrate assimilatory genes and their transcriptional regulation in a unicellular red alga *Cyanidioschyzon merolae*: genetic evidence for nitrite reduction by a sulfite reductase-like enzyme. Plant Cell Physiol 51:707–717

Imamura S, Kawase Y et al (2015) Target of rapamycin (TOR) plays a critical role in triacylglycerol accumulation in microalgae. Plant Mol Biol 89:309–318

Imamura S, Kawase Y et al (2016) TOR (target of rapamycin) is a key regulator of triacylglycerol accumulation in microalgae. Plant Signal Behav 11(3):e1149285

Imoto Y, Kuroiwa H et al (2013) Single-membrane-bounded peroxisome division revealed by isolation of dynamin-based machinery. Proc Natl Acad Sci U S A 110:9583–8588

Matsuzaki M, Mitsumi O et al (2004) Genome sequence of the ultrasmall unicellular red alga *Cyanidioschyzon merolae* 10D. Nature 428:653–657

Minoda A, Sakagami R et al (2004) Improvement of culture conditions and evidence for nuclear transformation by homologous recombination in a red alga, *Cyanidioschyzon merolae* 10D. Plant Cell Physiol 45:667–671

Miyagishima SY, Fujiwara T et al (2014) Translation-independent circadian control of the cell cycle in a unicellular photosynthetic eukaryote. Nat Commun 5:3807

Moriyama T, Sakurai K et al (2014) Subcellular distribution of central carbohydrate metabolism pathways in the red alga *Cyanidioschyzon merolae*. Planta 240:585–598

Nozaki H, Takano H et al (2007) A 100%-complete sequence reveals unusually simple genomic features in the hot spring red alga *Cyanidioschyzon merolae*. BMC Biol 5:28

Ohnuma M, Yokoyama T et al (2008) Polyethylene glycol (PEG)-mediated transient gene expression in a red alga, *Cyanidioschyzon merolae* 10D. Plant Cell Physiol 49:117–120

Ohnuma M, Misumi O et al (2009) Transient gene suppression in a red alga, *Cyanidioschyzon merolae* 10D. Protoplasma 236:107–112

Sumiya N, Fujiwara T et al (2014) Development of a heat-shock inducible gene expression system in the red alga *Cyanidioschyzon merolae*. PLoS One 9(10):e111261

Sumiya N, Kawase Y et al (2015) Expression of Cyanobacterial acyl-ACP reductase elevates the triacylglycerol level in the red alga *Cyanidioschyzon merolae*. Plant Cell Physiol 56:1962–1980

Sumiya N, Fujiwara T et al (2016) Chloroplast division checkpoint in eukaryotic algae. Proc Natl Acad Sci U S A 113:E7629–E7638

Taki K, Sone T et al (2015) Construction of a URA5.3 deletion strain of the unicellular red alga *Cyanidioschyzon merolae*: a background less host strain for transformation experiments. J Gen Appl Microbiol 61:211–214

Watanabe S, Ohnuma M et al (2011) Utility of a GFP reporter system in the red alga *Cyanidioschyzon merolae*. J Gen Appl Microbiol 57:69–72

Yagisawa F, Nishida K et al (2009) Identification of novel proteins in isolated polyphosphate vacuoles in the primitive red alga *Cyanidioschyzon merolae*. Plant J 60:882–893

Yoshida Y, Kuroiwa H et al (2010) Chloroplasts divide by contraction of a bundle of nanofilaments consisting of polyglucan. Science 329:949–953

# Part IV
# Techniques for Morphological and Structural Biology

**Sachihiro Matsunaga**

Techniques for morphological and structural biology have been multifariously developed. Fluorescent microscopic techniques combined with fluorescent reagents, fluorescent proteins and immunostaining have several advantages to visualize subcellular organelles. Electron microscopic techniques are very available to analyze subcellular structures at high magnification. The red alga *Cyanidioschyzon merolae* has become a promising model organism for morphological and structural biology. *C. molorae* has several advantages to reveal morphological characters of organelles and subcellular structures because it has a simple subcellular structure with a nucleus, a single mitochondrion, a single chloroplast and a minimal set of single-membrane-bound organelles.

Fluorescence and electron microscopic techniques using *C. molorae* are described in Chap. 8. Staining with fluorescent dyes and immunoistaining are basic methods to visualize subcellular structures and organelles of *C. molorae*. The electron microscopic techniques include transmission electron microscopy (TEM), immunoelectron microscopy and scanning electron microscopy (SEM). The applications of cryo-TEM, focused ion beam-SEM (FIB-SEM) and ultra-high voltage electron microscopy (UHVEM) to *C. molorae* are introduced in Chap. 9. These advanced electron microscopic techniques contribute to 3D structural modeling of *C. molorae*. In Chap. 10, cryogenic coherent X-ray diffraction imaging (CXDI) is also available for structural analyses of the whole cell and the organelles. CXDI enables to reveal the internal structure of a whole frozen-hydrated chloroplast of *C. merolae*.

# Chapter 8
# Basic Techniques in Fluorescence and Electron Microscopy

**Haruko Kuroiwa**

**Abstract** The primitive red alga *Cyanidioschyzon merolae* is a single cell with a very small size (1.5–2 μm) and has no thick cell wall; therefore, preparation of samples for fluorescence and electron microscopy requires different procedures compared to those for green algae or higher plants. Here, fluorescence microscopy methods, such as cell staining with fluorescent dye and the setup of the immunofluorescence microscope, are described. For electron microscopy, methods for transmission electron microscopy, immunoelectron microscopy, and negative staining methods are described. In addition, scanning electron microscopy methods for visualizing cells or organelles are presented.

**Keywords** Fluorescence microscopy • DAPI • Electron microscopy • $DiOC_6$ • Negative staining • Immunoelectron microscopy • Scanning microscopy

## 8.1 Introduction

*Cyanidioschyzon merolae* is a very small red alga with single copies of organelles and only a few lysosomes. Improved fluorescence microscopy (Kuroiwa and Suzuki 1980) and staining with the dye DAPI (4′,6-diamidino-2-phenylindole) has enabled the visualization of small algae and DNA in intracellular organelles (Kuroiwa et al. 1981). Moreover, by the establishment of cell culture method using light and dark, the cell cycles of *C. merolae* cultures can be highly synchronized (Suzuki et al. 1994). It thus became possible to follow the dynamics of all the intracellular organelles (Miyagishima et al. 1998c, 1999a). Furthermore, by establishing the methods of immunofluorescence microscopy, the localization of intracellular substance can be clearly shown (Miyagishima et al. 2001c; Nishida et al. 2003, 2004).

Various electron microscopy methods, such as optimized fixation and embedding protocols, have been developed for imaging detailed cellular features.

H. Kuroiwa (✉)
Department of Chemical and Biological Science, Faculty of Science, Japan Women's University, 2-8-1 Mejirodai, Bunkyo-ku, Tokyo 112-8681, Japan
e-mail: hkuroiwa@fc.jwu.ac.jp



T. Kuroiwa et al. (eds.), *Cyanidioschyzon merolae*,
https://doi.org/10.1007/978-981-10-6101-1_8

Chemical, high-pressure freezing, and, particularly, rapid-freezing fixation methods were applied to obtain convincing images (Miyagishima et al. 1998b, c, 2001a; Fujiwara et al. 2010; Yagisawa et al. 2012, 2016). Fractionation of cellular organelles and cell constituents enabled imaging with scanning electron microscopy (Miyagishima et al. 1999c; Yoshida et al. 2012). Furthermore, decoded genomic information (Matsuzaki et al. 2004; Nozaki et al. 2007) made it possible to elucidate mechanisms of organelle division, such as those for chloroplasts, mitochondria, and peroxisomes (Miyagishima et al. 2003; Nishida et al. 2003; Yoshida et al. 2006, 2010; Imoto et al. 2012, 2017). Negative staining immunoelectron microscopy has enabled immune detection of fractionated intracellular components and double labeling for clear robust information (Miyagishima et al. 1999c; Yoshida et al. 2012; Imoto et al. 2017).

In this section, the focus will be on (1) a staining method for nuclei, mitochondria, and peroxisomes, and a fluorescent antibody reaction method for cell and cell fractions for use with fluorescence and phase-contrast fluorescence microscopy. (2) For electron microscopy, the methods of chemical fixation, rapid-freezing substitution using liquid propane, and immunoelectron microscopy are described. (3) Also described is a negative staining method for observing cellular or fractionated intracellular organelle structures. Lastly, (4) a scanning electron microscopy method for observing cells or fractionated organelles is described.

## 8.2 Observation Techniques for Fluorescence Microscopy

### *8.2.1 Fluorescence Staining*

#### 8.2.1.1 Fluorescence Staining of DNA

Cells are collected by centrifugation at 2000 × g for 5 min. When the supernatant is removed, the cells are placed on a glass slide in a drop of 1% glutaraldehyde dissolved in distilled water. They are then stained with a drop of DAPI (4′, 6-diamidino-2-phenylindole) at a concentration of 1 μg/ml dissolved in NS buffer (1 mM EDTA, 1 mM $MgCl_2$, 0.1 mM $CaCl_2$, 0.1 mM $ZnSO_4$, 0.8 mM phenylmethylsulfonyl fluoride, 0.05% 2-mercaptoethanol, 0.5 M sucrose, 20 mM Tris-HCl at pH 7.6). The stained samples are squashed by pressing on a cover glass and are observed with an epifluorescence or an epifluorescence/phase-contrast microscope. DAPI fluorescence is excited by UV-light excitation and shows nuclear, chloroplast, and mitochondrial DNA emit *white-blue fluorescence*. Chloroplasts emit *red autofluorescence* by the chlorophyll a when excited with green light (Kuroiwa et al. 1981, 1995; Nishibayashi and Kuroiwa 1985; Suzuki et al. 1994; Fujiwara et al. 2010; Imoto et al. 2010) (Fig. 8.1).

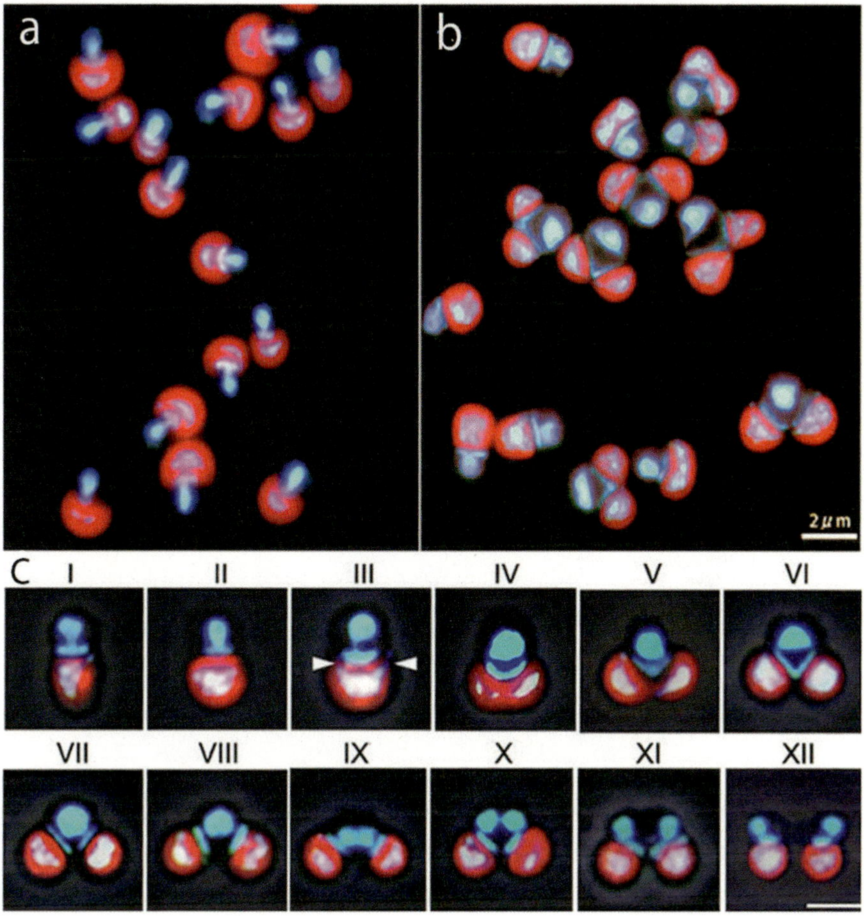


**Fig. 8.1** Synchronous culturing and definition of cell-cycle phases of *C. merolae*. (**a**, **b**) Images of synchronized interphase cells (2 h) and mitotic cells (18 h) stained with DAPI. Under UV light, nuclei, chloroplasts, and mitochondrial DNA emit *white-blue fluorescence*, while chloroplasts emit *red autofluorescence*. (**c**) Representative DAPI-stained micrographs of type I–XII cells observed through mitotic phases. The arrow indicates the edge of the plastid nucleus structure stretching toward both poles. Scale bar = 2 μm (Reproduced and modified from Imoto et al. 2010)

#### 8.2.1.2 Fluorescence Staining of Mitochondria with $DiOC_6$

Cells are harvested by centrifugation at 3000 × g for 5 min. $DiOC_6$ (3,3′-dihexyloxacarbocyanine iodide; 100 mg/ml stock in ethanol) stain is added to the culture medium to a final concentration of 0.1 mg/ml. After 10 min, the cells are washed twice with fresh medium by centrifugation and resuspended in medium.

A drop of this suspension is placed on a slide glass and is squashed by pressing on a cover glass. The stained samples are examined with an epifluorescence or an epifluorescence/phase-contrast microscope with blue excitation light to reveal each organelle and mitochondria emit *green fluorescence*. Under green excitation light, chloroplasts emit chlorophyll a *red autofluorescence* in the same field (Suzuki et al. 1994; Miyagishima et al. 2001a) (Fig. 8.2).

#### 8.2.1.3 Fluorescence Staining of Microbodies with Brilliant Sulfoflavine (BSF)

Cells are harvested by centrifugation at 2000 × g for 5 min. The supernatant is removed, and the cells are fixed in 1% glutaraldehyde dissolved in Allen's medium.

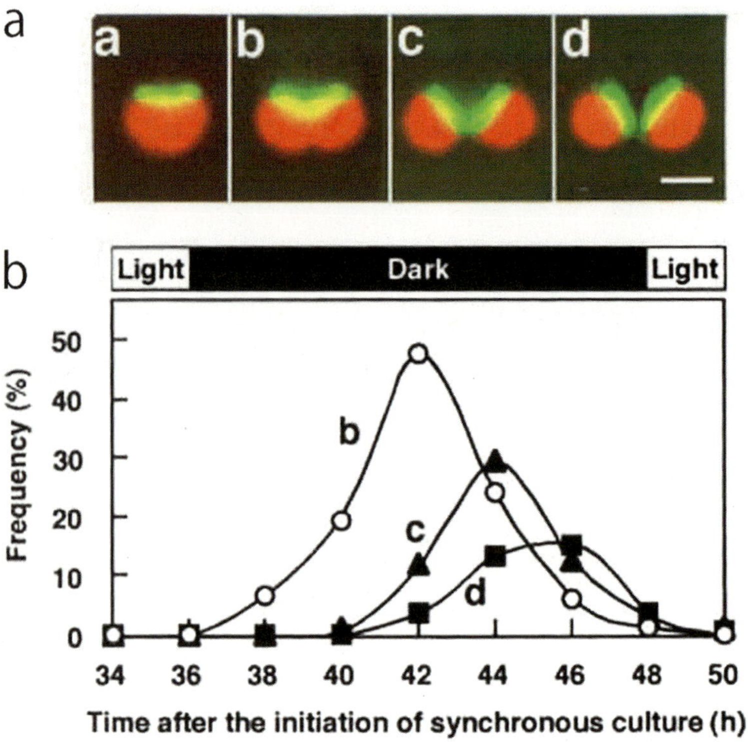


**Fig. 8.2** Progression of division of chloroplasts and mitochondria in a synchronized culture of *C. merolae*. (**a**) The mitochondria are imaged with $DiOC_6$ *green fluorescence*, while the chloroplasts emit *red autofluorescence* from chlorophyll a. (*a*) A cell with a non-dividing chloroplast and mitochondrion in interphase, (*b*) a dividing chloroplast and bent mitochondrion, (*c*) two divided chloroplasts and a dividing mitochondrion, and (*d*) two divided chloroplasts and two divided mitochondria. (**b**) Change in the frequencies of cell types progressing in (**a**). Scale bars = 1 μm (Reproduced and modified from Miyagishima et al. 2001a)

After washing the cells twice with 100 mM citric acid-sodium citrate buffer at pH 3.0, the cell suspension is stained with BSF (sodium 4-amino-N- (p-tolyl) -naphthalimide-3- sulfonate, $C_{19}H_{13}N_2O_5SNa$) at a concentration of 30 μg/ml. The BSF fluorescence is observed under blue excitation light with an epifluorescence or an epifluorescence/phase-contrast microscope (Yamada et al. 1985; Elrod et al. 1991; Miyagishima et al. 1998a).

## 8.2.2 *Immunofluorescence Microscopy*

### 8.2.2.1 Immunofluorescence Observation of Cells, Chloroplasts, and the Membrane Fraction of Chloroplasts

Pellets of cells, chloroplasts, or fractionated membrane of chloroplasts are suspended and fixed in paraformaldehyde (3% w/v) dissolved in 50 mM pipes, pH 6.8, 0.15 M sucrose, 10 mM EGTA, and 5 mM $MgSO_4$ for 1 h at room temperature. The suspension of samples is centrifuged and washed twice with phosphate-buffered saline (PBS). For immunofluorescence microscopy, the cover slips are coated with poly-D-lysine (*M*r 15,000 to 30,000; Sigma), and samples are affixed on them. They are squashed onto slide glasses, then, the cover glasses are peeled off the slides. For immune staining, cover glasses with samples are pretreated with PBS containing 0.05% Triton X-100 for 15 min, then blocked with 5% bovine serum albumin (BSA) in PBS for 30 min at 37 °C. Then, the reaction of the primary antibody (1:100 dilution) is performed to the samples affixed to the cover glasses for 2 h at 37 °C. After washing with 1% BSA in PBS for 30 min and blocking a second time with 5% BSA in PBS for 30 min, the secondary immune reaction with a fluorescein-conjugated secondary IgG antibody (1:200 dilution) is performed for 2 h at 37 °C. After washing with 1% BSA in PBS for 30 min, the samples are mounted with anti-fade (Slowfade, Molecular Probes, Eugene, OR, USA) and are examined under an epifluorescence microscope. The nucleus and chloroplasts can be observed in the same field after staining with a drop of DAPI at a concentration of l μg/ml. Observation under UV light reveals nuclear, chloroplast, and mitochondrial DNA with *white-blue fluorescence*. Chloroplasts contain chlorophyll a, which emits *red autofluorescence* when excited with green excitation light (Miyagishima et al. 2001b; Yoshida et al. 2006) (Fig. 8.3).

### 8.2.2.2 Immunofluorescence Microscopy

*C. merolae* cells are harvested by centrifugation for 5 min at 800 × g, then prefixed in 2% (w/v) paraformaldehyde in 10 mM phosphate buffer (pH 7.5), 30 mM NaCl, and 0.5 mM KCl for 3 min. Then, cells are centrifuged for 3 min at 2000 × g, and the supernatant discarded. The cell pellets are then suspended in an equal volume of prefixation buffer and are immediately fixed with 20 volumes of chilled solutions:

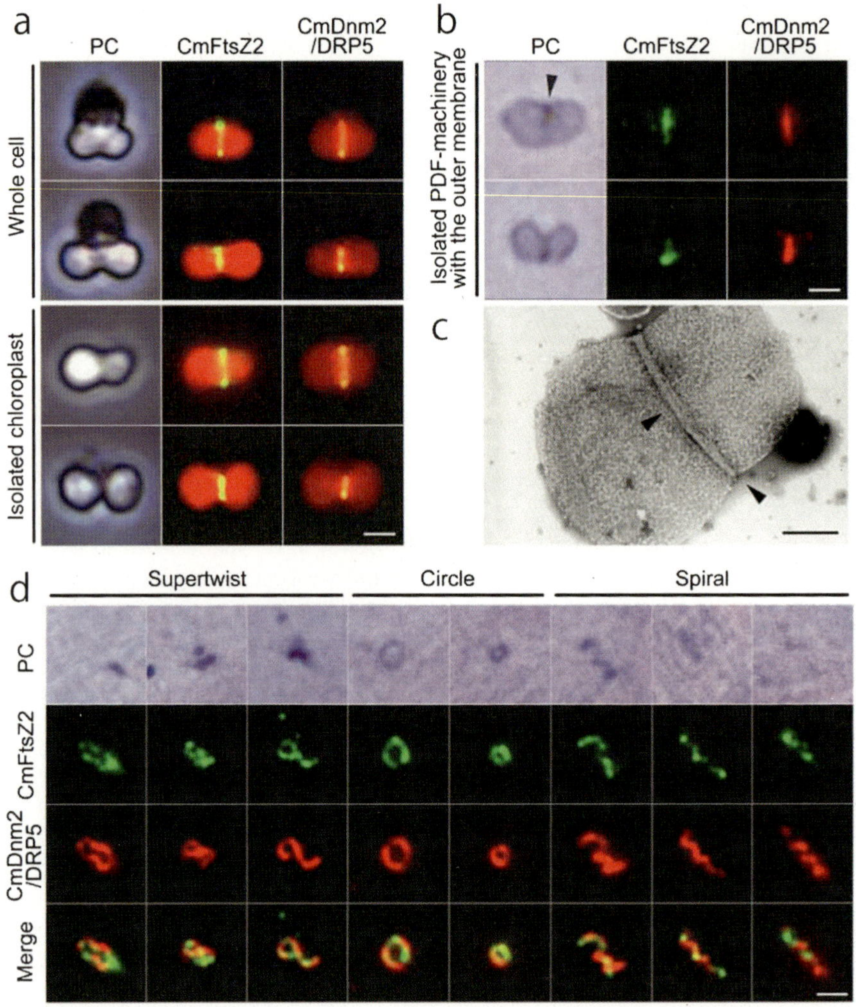


**Fig. 8.3** Isolation and immune staining of chloroplast division machinery in *C. merolae*. (**a**) Phase-contrast and immunofluorescence images of the FtsZ (CmFtsZ2; *yellow green*) and dynamin (CmDnm2/DRP5; *orange*) rings of whole cells and isolated chloroplasts at early and late phases of chloroplast division. (**b**) Phase-contrast and immunofluorescence images of PD (*arrowhead*), FtsZ, and dynamin rings of isolated chloroplast division machineries with the outer membrane at early and late phases of chloroplast division. (**c**) Electron micrograph of an isolated chloroplast with PDF machinery (*arrowheads*) on the outer membrane. (**d**) Phase-contrast and immunofluorescence images of the PD, FtsZ, and dynamin rings and merged images of isolated intact PDF-machineries showing supertwisted, circular, and spiral structures. Scale bars = 1 μm (**a**, **b**, **d**), 500 nm (**c**). (Reproduced and modified from Yoshida et al. (2006). Reprinted with permission from AAAS)

2% (w/v) paraformaldehyde, 85% (v/v) methanol, 15% (v/v) dimethylsulfoxide, and 1.5 mM NaOH for 5 min at −20 °C. After that, cells are washed with methanol and rehydrated with a solution of PBS, 137 mM NaCl, 2.7 mM KCl, 1 mM $Na_2HPO_4$, and 1.4 mM $K_2HPO_4$. Cells are then stored in PBS containing 50% (v/v) glycerol at −20 °C. For the immunofluorescence reactions, the prepared cells are incubated with primary antibody (1: 100 dilution) in blocking buffer (5% BSA in PBS) for 2 h at 30 °C. The secondary immunoreaction is performed with the secondary antibodies, Alexa Fluor 488 goat anti-rabbit IgG or goat anti-mouse IgG (Molecular Probes, Eugene, OR, USA) (1: 400 dilution) for 1 h at 30 °C, and cells are observed under an epifluorescence microscope with blue excitation light. In the same field, staining with DAPI (1 μg/ml) shows the nuclear, chloroplast, and mitochondrial DNA under UV excitation. Chloroplasts emit red autofluorescence by the chlorophyll a when excited with green excitation light. As a negative control, pre-immune antisera are used for antibody labeling and should produce no signal (Nishida et al. 2003, 2004; Miyagishima et al. 2001b, 2003) (Fig. 8.4).

## 8.3 Observation Techniques for Electron Microscopy

### *8.3.1 Observation of Ultrathin Sections with Transmission Electron Microscopy*

#### 8.3.1.1 Observation of Ultrathin Sections Using Chemical Fixation

*C. merolae* cells are suspended in their culture medium. To fix them, glutaraldehyde in 10 mM sodium cacodylate (pH 7.2) is added to the culture at a ratio of 1: 1 (final concentration 0.5% (w/v)) and incubated for 3 h. The cells are then washed twice with sodium cacodylate and fixed with 1% $OsO_4$ (w/v) overnight at 4 °C. Cells are harvested by centrifugation and mixed with a small volume of 5% low-melting agarose (type IX, ultralow-gelling temperature; Sigma). Solid agarose containing the samples is cut into 1.0 mm cubes and dehydrated in a graded ethanol series (30, 50, 70, 80, 90, and 100% (v/v)). The ethanol is then substituted with 100% (v/v) propylene oxide, and the samples are embedded in Spurr's resin (Polyscience, Warrington, PA, USA). They are then cut into 85-nm-thick sections and mounted on copper grids coated with a polyvinyl formal film. They are then stained with 3% (w/v) uranyl acetate and 1% (w/v) lead citrate and examined by transmission electron microscopy (Yagisawa et al. 2012) (Fig. 8.5).

#### 8.3.1.2 Observation of Ultrathin Sections Using Rapid-Freezing Fixation

*C. merolae* cells are collected by centrifugation and sandwiched between two ultrathin copper plates (1 mm × 4 mm × 0.2 mm). Propane gas is liquefied with

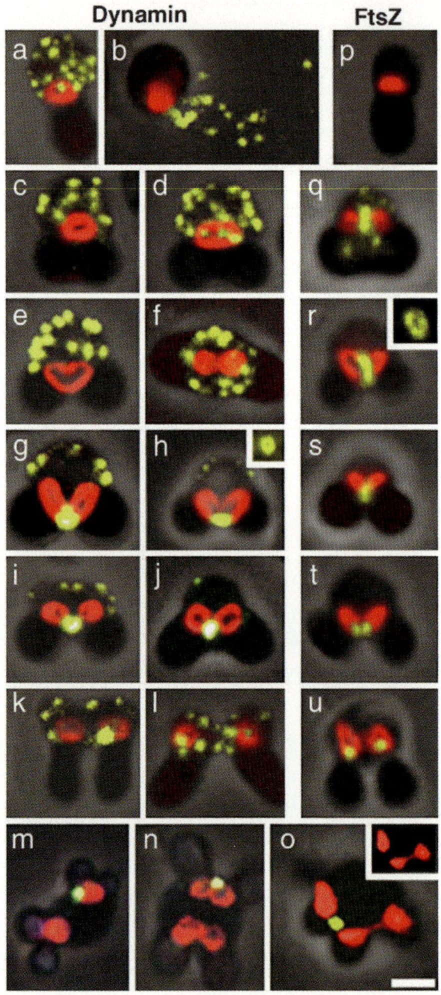


**Fig. 8.4** Confocal laser scanning images of *C. merolae*, and immune staining with CmDnm1 and CmFtsZ1. Cell outlines are shown with overlaying phase-contrast images. Mitochondria stained with Mito-Tracker CMXRos (Molecular Probes, Eugene, OR, USA) are shown in *red*. The CmDnm1 and CmFtsZ1 signals are detected with fluorescein-conjugated secondary antibodies shown in *yellow*. (**a**–**l**) Localization of CmDnm1 during the sequence of cell division. (**p**–**u**) Localization of CmFtsZ1 during the sequence of cell division. (**m**–**o**) Localization of CmDnm1 in cells with defective cell division. Insets in ***h*** and **r** show immunofluorescence of dynamin and FtsZ rings, respectively. Scale bar = 1 μm (Reproduced from Nishida et al. 2003, Copyright (2003) National Academy of Science, U.S.A.)

liquid nitrogen at −195 °C, and the copper plates are then rapidly dropped into the liquid propane. Frozen samples are separated from the copper plates and transferred to 1% (w/v) $OsO_4$ in dry acetone for 48 h at −80 °C. The samples are gradually warmed for 2 h at −20 °C, then for 2 h at 0 °C, and finally for 1 h at room temperature. They are dehydrated with 100% acetone and embedded in Spurr's resin (Polyscience, Warrington, PA, USA). They are then cut into 85-nm-thick sections and placed on copper grids coated with a polyvinyl formal film. They are then stained with 3% (w/v) uranyl acetate and 1% (w/v) lead citrate and examined by transmission electron microscopy (Miyagishima et al. 1998c, 1999b, 2001a) (Fig. 8.6).

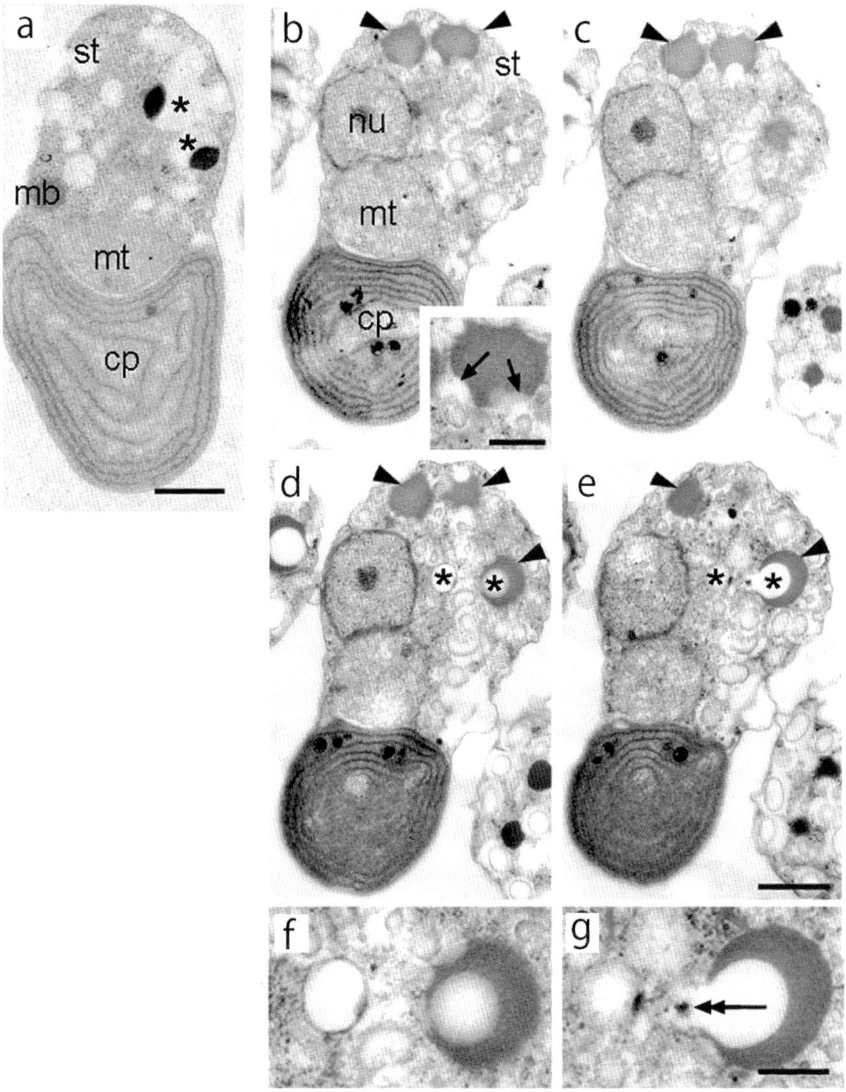


**Fig. 8.5** Electron micrographs of *C. merolae* cells embedded by chemical fixation and deprived of phosphate for 15 days. (**a**) A control cell cultured in 2× Allen's medium. Electron-dense materials are seen in the vacuoles (*). (**b**, **c**, **d**, **e**) Serial sections of a cell deprived of phosphate for 15 days. Lipid bodies (*arrowheads*) in close contact with starch grains [*arrows*, inset in (**b**)]; the vacuoles and the electron-dense regions decrease. (**f**, **g**) Magnified images of vacuoles and lipid bodies in (**d**, **e**). The vacuole with an electron-dense material (*double arrow*) is taken into the lipid bodies. *st* starch grains, *po* peroxisome, *mt* mitochondrion, *cp* chloroplast, *n* nucleus. Scale bars = 500 nm (A, E), 200 nm (inset in **b**, **g**) (Reproduced and modified from Yagisawa et al. 2016. Reprinted with permission from CYTOLOGIA Japan Mendel Society)

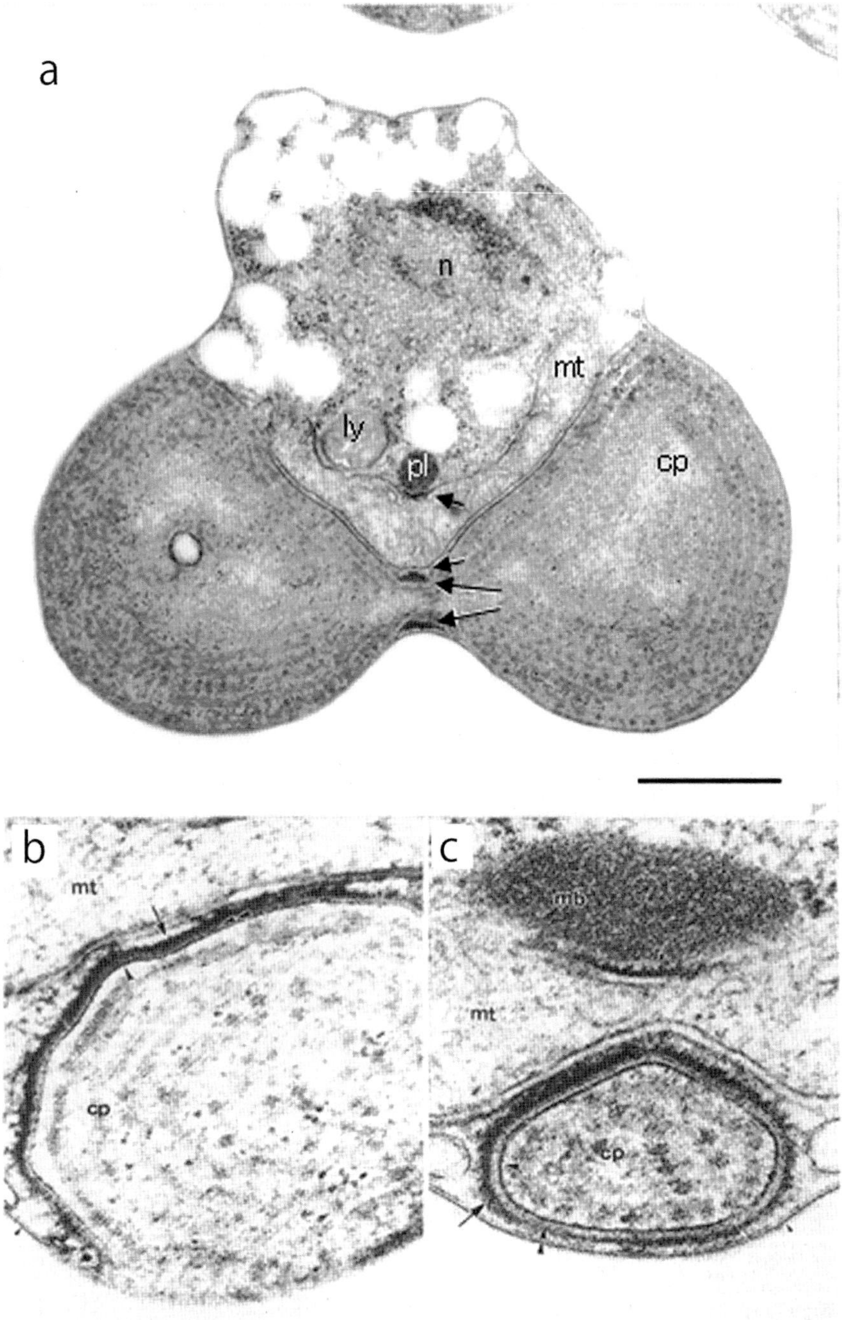


**Fig. 8.6** Electron micrographs of thin sections of M-phase *C. merolae* cells cut perpendicular to the divisional plane at the center of the cell. (**a**) Chloroplast with PD rings (*large arrows*) and mitochondrion with MD rings (*small arrows*) showing the early stages of chloroplast division. (**b**, **c**)

#### 8.3.1.3 Observation of Isolated Organelles in Ultrathin Sections

Samples as pellets of isolated chloroplasts or fractionated membrane of chloroplasts are fixed in solutions of 1% glutaraldehyde, 300 mM sucrose, and 20 mM sodium cacodylate (pH 7.2), for 1 h at 0 °C. The samples are then washed twice with 20 mM sodium cacodylate and postfixed in 1% (w/v) $OsO_4$ in 20 mM sodium cacodylate for 2 h at 0 °C. Samples are harvested by centrifugation and mixed with a small volume of 5% low-melting agarose (type IX, ultralow-gelling temperature; Sigma). The solid agarose with the samples is cut into 1.0 mm cubes and dehydrated in a graded ethanol series (30, 50, 70, 80, 90, and 100% (v/v)). The ethanol is then substituted with 100% (v/v) propylene oxide, and the samples are embedded in Spurr's resin (Polyscience, Warrington, PA, USA). Serial 85-nm-thick sections are mounted on copper grids coated with a polyvinyl formal film. They are stained with uranyl acetate and lead citrate and examined by transmission electron microscopy (Miyagishima et al. 2001b, c) (Fig. 8.7).

### *8.3.2 Immunoelectron Transmission Microscopy of Ultrathin Sections*

#### 8.3.2.1 Immunoelectron Microscopy of Ultrathin Sections Using Chemical Fixation

*C. merolae* cells are harvested by centrifugation and then fixed in 0.5% glutaraldehyde dissolved in 10 mM sodium cacodylate (pH 7.2) at room temperature for 4 h. Fixed cells are centrifuged and mixed with a small volume of 5% low-melting agarose (type IX, ultralow-gelling temperature; Sigma). Solid agarose with cells is cut into 1.0 mm cubes and dehydrated in a graded ethanol series. The ethanol is then substituted with 100% (v/v) propylene oxide, and the samples are embedded in LR white resin (London Resin Co). The resin block is then cut into 85-nm-thick sections and placed on nickel grids coated with a polyvinyl formal film. For the immune reactions, samples on the grids are blocked with 5% BSA in PBS for 30 min at 37 °C, and the primary antibody reaction is performed (at a dilution 1:10 ~ 1:100) at 37 °C for 2 h. After washing the grids with PBS, the secondary immune reactions are performed with the antibodies: goat anti-rabbit or anti-mouse

**Fig. 8.6** (continued) Sections of cells cut parallel to the plane of division at the constricting neck of dividing chloroplast and mitochondria. Three PD rings, that is, outer, middle, and inner rings are recognized closely contact to each envelope. The outer PD ring (*large arrow*) lies on the cytoplasmic face of the outer envelope membrane, the middle PD ring (*double arrowhead*) is located between the inner and outer envelopes, and the inner PD ring (*arrowhead*) is present on the stromal face of the inner envelope membrane. *cp* chloroplast, *mt* mitochondrion, *n* nucleus, *mb* microbody, *ly* lysosome. Scale bars = 500 nm (Reproduced and modified from Miyagishima et al. 1998c)

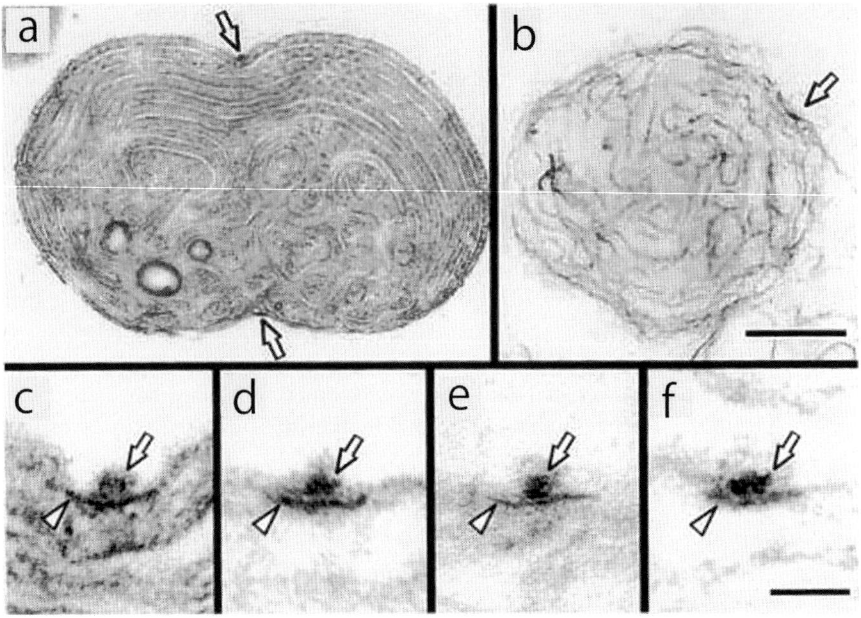


**Fig. 8.7** Effect of pH on the stability of the PD ring. (**a**, **b**) Electron microscope image of whole isolated chloroplasts before (**a**) and after (**b**) being burst in medium at pH 7.6. (**c**, **d**, **e**, **f**) Magnified cross-sectional images of the PD ring before (**c**) and after being burst in medium at pH 5.5 (**d**), 6.5 (**e**), or 7.5 (**f**). *Arrows* and *arrowheads* indicate the outer and inner PD rings, respectively. Scale bar = 500 nm (**a**, **b**), 50 nm (**c**, **d**, **e**, **f**) (Reproduced and modified from Miyagishima et al. 2001b. Copyright American Society of Plant Biologists)

IgG conjugated with 10 nm or 15 nm colloidal gold (at a dilution 1: 50 ~ 1: 80) at 37 °C for 2 h. After being washed with PBS and milliQ water, the grids are then stained with uranyl acetate and examined by transmission electron microscopy (Yagisawa et al. 2009, 2012) (Fig. 8.8).

#### 8.3.2.2 Immunoelectron Microscopy of Ultrathin Sections via Rapid-Freezing Fixation

*C. merolae* cells are collected by centrifugation and sandwiched between two ultrathin copper plates (1 mm × 4 mm × 0.2 mm). Propane is liquefied with liquid nitrogen at −195 °C. The copper plates with samples are rapidly dropped into the liquid propane, and the frozen cells are then transferred to dry acetone (−80 °C) for 48 h. Then, the samples are gradually warmed for 2 h at −20 °C, for 2 h at 0 °C, and for 1 h at room temperature. They are then dehydrated with 100% acetone and embedded in LR white resin (London Resin Co). After being cut into 85-nm-thick

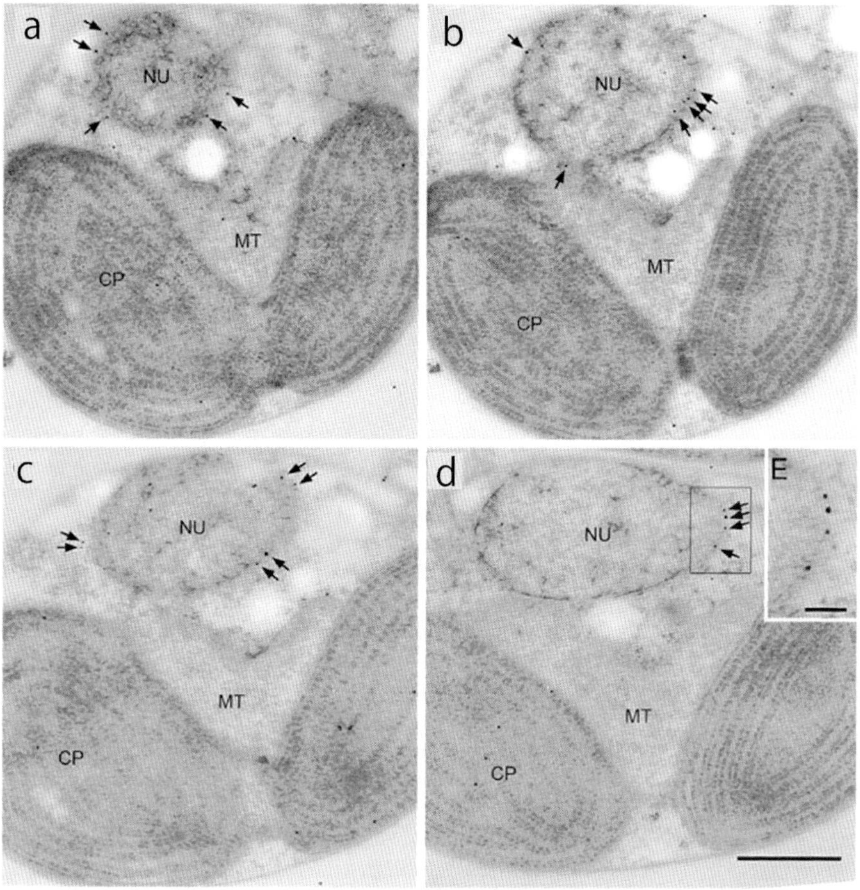


**Fig. 8.8** Immunoelectron microscopy using antibodies against *C. merolae* calnexin. (**a**, **b**, **c**, **d**) Serial sections of a cell and gold particles of the secondary immune-reaction (*arrows*) appear on the nuclear ER or on the edges of the cell nucleus. (**e**) Enlarged image of boxed area in (**d**). Scale bars = 500 nm (**d**), 100 nm (**e**) (Reproduced and modified from Yagisawa et al. 2012)

sections, they are placed on nickel grids coated with a polyvinyl formal film. For immune reactions, the samples on the grids are blocked with 5% BSA in PBS for 30 min at 37 °C and then labeled with primary antibody (at a dilution of 1: 10 ~ 1: 100) for 2 h at 37 °C. After washing the grids with PBS, the secondary immune reactions are performed with secondary antibody: goat anti-rabbit or anti-mouse IgG conjugated with 10 nm or 15 nm colloidal gold (at a dilution of 1: 50) for 2 h at 37 °C. The grids are then stained with uranyl acetate and examined by transmission electron microscopy (Miyagishima et al. 1999c; Nishida et al. 2003, 2007; Fujiwara et al. 2010) (Fig. 8.9).

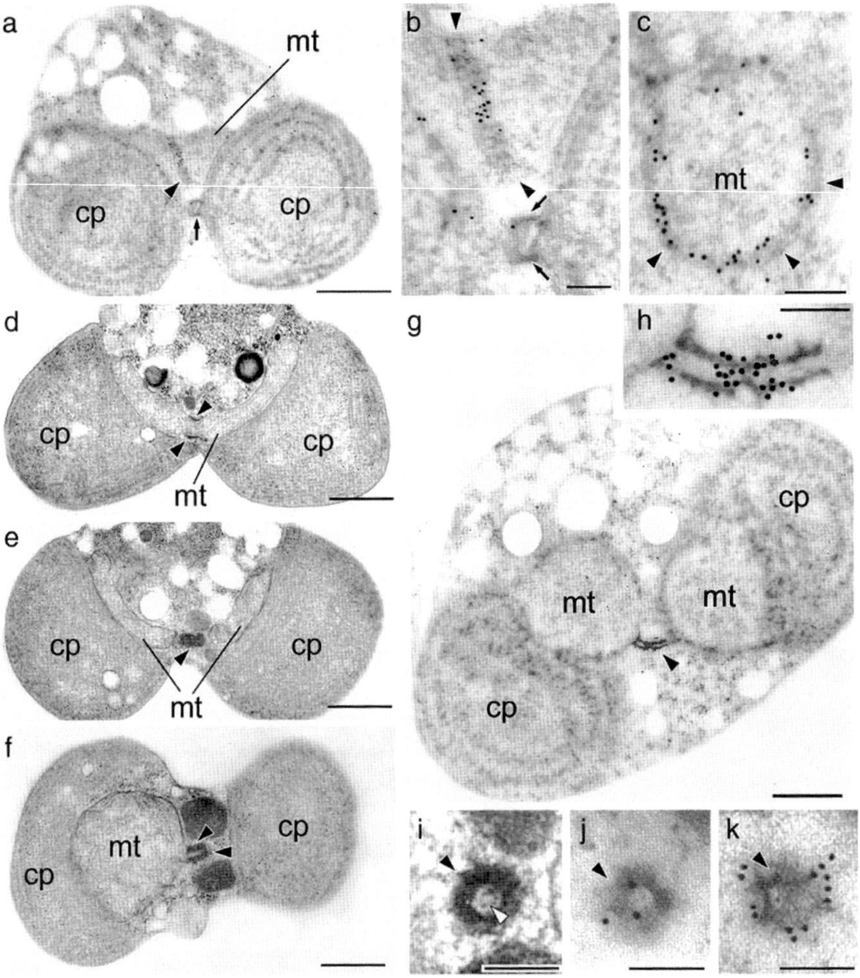


**Fig. 8.9** Immunoelectron micrographs against CmDnm1 of a cell at late phase of mitochondrial division. (**a**) Sections grazing the MD and PD rings and gold particles of the secondary immune reaction of CmDnm1 appear on the MD ring. (**b**) A threefold magnified image of **a**. (**c**) A round section of a MD ring with gold particles. (**d**, **e**, **f**) Electron micrographs of cells with MD ring bridging dividing mitochondria. (**g**) Immunoelectron micrographs of a MD ring bridging dividing mitochondrion at the latest phase. (**h**) A fivefold magnified image of **g**. (**i**, **j**, **k**) Vertical section of the constricted MD ring site at the late phase, which is predicted to correspond to the midpoint of the MD ring in **h**. (**i**) Thin-section electron microscopy showing higher-contrast images. (**j** and **k**) Immunoelectron micrographs of two serial sections. The inner membrane (*open arrowhead*) can be observed in **i** but is invisible in **j**. *pt* plastid, *mt* mitochondrion, *arrow* indicate PD ring; *filled arrowhead* indicates MD ring. Scale bars = 500 nm (**a**, **d**, **e**, **f**), 100 nm (**b**, **c**, and **h**, **i**, **j**, **k**) (Reproduced from Nishida et al. 2003. Copyright (2003) National Academy of Science, USA)

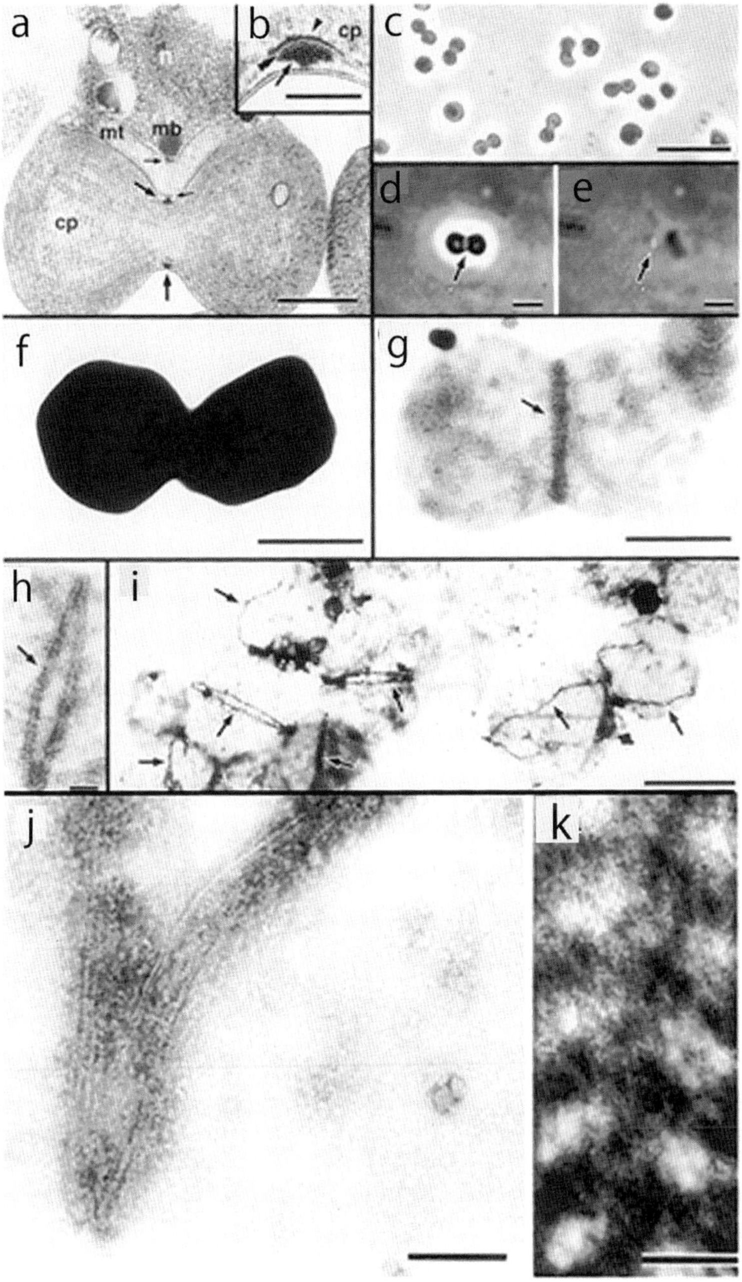


**Fig. 8.10** Visualization of the outer PD ring by negative staining after treatment with Nonidet P-40. (**a**) Thin section of a *C. merolae* cell containing a dividing chloroplast and mitochondrion showing the PD ring and the mitochondrion-dividing ring. The section cut the cell at the center perpendicular to the equator. (**b**) Magnified cross-sectional view of the PD ring showing that the

## *8.3.3 Observation with Negative Staining*

### 8.3.3.1 Observation of Extracted Materials with Negative Staining

Dividing chloroplasts are isolated from *C. merolae* cells and lysed in isolation medium and 0.1% Nonidet P-40 (at a concentration of 1 mg total protein/mL) for 1 h at 0 °C. The insoluble fraction is collected by centrifugation at 10,000 × g for 15 min at 4 °C. The pellets are fixed in a 1% glutaraldehyde, 300 mM sucrose, and 20 mM sodium cacodylate solution (pH 7.2) for 30 min at 0 °C. Samples are then washed twice with distilled water, placed on copper grids, and negatively stained with 0.5% phosphotungstic acid (pH 7.0). The grids are examined by transmission electron microscopy (Miyagishima et al. 2001b, c) (Fig. 8.10).

### 8.3.3.2 Immunoelectron Microscopy of Extracted Materials Via Negative Staining

Isolated chloroplasts, which are insoluble after the 0.25% Nonidet P-40 treatment, are not fixed. For immune reactions, samples are blocked with 5% BSA in PBS for 30 min at 4 °C. Primary antibody is diluted in Can Get Signal Immunostain Solution B (TOYOBO, Japan) (at a dilution 1:100), and primary immune reactions are performed for 1 h at 4 °C. The secondary immune reactions are proceeded with secondary antibody conjugated with 10 nm or 15 nm gold particles (at a dilution 1:20). Samples are placed on nickel grids, negatively stained with 0.5% phosphotungstic acid (pH 7.0), and examined by transmission electron microscopy (Miyagishima et al. 2001b; Yoshida et al. 2006, 2009, 2010, 2012) (Fig. 8.11).

**Fig. 8.10** (continued) PD ring is composed of three rings. (**c**) Phase-contrast image of isolated dividing chloroplasts from a synchronized culture. (**d**) The outer PD ring of an isolated dividing chloroplast was visualized as *green fluorescence* by labeling surface proteins of the chloroplast with NHS-biotin and detecting them with FITC-avidin. (**e**) Chloroplasts shown in (**d**) are lysed by the addition of a solution containing 0.5% Nonidet P-40 from the edge of the cover glass. The outer PD ring remained insoluble. (**f** and **g**) Negative staining images of isolated chloroplasts before (**f**) and after 0.1% Nonidet P-40 treatment (**g**). (**h**) The outer PD ring was clearly observed as a closed ring by negative staining after Nonidet P-40 treatment in oblique samples. (**i**) Low magnification images frequently showed that the outer PD rings are attached to the outer envelopes of chloroplasts. (**j**) The outer ring is composed of a number of 5nm filaments. (**k**) Higher magnification of two adjacent filaments showing straight arrays of globular proteins along the filaments. *Small arrows* indicate the mitochondria-dividing ring. *Large arrows*, *arrowhead*, and *double arrowhead* indicate the outer, inner, and middle PD rings, respectively. *cp* chloroplast, *mb* microbody, *mt* mitochondrion, *n* nucleus. Scale bars = 500 nm (**a**, **f**, **g**), 100 nm (**b**, **h**, **j**), 5 μm (**c**), 2 μm (**d**, **e**), 1 μm (**i**), and 5 nm (**k**) (Reproduced and modified from Miyagishima et al. 2001b. Copyright American Society of Plant Biologists)

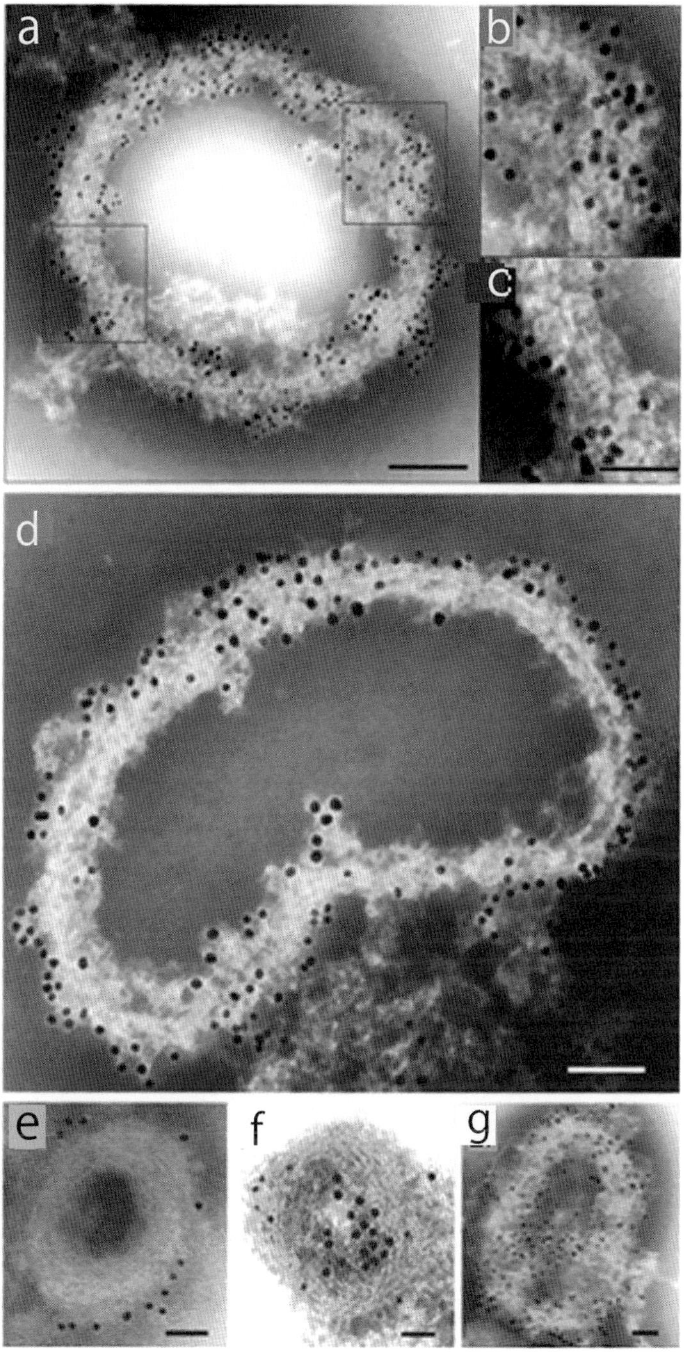


**Fig. 8.11** Construction of plastid-dividing machinery demonstrated by immunoelectron microscopy. (**a**, **b**, **c**) Immuno-EM micrographs of PDR1 and (**d**, **e**, **f**, **g**) FtsZ, dynamin, and PDR1 in the plastid-dividing (PD) machinery. Immunogold particles labeling PDR1 localized on the whole PD ring. (**b**, **c**) Enlarged micrographs of the boxed areas in (**a**). More immunogold particles appeared

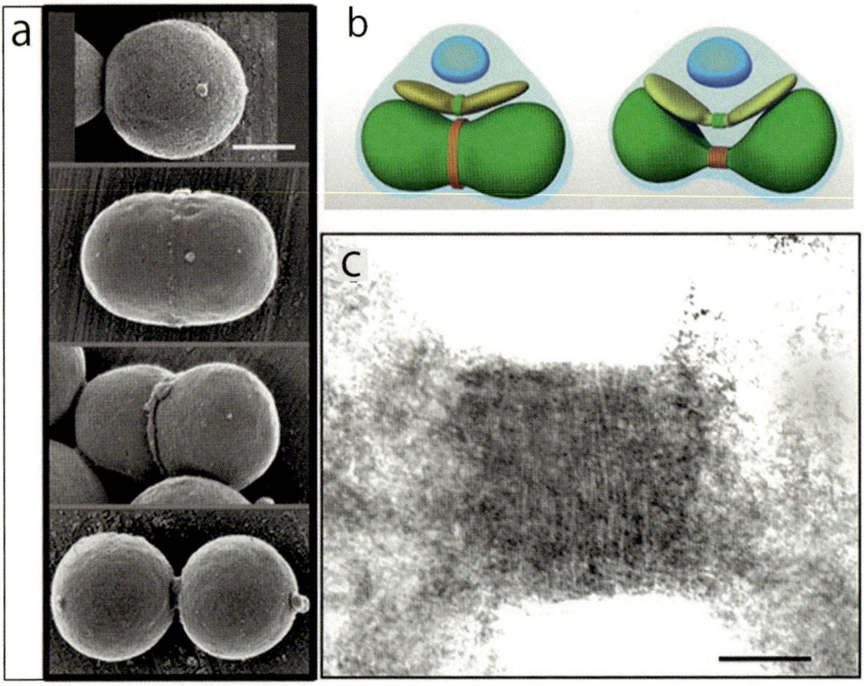


**Fig. 8.12** Micrographs and schema of dividing plastids in the red alga *C. merolae*.(**a**) Field-emission scanning electron microscopy images showing formation and constriction of the outer PD machineries at the plastid division site (from *top* to *bottom*). At the early phase of division, small granules appear along the division site. (**b**) Schematic illustration of the PD machinery (*red* ring) and the MD machinery (*green* ring) in early (*left*) and late (*right*) division phases. The outer PD machinery and MD machinery consist of a bundle of filaments, respectively. Nucleus, *blue*; mitochondrion, *yellow*; plastid, *green*. (**c**) Electron micrograph of the outer PD ring obtained by treating isolated plastids with detergent and negative staining. The image during the late stage of constriction reveals that the ring is composed of filaments that are 5–7 nm in diameter. Scale bars = 500 nm (**a**), 50 nm (**c**). (Reproduced and modified with permission from Yoshida et al. 2012)

**Fig. 8.11** (continued) in the less condensed F-PDR region (**b**) than in the solid F-PDR region (**c**). The PD machineries during middle (**d**) and late phases (**e**, **f**, **g**) of constriction are shown. (**d**) Dynamin labeled with 10 nm gold particles localizes on the outer (cytosolic) side of the machinery, while PDR1 labeled with 15 nm gold particles appears randomly on the machinery. (**e**) Dynamin labeled with 10 nm gold particles localizes on the outer side of the machinery. (**f**) In contrast to dynamin labeled with 10 nm gold particles, FtsZ labeled with 15 nm gold particles localizes at the inner (stromal) side of the PD machinery. The machinery appears to be a bundle of fine filaments. (**g**) PDR1 labeled with 10 nm gold particles localizes throughout the outer PD ring filaments. Scale bars = 50 nm (**a**, **e**, **f**, **g**), 500 nm (**b**, **c**), and 100 nm (**d**). [**a**, **b**, **c**, **d**: Reproduced and modified from Yoshida et al. 2010. Reprinted with permission from AAAS], (**e**, **f**, **g**: Reproduced and modified with permission from Yoshida et al. 2012)

### 8.3.4 Scanning Electron Microscopy

Cells or isolated chloroplasts are collected by centrifugation and then fixed with 1% glutaraldehyde in 20 mM sodium cacodylate (pH 7.2) for 2 h at room temperature. Samples are washed with PBS (pH 7.2) for 20 min three times and then dehydrated in a graded ethanol series. They are then dried to the critical point, mounted, and then sputter coated with platinum. The samples are examined by a field-emission scanning electron microscope (Miyagishima et al. 1999c; Yoshida et al. 2012) (Fig. 8.12).

## References

Elrod VA, Johnson KS, Coale KH (1991) Determination of subnanomolar levels of iron (II) and total dissolved iron in seawater by flow injection analysis with chemiluminescence detection. Anal Chem 63:893–898

Fujiwara T, Kuroiwa H, Yagisawa F, Ohnuma M, Yoshida Y, Yoshida M, Nishida K, Misumi O, Watanabe S, Tanaka K, Kuroiwa T (2010) The coiled-coil protein VIG1 is essential for tethering vacuoles to mitochondria during vacuole inheritance of *Cyanidioschyzon merolae*. Plant Cell 22:772–781. https://doi.org/10.1105/tpc.109.070227

Imoto Y, Fujiwara T, Yoshida Y, Kuroiwa H, Maruyama S, Kuroiwa T (2010) Division of cell nuclei, mitochondria, plastids, and microbodies mediated by mitotic spindle poles in the primitive red alga *Cyanidioschyzon merolae*. Protoplasma 241:63–74. https://doi.org/10.1007/s00709-010-0107-y

Imoto Y, Kuroiwa H, Ohnuma M, Kawano S, Kuroiwa T (2012) Identification of peroxisome-dividing ring in *Cyanidioschyzon merolae* based on organelle partner hypothesis. Cytologia 77(4):515–522. https://doi.org/10.1508/cytologia 77.515

Imoto Y, Abe Y, Okumoto K, Honsho M, Kuroiwa H, Kuroiwa T, Fujiki Y (2017) Defining the dynamin-based ring organizing center on the peroxisome-dividing machinery isolated from *Cyanidioschyzon merolae*. J Cell Sci 130:853–867. https://doi.org/10.1242/jcs.199182

Kuroiwa T, Suzuki T (1980) An improved method for the demonstration of the in situ chloroplast nuclei in higher plants. Cell Struct Funct 5:195–197

Kuroiwa T, Suzuki T, Ogawa K, Kawano S (1981) The chloroplast nucleus: distribution, number, size and shape, and a model for the multiplication of the chloroplast genome during chloroplast development. Plant Cell Physiol 22:381–396

Kuroiwa T, Suzuki T, Itoh R, Toda K, O'Keefe TC, Kuroiwa H (1995) Mitochondria-dividing ring : ultrastructural basis for the mechanism of mitochondrial division in *Cyanidioschyzon merolae*. Protoplasma 186:12–23. https://doi.org/10.1007/BF01280718

Matsuzaki M, Misumi O, Shin-i T, Maruyama S, Takahara M, Miyagishima S, Mori T, Nishida K, Yagisawa F, Nishida K, Yoshida Y, Nishimura Y, Nakao S, Kobayashi T, Momoyama Y, Higashiyama T, Minoda A, Sano M, Nomoto H, Oishi K, Hayashi H, Ohta F, Nishizaka S, Haga S, Miura S, Morishita T, Kabeya Y, Terasawa K, Suzuki Y, Ishii Y, Asakawa S, Takano H, Ohta N, Kuroiwa H, Tanaka K, Shimizu N, Sugano S, Sato N, Nozaki H, Ogasawara N, Kohara Y, Kuroiwa T (2004) Genome sequence of the ultrasmall unicellular red alga *Cyanidioschyzon merolae* 10D. Nature 428:653–657. https://doi.org/10.1038/nature02398

Miyagishima S, Itoh R, Toda H, Takahashi H, Kuroiwa H, Kuroiwa T (1998a) Visualization of the microbody division in *Cyanidioschyzon merolae* with the fluorochrome brilliant sulfoflavin. Protoplasma 201:115–119. https://doi.org/10.1007/BF01280718

Miyagishima S, Itoh R, Toda K, Takahashi H, Kuroiwa H, Kuroiwa T (1998b) Orderly formation of the double ring structures for plastid and mitochondrial division in the unicellular red alga *Cyanidioschyzon merolae*. Planta 206:551–560. https://doi.org/10.1007/s004250050432

Miyagishima S, Itoh R, Toda K, Takahashi H, Kuroiwa H, Kuroiwa T (1998c) Identification of a triple ring structure involved in plastid division in the primitive red alga *Cyanidioschyzon merolae*. J Electron Microsc 47(3):269–272

Miyagishima S, Itoh R, Aita S, Kuroiwa H, Kuroiwa T (1999a) Isolation of dividing chloroplasts with intact plastid-dividing rings from a synchronous culture of the unicellular red alga *Cyanidioschyzon merolae*. Planta 209:371–375. https://doi.org/10.1007/s004250050645

Miyagishima S, Itoh R, Toda K, Kuroiwa H, Kuroiwa T (1999b) Real-time analyses of chloroplast and mitochondrial division and differences in the behavior of their dividing rings during contraction. Planta 207:343–353. https://doi.org/10.1007/s0042500491

Miyagishima S, Itoh R, Toda K, Kuroiwa H, Nishimura M, Kuroiwa T (1999c) Microbody proliferation and segregation cycle in the single-microbody alga *Cyanidioschyzon merolae*. Planta 208:326–336. https://doi.org/10.1007/s004250050566

Miyagishima S, Kuroiwa H, Kuroiwa T (2001a) The timing and manner of disassembly of the apparatuses for chloroplast and mitochondrial division in the red alga *Cyanidioschyzon merolae*. Planta 212:517–528. https://doi.org/10.1007/s004250000426

Miyagishima S, Takahara M, Kuroiwa T (2001b) Novel filaments 5 nm in diameter constitute the cytosolic ring of the plastid division apparatus. Plant Cell 13:707–721. https://doi.org/10.1105/tpc.13.3.707

Miyagishima S, Takahara M, Mori T, Kuroiwa H, Higashiyama T, Kuroiwa T (2001c) Plastid division is driven by a complex mechanism that involves differential transition of the bacterial and eukaryotic division rings. Plant Cell 13:2258–2268

Miyagishima S, Nishida K, Mori T, Matsuzaki M, Higashiyama T, Kuroiwa H, Kuroiwa T (2003) A plant-specific Dynamin-related protein forms a ring at the chloroplast division site. Plant Cell 15:655–665. https://doi.org/10.1105/tpc.009373

Nishibayashi S, Kuroiwa T (1985) Division of mitochondrial nuclei in protozoa, a green alga and a higher plant. Cytologia 50:75–82

Nishida K, Takahara M, Miyagishima S, Kuroiwa H, Matsuzaki M, Kuroiwa T (2003) Dynamic recruitment of dynamin for final mitochondrial severance in a primitive red alga. Proc Natl Acad Sci 100:2146–2151. https://doi.org/10.1073/pnas.0436886100

Nishida K, Misumi O, Yagisawa F, Kuroiwa H, Nagata T, Kuroiwa T (2004) Triple Immunofluorescent labeling of FtsZ, Dynamin, and EF-Tu reveals a loose association between the inner and outer membrane mitochondrial division machinery in the red alga *Cyanidioschyzon merolae*. J Histochem Cytochem 52:1–7. https://doi.org/10.1369/jhc.3C6315.2004

Nishida K, Yagisawa F, Kuroiwa H, Yoshida Y, Kuroiwa T (2007) WD40 protein Mda1 is purified with Dnm1 and forms a dividing ring for mitochondria before Dnm1 in *Cyanidioschyzon merolae*. Proc Natl Acad Sci 104(11):4736–4741. https://doi.org/10.1073/pnas.0609364104

Nozaki H, Takano H, Misumi O, Terasawa K, Matsuzaki M, Maruyama S, Nishida K, Yagisawa F, Yoshida Y, Fujiwara T, Takio S, Tamura K, Chung SJ, Nakamura S, Kuroiwa H, Tanaka K, Sato N, Kuroiwa T (2007) A 100%-complete sequence reveals unusually simple genomic features in the hot-spring red alga Cyanidioschyzon. BMC Biol 5:28–37. https://doi.org/10.1186/1741-7007-5-28

Suzuki K, Ehara T, Osafune T, Kuroiwa H, Kawano S, Kuroiwa T (1994) Behavior of mitochondria, chloroplasts and their nuclei during the mitotic cycle in the ultramicroalga *Cyanidioschyzon merolae*. Eur J Cell Biol 63:280–288

Yagisawa F, Nishida K, Yoshida M, Ohnuma M, Shimada T, Fujiwara T, Yoshida Y, Misumi M, Kuroiwa H, Kuroiwa T (2009) Identification of novel proteins in isolated polyphosphate vacuoles in the primitive red alga *Cyanidioschyzon merolae*. Plant J 60:882–893. https://doi.org/10.1111/j. 1365-313X.2009.04008.x

Yagisawa F, Fujiwara T, Kuroiwa H, Nishida K, Imoto Y, Kuroiwa T (2012) Mitotic inheritance of endoplasmic reticulum in the primitive red alga *Cyanidioschyzon merolae*. Protoplasma 249:1129–1135. https://doi.org/10.1007/s00709-011-0359-1

Yagisawa F, Kuroiwa H, Fujiwara T, Kuroiwa T (2016) Intracellular structure of the unicellular red alga, *Cyanidioschyzon merolae*, in response to phosphate depletion and resupplementation. Cytologia 81(3):341–347. https://doi.org/10.1508/cytologia.81.341

Yamada M, Sudo A, Suzuki S (1985) Chemiluminescence method for selective determination of iron (II) and chromium (II) with single reaction system. Che Lett Chem Soc Jpn:801–804

Yoshida Y, Kuroiwa H, Misumi O, Nishida K, Yagisawa F, Fujiwara T, Nanamiya H, Kawamura F, Kuroiwa T (2006) Isolated chloroplast division machinery can actively constrict after stretching. Science 313:1435–1438. https://doi.org/10.1126/science.1129689

Yoshida Y, Kuroiwa H, Hirooka S, Fujiwara T, Ohnuma M, Yoshida M, Misumi O, Kawano S, Kuroiwa T (2009) The bacterial ZapA-like protein ZED is required for mitochondrial division. Curr Biol 19:1–7. doi: 101016/j.cub.2009.07.035

Yoshida Y, Kuroiwa K, Misumi O, Yoshida M, Ohnuma M, Fujiwara T, Yagisawa F, Hirooka S, Imoto Y, Matsushita K, Kawano S, Kuroiwa T (2010) Chloroplasts divide by contraction of a bundle of Nanofilaments consisting of Polyglucan. Science 329:949–953. https://doi.org/10.1126/science.1190791

Yoshida Y, Miyagishima S, Kuroiwa H, Kuroiwa T (2012) The plastid-dividing machinery: formation, constriction and fission. Curr Opin Plant Biol 15:1–8. https://doi.org/10.1242/jcs.116798

# Chapter 9
# Cytological Analyses by Advanced Electron Microscopy

**Takako M. Ichinose and Atsuko H. Iwane**

**Abstract** Recently several microscopy techniques have revealed the dynamics and structure of intracellular material such as organelles and supramolecular proteins for a better understanding of the fundamental properties of a cell. Electron microscopy provides information on extremely small biological specimens, providing images of several target molecules and organelles, including mitochondria and plastids, and their surrounding intracellular environments. Technical development using several advanced electron microscopes has been advanced to obtain 3D structural modeling. Among them, cryo-TEM and tomography techniques are useful for visualization of fine structures, but the observation area is limited. Conversely, "FIB-SEM" theoretically is not affected by the thickness of the sample, and FIB-SEM and reconstruction are effective for 3D structural modeling of *C. merolae* cells (2–5 microns) during cell division at the whole cell level. Another technique, UHVEM tomography, visualizes a detailed 3D structure of the light-harvesting complex on the surface of the thylakoid membrane in the plastid and the micro-compartments (cristae) in mitochondria. Furthermore, fine whole cell 3D structural model using cryo-STEM tomography, which is obtained under physical fixation and without staining, instead of chemical fixation, is compared with the models of FIB-SEM and 3D reconstruction.

**Keywords** FIB-SEM • SBF-SEM • EM-tomography • EM (electron microscopy) • Block-face imaging • Cryo-TEM • Cryo-TEM tomography • *Cyanidioschyzon merolae* • Physical fixation

## Abbreviations

| | |
|---|---|
| 3D | Three-dimensional |
| ATUM-SEM | Automated tape-collecting ultramicrotome |

T.M. Ichinose • A.H. Iwane (✉)
Cell Field Structure, Quantitative Biology Center, Riken, 3-10-23, Kagamiyama, 739-0046 Higashi-Hiroshima, Hiroshima, Japan

Special Research Promotion Group, Graduate School of Frontier Biosciences, Osaka University, 1-3, Yamadaoka, 565-0871 Suita, Osaka, Japan
e-mail: atsuko.iwane@riken.jp



T. Kuroiwa et al. (eds.), *Cyanidioschyzon merolae*,
https://doi.org/10.1007/978-981-10-6101-1_9

| | |
|---|---|
| BSE | Backscattered electron |
| CMOS | Complementary metal oxide semiconductor |
| EM | Electron microscopy |
| FIB-SEM | Focused ion beam SEM |
| HAADF | High-angle annular dark field |
| HPF | High-pressure freezing |
| LM | Light microscopy |
| SBF-SEM | Serial block-face scanning electron microscopy |
| SBF-SEM | Serial block-face SEM |
| SEM | Scanning electron microscopy |
| STEM | Scanning transmission electron microscopy |
| TEM | Transmission electron microscopy |
| UHVEM | Ultra-high-voltage electron microscopy |

## 9.1 Introduction

Electron microscopy offers information on extremely small biological specimens, providing images of various target molecules and organelles, including mitochondria and plastids, and their surrounding intracellular environments. EM structural analysis of *C. merolae* has been mainly achieved using conventional embedding, ultrathin sectioning, and TEM observation (Fig. 9.1, Itoh et al. 1999). Using this technique, useful structural information that includes the position of each organelle during mitosis and the identification of the division rings has been reported (Kuroiwa et al. 2006; Yoshida et al. 2010). Furthermore, Miyagishima et al. reported the morphological changes of several organelles, especially the peroxisome 3D model during mitosis from photographic prints of serial thin sections

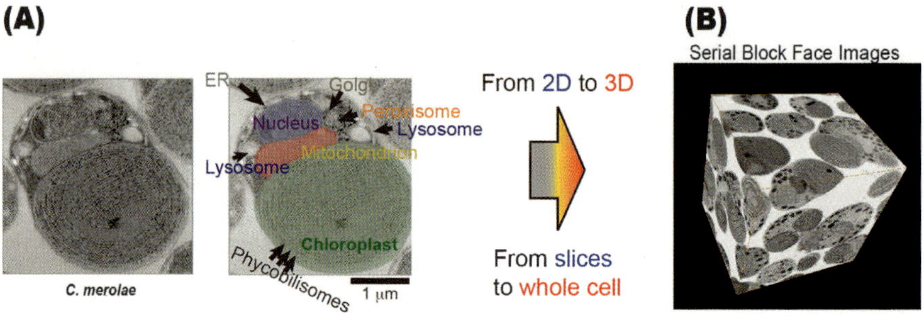


**Fig. 9.1** The difference between electron microscope modalities
Conventional biological TEM image has been applied for fine (2D) imaging of structures thousands of times smaller than the smallest resolvable object with a light microscope. TEM is capable of expressing shape and positional relationship on the same plane and has been positively performed as a representation method for structural analysis (**a**). On the other hand, serial block-face scanning electron microscopy (SBF-SEM) is one example of the methods which generate at high-resolution three-dimensional images from small samples, often biological samples. This method is particularly good at imaging large fields of view in X, Y, Z at nanometer resolution (**b**)

using a diamond knife and 3D reconstruction technique (Miyagishima et al. 1998, 1999). This method requires very advanced skills, dexterity, and time. Sections can be created by ultramicrotome, but picking up sections sequentially, without missing contiguous pieces floating on the water surface, has required a fairly advanced technique. On the other hand, conventionally stained cell slices with thickness of about 100–150 nanometers visualized with the TEM tomography might retain specific information at partial part on cells. However, the resolution of TEM is better than that of SEM, which can obtain contrast information about the sample surface from the backscattered electrons generated by the applied electron beam. Now, it is possible to automatically obtain a sequential two-dimensional image without physically picking up the sections. Typical devices are FIB-SEM and SBF-SEM. The difference between these typical devices is that the cutting interval and the range of observational field of view are different in on-block sectioning with gallium ion beams or diamond knives. Generally, gallium ions can mill off the block surface at intervals of several nanometers to 10 nanometers, but in the case of diamond knives attached to the SEM, it is 30–50 nm sections (Peddie and Collinson 2014). Because *C. merolae* cells are small, some intra-structure information about the cells might be overlooked, even if sequential sections are used. Although the observational field of view of FIB-SEM is smaller than that of SBF-SEM, it is not very important for *C. merolae* as the size of the cells is small. So, in the case of *C. merolae* cells at whole cell level, FIB-SEM is thought to be suitable.

For 3D EM structural analysis, several new advantageous techniques (Fig. 9.1b), "large volume analysis" as generally speaking including FIB-SEM, SBF-SEM, ATUM-SEM, FIB tomography, array tomography, and TEM tomography, have recently been developed, mainly classified into two groups. First are the machines and/or techniques which obtain "sequential x-y slices" with the introduction of high-performance field emission scanning electron microscopy (SEM): FIB-SEM, SBF-SEM, and ATUM-SEM (FIB-SEM, Giannuzzi et al. 1999, Denk and Horstmann 2004, Knott et al. 2008, Ohta et al. 2012; SBF-SEM, Pinali and Kitmitto 2014; ATUM-SEM, Schalek et al. 2011, Hayworth et al. 2014, Wanner et al. 2015, Peddie and Collinson 2014, Ichimura et al. 2015, Titze and Genoud 2016). Second is "TEM tomography by UHVEM" (Takaoka et al. 2008; Kunimoto et al. 2012; Takahashi et al. 2015). SEM observation can be acquired from the surface of ultrathin sections thereby removing the need to mount sections on an electron-transmission substrate. Although TEM is expected to be of higher resolution than SEM, sample thickness is limited. The character of these two techniques is shown as follows (Fig. 9.2).

*C. merolae* is a primitive unicellular red alga, and its genome is covered with nuclear membrane. So *C. merolae* is considered to be the oldest eukaryotic organism with mitochondria. The size of a *C. merolae* cell is 2–5 microns. The cell has simple individual double-membrane organelles such as nucleus, plastid, and mitochondrion and single-membrane organelles such as the ER, Golgi, and peroxisome. Each parent cell divides into two daughter cells. Cell division is an essential life-sustaining process, and because the number of organelles is few and the cooperation

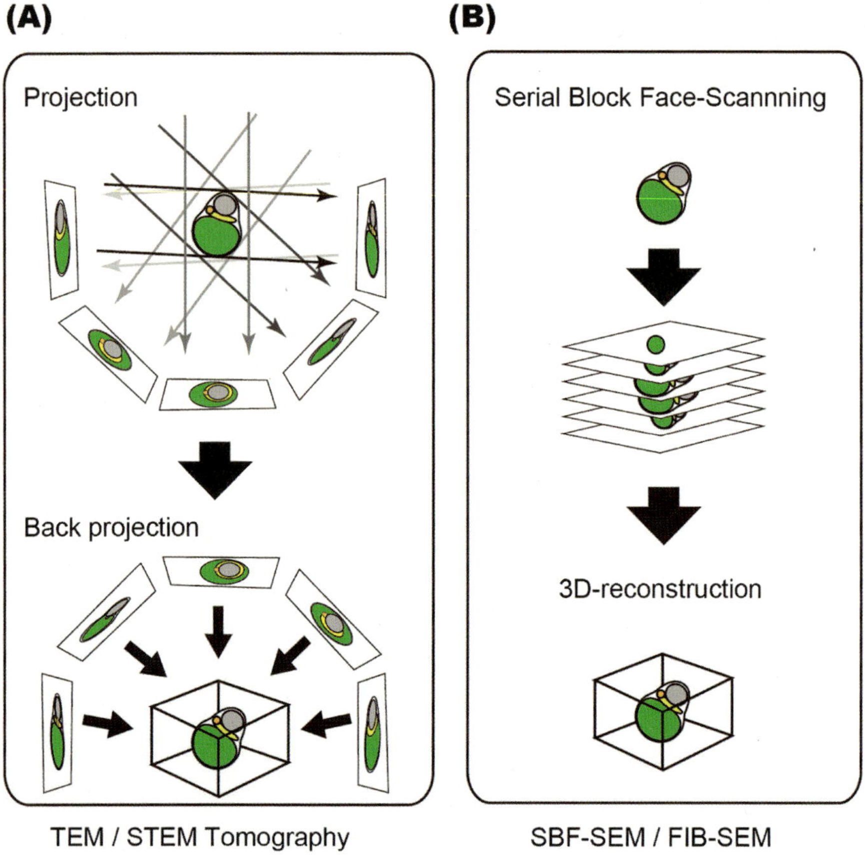


**Fig. 9.2** Two types of 3D reconstruction methods
Electron tomography (ET) of biological samples is used to study the organization and the structure of the whole cell and subcellular complexes in great detail. The illustration on the *left* (**a**) shows the principles of tomography. The projection from various angles was mounted against the specimen (*upper*). Projections cannot be acquired over a full tilt angle range (usually limited to −70°−+70 ° at 2x cosθ degrees tilt interval) with biological samples in electron microscopy. Then the 3D model was reconstructed by Fourier synthesis transformation from all sections (*lower*). Electron tomography image reconstruction is considered to contain incompleteness and resolution problems because of the missing edge for angle dependency. Sequential x-y slice images which are obtained using FIB-SEM/SBF-SEM are reconstructed for 3D model (**b**). Although there is no theoretical limit in the thickness direction (z-axis), the resolution of SEM is some inferior to that of TEM

between organelles can be shown relatively clearly, *C. merolae* was chosen for 3D structural analysis at the whole organism level including the complete intracellular environment.

In the first half of this chapter, the division process of *C. merolae* at the whole cell level, which was obtained mainly from FIB-SEM sequential x-y slice images with 3D reconstruction of the form change and interaction of the main organelles, and the thylakoid membrane model inside the plastid is shown. In the latter half, the model of crista in mitochondrion, the model of phycobilisome (photosynthesis-related protein) complex on the thylakoid membrane in the plastid, and the plastid division ring are visualized with UHVEM using the 200 MV; super high-voltage TEM tomography method, as they are somewhat difficult to obtain from FIB-SEM observation. Finally, the preliminary 3D model of the whole *C. merolae* cell is shown using the cryo-tomography technology which can image the true structure without chemical fixation and staining. Through these advanced electron microscope analyses, the localization and interaction of target molecules and organelles observed by optical microscopy can be displayed in a more detailed 3D model, and new technologies that can approach the functions and behaviors of target molecules and organelles from the structural point of view are applied.

## 9.2 Electron Microscopy

### *9.2.1 General Biological Sample Preparation for EM Analysis*

Figure 9.3 shows a flowchart of several types of sample preparation of cells. The gray line of arrows is a conventional method for TEM observation. Chemical fixation by glutaraldehyde or paraformaldehyde and osmium tetroxide, and dehydration by ethanol. After EPON embedding and ultrathin sectioning by ultramicrotome (with optional restaining with uranyl acetate and/or a suitable heavy metal), the ultrathin section is observed using conventional TEM (e.g., JEOL: JEM1011). This process takes about a week as a series of steps. Another TEM observing method using a high-pressure freezing instead of chemical fixation is the line of blue arrows (I)*. Chemical fixation remains the gold standard in biological research and histopathology. However, using this approach the specimen is taken from an aqueous environment and placed in a nonaqueous milieu with changes to the osmolarity, pH, and ionic strength as a result of chemical fixation, followed by dehydration, embedding, and sectioning processes (Maunsbach 1966). Cryogenic methods have been developed in attempts to preserve specimens in as close to the native state as possible and avoid chemical fixation (Dubochet et al. 1981). Physical fixation is achieved mainly by three different techniques: cold metal block freezing, plunge freezing, and high-pressure freezing (Pinali and Kitmitto 2014). *C. merolae* does not have a wall, and it changes easily the shape of the cell by temperature, pH, etc. High-pressure freezing (HPF) as presented by Moor and Riehle (1968) is a technique that allows for a depth of 100–300 μm from the specimen surface to be

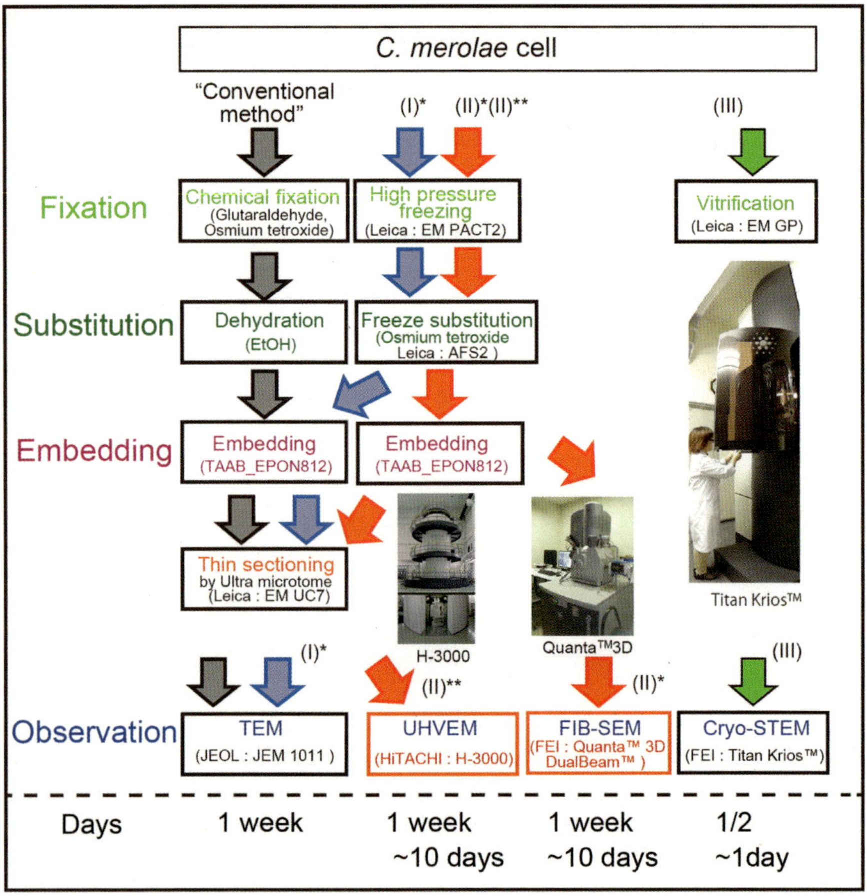


**Fig. 9.3** Flowchart for sample preparation for EM analysis
*Gray line of arrows*: Conventional TEM observation methods using chemical fixing. (I)* *Blue line of arrows*: Conventional TEM observation methods using physical fixing. (II)* *Red line of arrows*: FIB-SEM observation methods using physical fixing. (II)** *Red line of arrows*: UHVEM tomography observation methods using physical fixing. (III)* *Green line of arrows*: cryo-STEM observation methods using physical fixing and no staining with heavy metals. Sample preparation period for each observations are shown on the bottom column. Each of the displayed devices is an example

frozen without detectable ice crystallization damage (Dahl and Staehelin 1989; Studer et al. 1989; Vanhecke et al. 2008; Leunissen and Yi 2009), which could damage the organelle structure with the release/leakage of soluble elements into the cell. Fixation of *C. merolae* in which cell walls are absent is easily damaged by chemical fixatives and is considered that the form of organelles may be affected, so physical fixation by HPF is suitable.

FIB-SEM is the red line of arrows (II)*. High-pressure freezing and freeze substitution and EPON embedding. After smoothing surface of the embedding by ultramicrotome, the obtained specimen is put on FIB-SEM machine. Sequential SEM images which are obtained by FIB-SEM machine are aligned and used to create a 3D model using the Amira 3D analysis software. It takes 1 week to 10 days in a series of steps depending on the depth of FIB milling. (Details are described later, Sect. 9.3.1.)

UHVEM is the red line of arrows (II)**, and until EPON embedding is the same as (II)*. However, after ultrathin sectioning by ultramicrotome, there is restaining by uranyl acetate and lead.

Cryo-STEM tomography observation is the green line of arrows (III). After just physical fixation, by vitrification, the intracellular architecture is obtained using cryo-scanning transmission electron tomography. This process takes about 1 day.

## 9.3 Advanced Electron Microscopy I (FIB-SEM)

### *9.3.1 Introduction*

Three-dimensional (3D) structural information on multiple scales is of central importance in biological research. Electron microscopy (EM) offers the best resolution of extremely small biological specimens, providing images of the minutest of organelles and molecules that are responsible for phenomena fundamental to life including cell division, differentiation, and proliferation. Obtaining such volumetric reconstructions is especially useful for analysis of *C. merolae* cells at whole cell level. FIB-SEM is a microscope that has already contributed to surface processing and observation of metals and ceramics (Li et al. 1998; Dunn and Hull 1999). Observation with FIB-SEM machine does not affect the thickness of the spatial resolution, so this technology might be applicable to biological specimens. For the 3D analysis at whole cell level, FIB-SEM is suitable for mesoscale imaging just between electron beam tomography and X-ray CT (Fig. 9.4). It has a suitable resolution for observing intracellular architecture and intercellular interactions. Figure 9.5 shows a cartoon illustration of essential part of FIB-SEM. Sequential 2D images of cells are obtained by taking a SEM image for each FIB ablation of an ultrathin section.

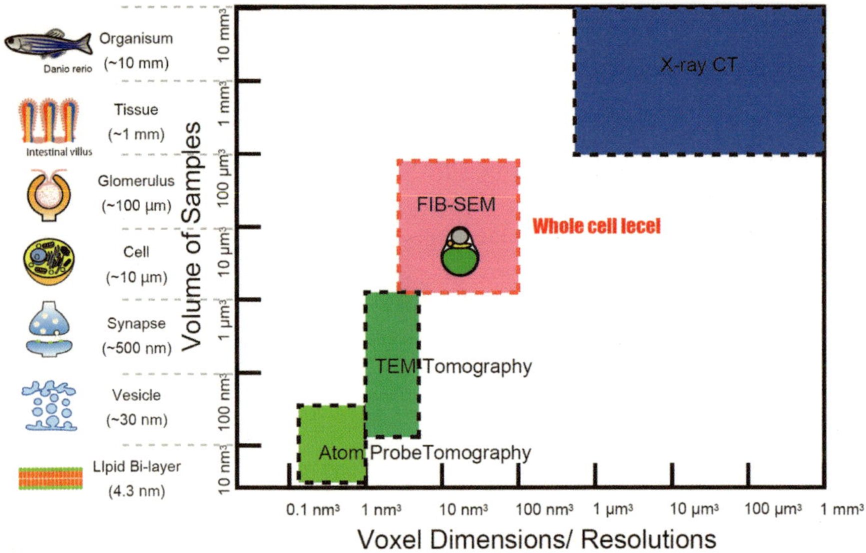


**Fig. 9.4** 3D structural analysis
Characteristic information of the structural analysis tools are presented. Based on the size of the biological material and the observational field, selection of an appropriate measurement system is required. The horizontal axis shows voxel dimensions/resolution, and the vertical axis shows the sample volume. For the 3D analysis of *C. merolae* cells at whole cell level, FIB-SEM is suitable for mesoscale imaging just between electron beam tomography and X-ray tomography

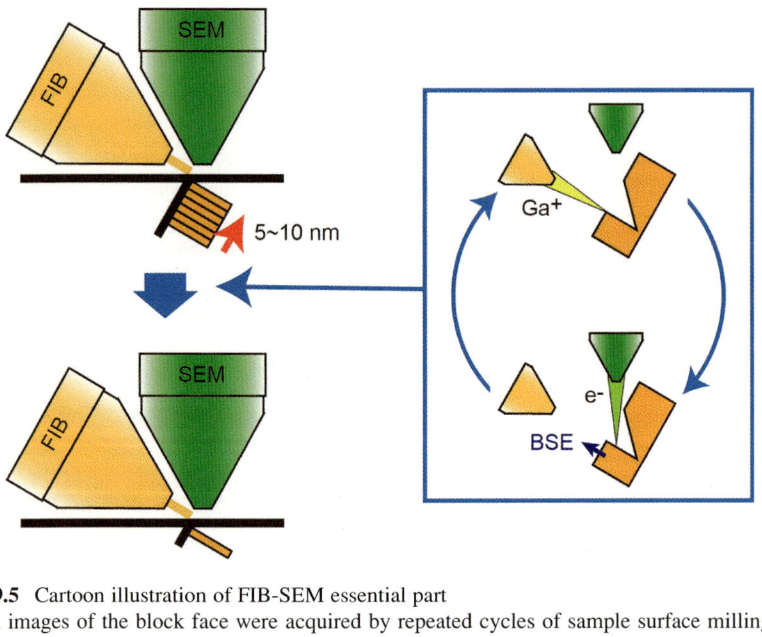


**Fig. 9.5** Cartoon illustration of FIB-SEM essential part
Serial images of the block face were acquired by repeated cycles of sample surface milling and imaging

### 9.3.2 *Preparation of* C. merolae *Cells for FIB-SEM Analysis (Fig. 9.3)*

*C. merolae* cells (NIES-1804) were obtained from the National Institute for Environmental Studies and cultured in 2x Allen's medium under continuous light (50 W/m$^2$ square meter) at 42°. Cells were harvested at the log phase. For keeping the delicate cell structure of the *C. merolae*, the physical fixation by high-pressure freezing is suitable for vitrifying samples up to 200 μm in thickness without the artifacts of chemical fixation (Vanhecke et al. 2008). High-pressure freezing and freeze substitution of *C. merolae* cells were processed according to the modified Leica microsystems manuals. After centrifugation of 100 ml cell culture at 1,800xg for 2 min at room temperature, the culture medium is largely removed, and cells are kept at 40°. Physical fixation of the obtained cells is carried out by dividing it into five parts using a flat specimen carrier (Leica microsystems parts: No.16706898). Freeze substitution is carried out in a 2% $OsO_4$/1% $H_2O$/0.1% potassium ferrocyanide/acetone solution over 72 h from −90° to room temperature. According to the OTO method (Willingham and Rutherford 1984), the *C. merolae* sample is stained in an acetone solution. Afterward, the *C. merolae* sample is post-stained with 1% uranyl acetate and 1% Kushida's lead in an acetone solution, and resin embedding was carried out with EPON 815 (6:4). After smoothing surface of the embedding by ultramicrotome, the obtained specimens (~1 mm × 1 mm × 1 mm) are put on the suitable specimen holder for FIB-SEM machine and then are coated with carbon to prevent charge-up.

### 9.3.3 *FIB-SEM Observation and 3D Reconstruction*

Sequential (7–15 nm intervals) SEM images which are obtained by a vCD detector (solid-state backscattered electron detector) in FIB-SEM machine using Slice & View G2 operating software (FEI) are aligned and created serial images of the block face using the "Amira" 3D analysis software (commercial program Ver5.6). Additionally, 3D reconstruction, modeling, and display programs are possible with an open-source software, "IMOD," by the University of Colorado, Home Page (refer to http://bio3d.colorado.edu/imod).

### 9.3.4 *Serial Block-Face Imaging*

A BSE (backscattered electron: reflection electron) signal from biological samples is collected as a SEM image with enhanced brightness. Sequential SEM image stack is processed and aligned using "Amira" 5.6 software (FEI). In addition, this is a sub-volume of a part of the image that was inverted black and white again

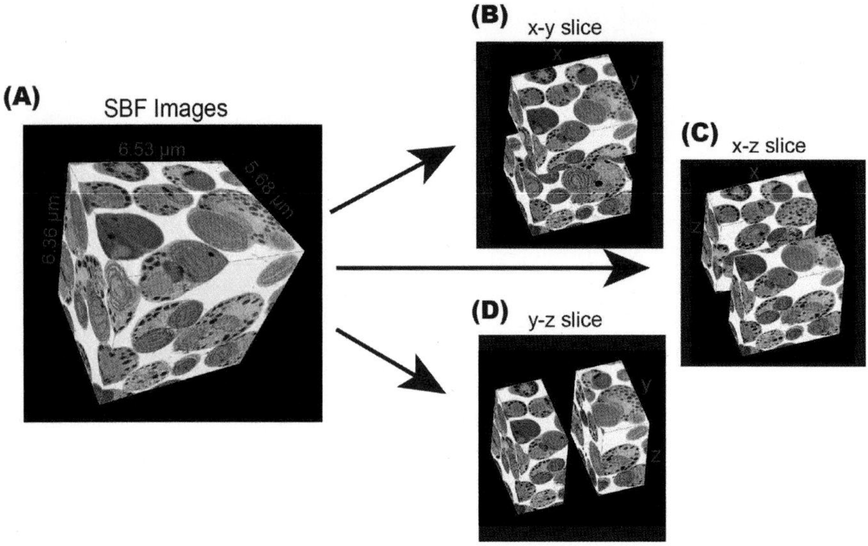


**Fig. 9.6** 3D view image of *C. merolae* cells from FIB-SEM and reconstruction
The milling was performed with a gallium ion beam with a milling pitch of 11.7 nm/step. Sequential 2D images which are collected from a FIB-SEM (FEI Quanta™ 3D DualBeam™) are inverted black and white, aligned and reconstructed using Amira (Ver.5.6) software. Size of SBF (serial block face) embedding view is 6.53 (x), 5.68 (y), and 6.36 (z) m (**a**). z-axis center block cut views (**b**), y-axis center block cut views (**c**), and x-axis center block cut views (**d**). The major parameters were as follows: beam current = 1.9 pA, dwell time = 6 μs/pixel, image size = 2048 × 1768 pixel (11.9 × 10.3 μm), pixel size = 5.8 nm/pixel

(Fig. 9.6a). This size of the SBF is x = 6.53 μm, y = 5.67 μm, z = 6.36 μm. In this three-dimensional block image, structural information of several *C. merolae* cells is included. Once an image of SBF is obtained, it is possible to obtain 2D information within the 3D structure of *C. merolae* cell from a free angle on the PC. As an example, a block cut in half in the z, y, and x directions can be indicated by Figs. 9.6b–d. A single-cell volume data from SBF is possible to be picked up from the image by cropping. Figure 9.7 illustrates this and the making of a 3D model (see Fig. 9.7 legend for details). After cropping of a single-cell volume data and segmentation of the suitable region, 3D structural model can present several types of visualization methods by adjusting the contrast and brightness of each section independently. One 3D visualization method is surface rendering (surface view), which displayed the region of segmentation. Another is volume rendering (Volren and Voltex), which is a method of reconstructing a 3D image using transparency information in addition to the luminance value of the voxel data. There is also a 3D visualization method called maximum intensity projection. Details of the 3D visualization method leave to the information of Amira (https://www.fei.com/software/amira-62-user-guide.pdf).

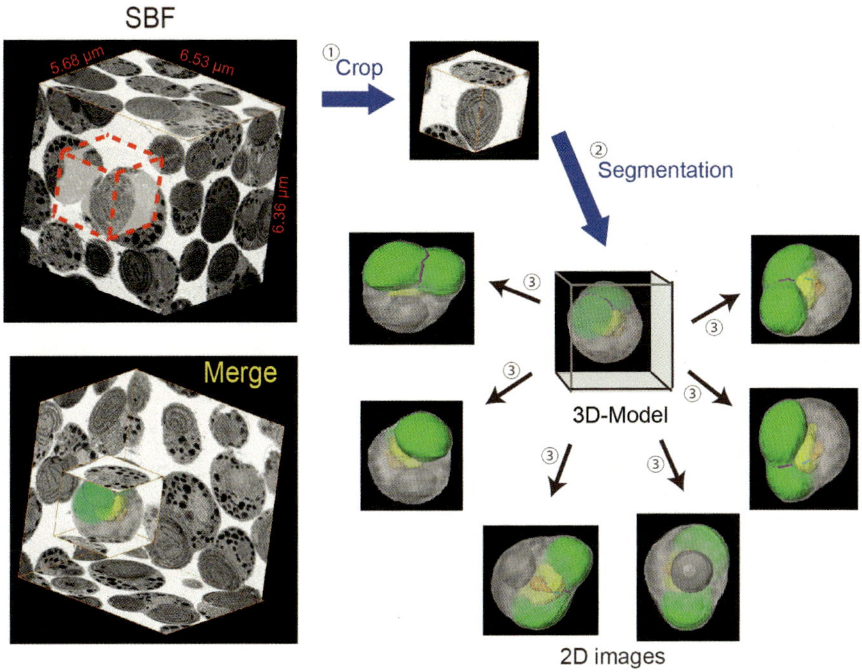


**Fig. 9.7** The converting from SBF image to 3D model
A stage II single-cell volume data from SBF images obtained by cropping and realignment. After image segmentation that is typically used to locate objects and boundaries (area, lines, curves, etc.) in sequential 2D-SEM images, for example, the surface membrane of organelles, 3D structural model is presented from several directions and several presentation methods (surface rendering, volume rendering, maximum intensity projection, and so on), and the threshold is adjusted or checked by eye. Plastid (*green*: surface view), mitochondrion (*yellow*: surface view), peroxisome (*orange*: surface view), nucleus (*gray*: surface view), PD ring (*magenta*: surface view), MD ring (*sky blue*: surface view), and cell body (*white gray*: Volren view)

### *9.3.5 Mitotic Cycle of* C. merolae *and Their Models*

There are several classification methods of *C. merolae* cell division, but as an example of this, the 3D models at whole cell level are shown comparing to one classification. Figure 9.8a is an illustration of the mitotic cycle of *C. merolae* cell which presented with reference to Imoto et al. (2010). The *C. merolae* cell division is classified into five stages. Types also known as stages I, II, III, IV, and V correspond to the G1-S, G2-M (prophase), M (prophase and metaphase), M (anaphase), and M (telophase and cytokinesis) of the mitotic cycle, respectively. Morphological features of three major organelles at each stage are as follows. Stage I: there is one cell nucleus, one mitochondrion, and one plastid. In the later stage, DNA replication of mitochondrion, plastid, and nucleus is performed. Plastid division is commenced. Stage II: mitochondrial division is commenced. Stage III:

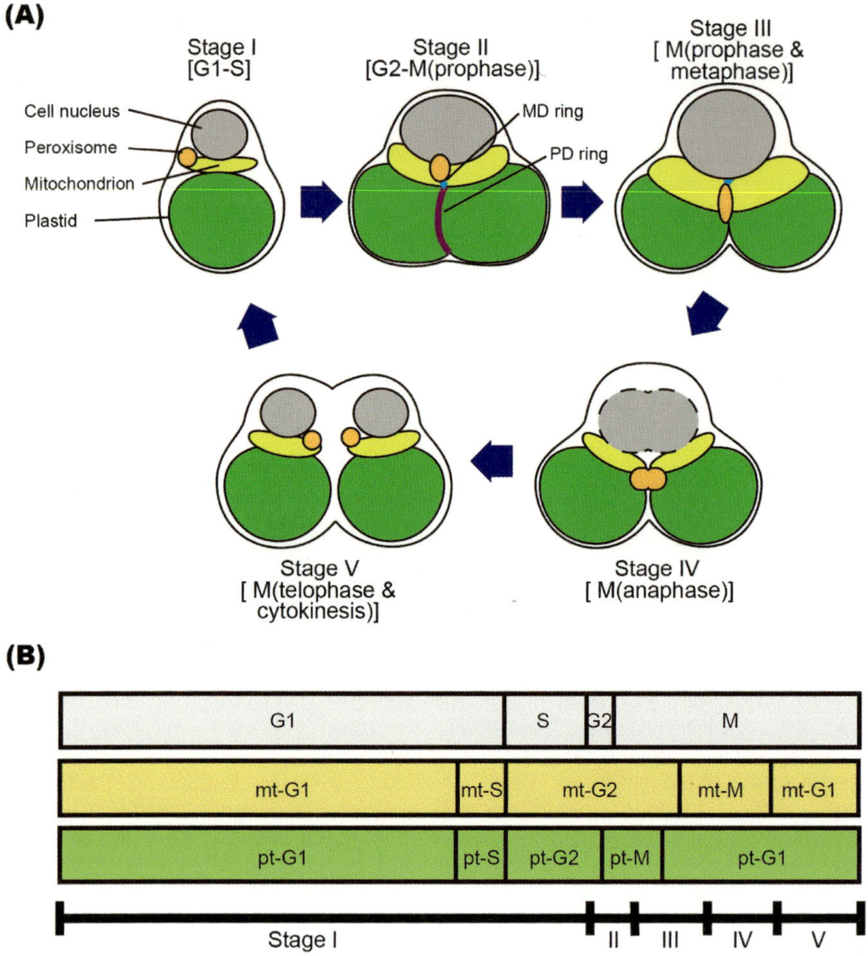


**Fig. 9.8** The mitotic cycle of *C. merolae*
Schematic representation of the mitotic cycle of *C. merolae* (**a**). However, there are several classifications of the cell division cycle of the *C. merolae*. (**a**) *C. merolae* cell has simple individual double-membrane organelles such as nucleus (*gray*), plastid (*green*), and mitochondrion (*yellow*) and single-membrane organelles such as the ER (not shown), Golgi complex (not shown), and peroxisome (*orange*). Plastid-dividing ring (PD ring: *purple*) and mitochondria-dividing ring (MD ring: *sky blue*) exist in stage II (G2–M(prophase)) and/or stage III (M(prophase and metaphase)). As the figure depicts, a single cell perfectly divides into two daughter cells during the mitotic cycle. Relationship between division of cell nucleus, mitochondrial nucleus, and plastid nucleus in the cell cycle (**b**)

plastid division is completed. Stage IV: mitochondrial division is completed, and cell nucleus division is initiated. Stage V: cell nucleus division is completed. One *C. merolae* cell contains two nuclei, two mitochondria, and two plastids. At the end of this stage, cytokinesis is performed.

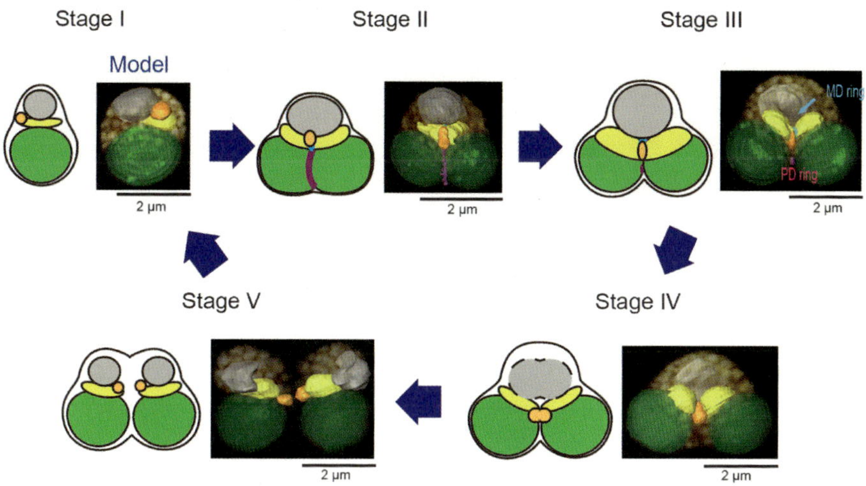


**Fig. 9.9** *C. merolae* cell models during cell division
*C. merolae* cells were presented in typical single x-y slice views after cropping the single 3D image of a whole cell from sequential images obtained from FIB-SEM and classifying each mitotic stage. For each type, the illustration on the left and the oblique 3D model on the right are in the same plane of view.
Models: plastid (*green*: Volren model), mitochondria (*yellow*: surface model), nucleus (*gray*: surface model), peroxisome (*orange*: surface model), and cell body (*volren Glow*: Volren model). Many starches exist in the cell body. Bars are 2 micrometer

Figure 9.9 shows the illustration and its 3D model of a *C. merolae* cell during the mitotic cycle by the volume rendering method that is a set of techniques used to display a 2D projection of a 3D discretely sampled data set, typically a 3D scalar field (Iwane and Ohta 2015). These models are shown as surface view for the main organelles, the nucleus (gray), mitochondria (yellow), plastid (green), and peroxisome (orange). The PD ring (magenta) and the MD ring exist in the stage II and stage III cells. Additionally, there is much starch in the cytoplasm. Cytoplasm is shown as volume rendering, in which displayed transparency information is added to luminance value. Starch grains supply energy for cell division. These starch grains decrease in dark conditions over several hours (data not shown). In this section, cell division was classified into only five stages and presented the representative examples of corresponding 3D structural models. As the change in the external shape of each organelle and the interaction among the organelles inside of a cell exist over a whole cell cycle, we are confident that we need to observe their change using more *C. merolae* cells.

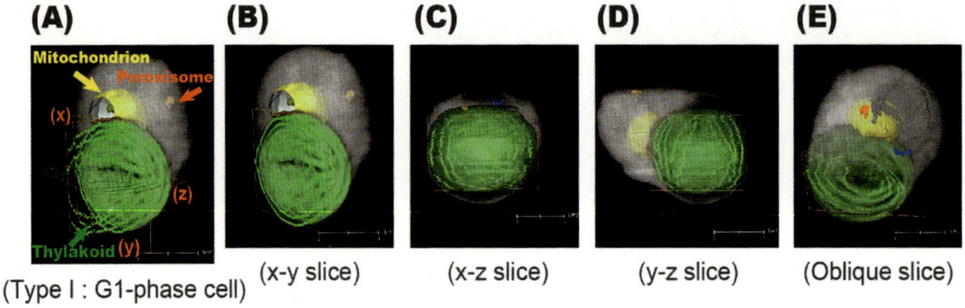


**Fig. 9.10** Inside 3D structure of organelle in a *C. merolae* cell at the G1 phase (**a**, **b**, **c**, **d**, **e**) Surface cut images of the thylakoid from several directions in the same plastid. x-y slice view: (**a**, **b**), x-z slice view: (**c**), y-z slice view (**d**), and oblique slice view (**e**). Oblique slice view is from imaging technology which enables examination and view of 3D volume data from any direction and plane desired. Sequential 2D images which are collected from a FIB-SEM (FEI Helios NanoLab™ DualBeam™)

## 9.3.6 Internal Structure of the Plastid

Although the outer shape of each organelle is expected to be observable with an optical microscope, depending on the size, EM has an advantage observing the 3D structure inside organelles. The plastid of *C. merolae* probably arose by endosymbiosis of a cyanobacterium, and it is the site of manufacture and storage of important chemical compounds in the cell. It has enormous phycobilisomes, the light-harvesting antennae of photosystem II complexes containing ATP synthesis anchored to thylakoid membranes in the plastid. The shape of phycobilisomes in cyanobacterium is hemi-discoidal, but the details including the distribution are unknown. This is also true for *C. merolae*. Until now, neither this nor even the connection of the thylakoid membrane had been elucidated. The connection of the thylakoid membrane in G1 phase cells was or has been clarified (Fig. 9.10) (Iwane and Ohta 2015). After 3D modeling construction of G1 phase cells, the images of the plastid from several directions are depicted. Thylakoid membranes in plastids are three to four layers using conventional TEM observation (Toda et al. 1998, see Fig. 9.1), but by using this method, it became possible to more clearly show the connection of these membranes. The plastid resembles an onion. This cell seems to be the early stage of G1 phase, especially the G1 phase after cytokinesis, because it is the same Plastid structure model as that of G2 phase and M phase cells. Furthermore, the number and morphology of thylakoid membranes seem to be affected by various factors, such as growth under the environment of nitrogen deficiency (Klotz et al. 2016). It will be expected to confirm its influence directly not only from the biochemical approach but also from the analysis of a 3D structural model.

## 9.4 Other Advanced Electron Microscopes

### *9.4.1 Ultra-High-Voltage TEM and Tomography*

UHVEM (H-3000, Hitachi) is a specific TEM of Osaka University with an accelerating voltage of 3MV; thick specimens can be observed due to increasing penetration power of electrons. Currently, UHVEM is called the EM with an accelerating voltage of over 1MV. All such electron microscopes are made in Japan (HITACHI, JEOL, etc.) and are installed in several places in Japan, Germany, and Korea. The development of the UHVEM of Osaka University started from around 1985 and was completed in 1996 (Fijita 1986). Since it is advantageous for observation of biological samples as thick as 2–3 microns, structural information in the fixed cells and tissues using the EM has been clarified (Takaoka et al. 2008; Kunimoto et al. 2012; Takahashi et al. 2015). Because *C. merolae* cell is about 2–5 microns in size, it may be possible to obtain structural information on the external shape of each organelle in the cell using this microscope. However, in order to obtain good information of the internal structure of organelle with higher resolution due to TEM, the sample is observed with processing to a thickness of 0.5 micron; in this section, it is not information on the whole cell though. In addition, the missing wedge from the limited view data is reduced by two-axis tomographic observation.

Using UHVEM TEM tomography, a more detailed information about the inside of the organelles in *C. merolae* cell can also be obtained (Iwane and Ohta 2016). Mitochondrial 3D structure models are obtained by UHVEM TEM tomography (Fig. 9.11a, b). The cell is at G1 phase. The internal structure information of mitochondria, cristae structure, is a bit difficult to obtain from our FIB-SEM analysis and also from sequential TEM analysis using even though 90 nm serial thin sections shown by Nishida et al. (2003). Despite of whole cell analysis might be a little bit difficult using 0.5 micron thick section, the detail structural information was expected to obtain from the samples using an UHVEM operating at 2 MeV (H-3000; Hitachi). Figure 9.11c shows an x-y slice plus a part of thylakoid surface view after 3D reconstruction of *C. merolae* cell at a G2-M phase. When viewing the part of thylakoid surface view magnified, there are regularly arranged structures on the surface of the thylakoid membrane (Fig. 9.11d). The structure present on both sides of the thylakoid membrane must be a photosynthesis-related complex including phycobilisomes. Is there any influence of the spacing between these complexes due to under light or dark conditions and light stress, etc.? These questions can be answered as the structure in the organelle, which is difficult to observe with a normal optical microscope, becomes clear. In addition, the division of plastids can be learned by an approach from 3D structural modeling. There are two types of plastid-dividing rings, an outer ring and an inner ring at this time (Miyagishima et al. 1998, 2001). Figure 9.12a shows a single x-y slice of *C. merolae* cell at M phase after reconstruction, which is going to be divided into two plastids. In addition to 3D structural models of the thylakoid membrane, mitochondrion and PD ring are added on the x-y slice image (Fig. 9.12b). The relationship of each

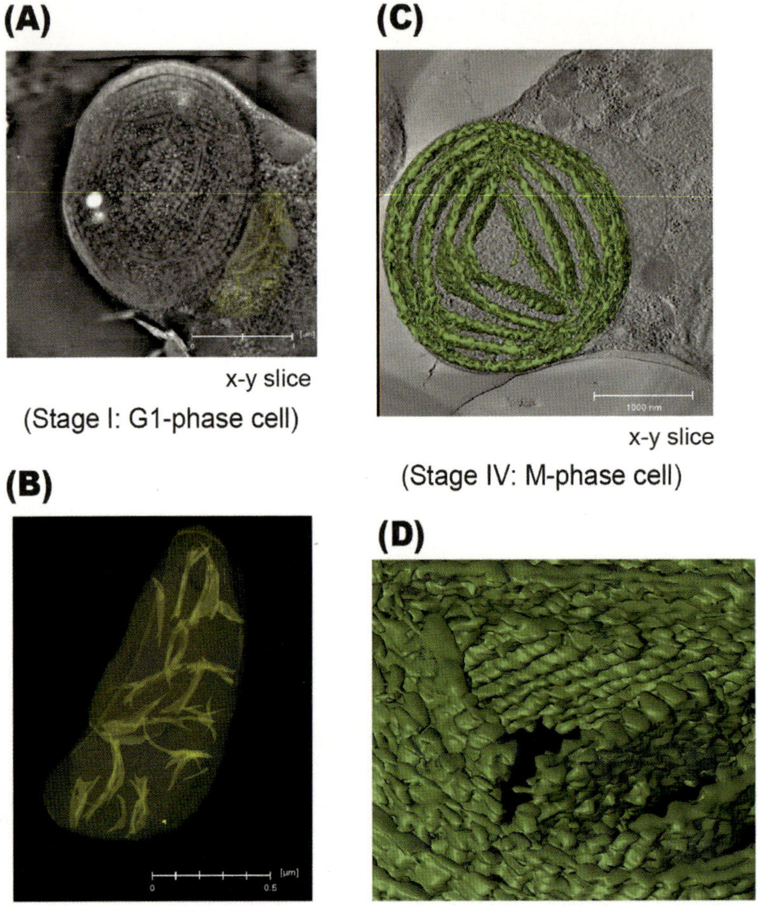


**Fig. 9.11** Selected tomographic slices and 3D models after 3D reconstruction of UHVEM
One x-y slice of part of a *C. merolae* type I cell (**a**) and its mitochondrion 3D model (*yellow*: surface model) (**b**), one x-y slice of a part of a *C. merolae* stage IV cell, thylakoid membrane in the plastid (green: surface model) (**c**), and its expanded surface model from different directions (**d**). For observation by UHVEM, 500 nm thick sections are cut using an ultramicrotome and mount on Formvar-coated copper grids. The thick sections are stained in 10% uranyl acetate in 70% methanol with 150 W microwave for 30 s and then incubate for 20 min. After washing and drying, the sections are stained in lead citrate with 150 W microwave for 30 s and then incubated for 10 min. 20 nm GNPs (Gold nanoparticles) are attached outside of the EM specimen grids as corrective 3D markers. To improve the resolution of the structure model, cells were dual-axially imaged at 4096 × 4096 pixels (0.92 nm/pixel) slow scan CCD camera (TVIPS) over an angular range from −60 to 60° at 2° intervals. Reconstruction of the tomographic and 3D modeling was performed with an open-source suite of modeling software (IMOD 4.7.11). Segmentation and 3D view are created just as in FIB-SEM analysis. Bars are 1 micrometer (**a**) (**c**) and 0.5 micrometer (**b**)

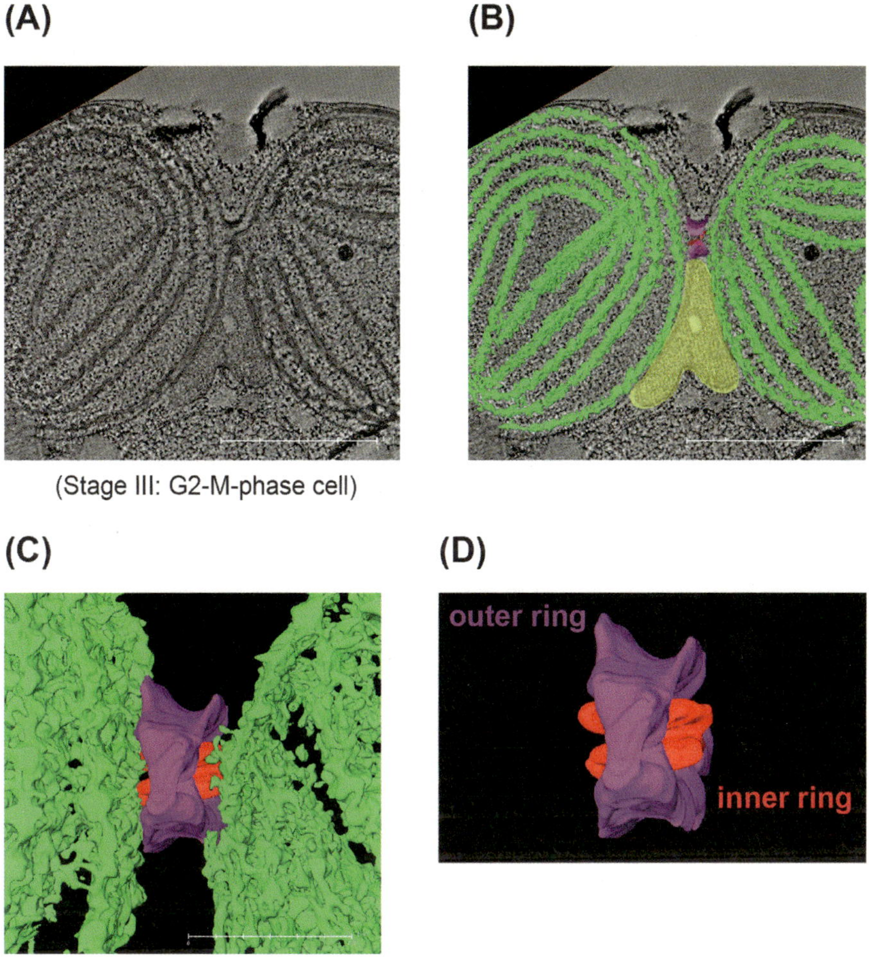


**Fig. 9.12** 3D model of plastid-dividing ring after 3D reconstruction of UHVEM
One x-y slice at stage III cell (**a**) and its thylakoid (*green*: surface model), mitochondrion (*yellow*: surface model with transmittance 50%), and plastid-dividing rings (outer ring, *purple*, and inner ring, *red*: surface model) (**b**), thylakoid (*green*: surface model) plus plastid-dividing rings (outer ring, *purple*, and inner ring, *red*: surface model) (**c**), extracted plastid-dividing rings from model C (**d**). Scales are 1 micrometer (**a** and **b**) and 0.3 micrometer (**c**)

organelle and inside the plastid is clarified although it is a limited range within the *C. merolae*. Expanding around the plastid division rings, the 3D structure model of the complete two plastid-dividing rings is revealed, and an outer ring seems to be bound to the membrane of the plastid, and inner ring seems to be bound to the thylakoid membrane on the inside of the plastid outer membrane (Fig. 9.12c, d) as 2D EM image which is reported by Miyagishima et al. (2003). Recently, in 3D modeling of the ellipsoidal unicellular alga *Glaucocystis* cells, this was reported by a single-axis UHVEM tomography (Takahashi et al. 2015).

### *9.4.2 Cryo-TEM Imaging of* C. merolae

Although many reports have provided structural models at superfine resolution using cryo-TEM tomography and cryo-TEM/direct detector (Kuijper et al. review, 2015), the fine 3D structural model of the membrane surface from specimens that are not purified using specific detergents is offered. The cryo-TEM is mainly equipped with two kinds of detectors, direct electron detectors which are represented by "Falcon (FEI)" and "K2 (Gatan)" and then "HAADEF (FEI)" detector for STEM observation. The HAADEF detector is suitable for observation of samples that are thick in the z direction, such as cells. *C. merolae* cells are directly cultured on a carbon-coated EM grid in a monolayer and are physically fixed into liquid ethane ($\sim -180°$) containing the culture medium. As the behavior of *C. merolae* cells on the carbon membrane is similar to general suspension culture, the physically fixed *C. merolae* cells are expected to have stable structural information. As cryo-tomography observation uses TEM machine, so there is a limitation on the size, and small size organisms like *Dictyostelium discoideum* have been taken as observation targets (Medalia et al. 2002). Although it is a preliminary result and it is more expected that the thickness of the *C. merolae* cell is affected, Fig. 9.13 shows an image of cryo-STEM tomography of stage IV *C. merolae* cell and after reconstruction. Sequential slice view images every 111.54.nm (a) and a single image after 3D reconstruction (b) are shown. Despite being unstained and uniaxial tomography observation, thylakoid membranes in plastids, nucleus, are clear. Especially, the nuclear membrane and thylakoid membrane complex appear to be more clear than those obtained from FIB-SEM. Starch in the cytoplasm is also well visualized. On the other hand, mitochondrion and peroxisome are somewhat unclear. It seems to be different from the structural information obtained from EM observation of a sample prepared using dehydrated and heavy metals using FIB-SEM or UHVEM. Proceeding with the dual-axial tomography system, more detailed structural model information of *C. merolae* cell might be obtained.

## 9.5 For the Future

Recently, several advanced EM observation techniques have been developed, and the development of the microscopes is rapidly advancing. Each electron microscope has unique characteristics, for example, range of observation field, thickness

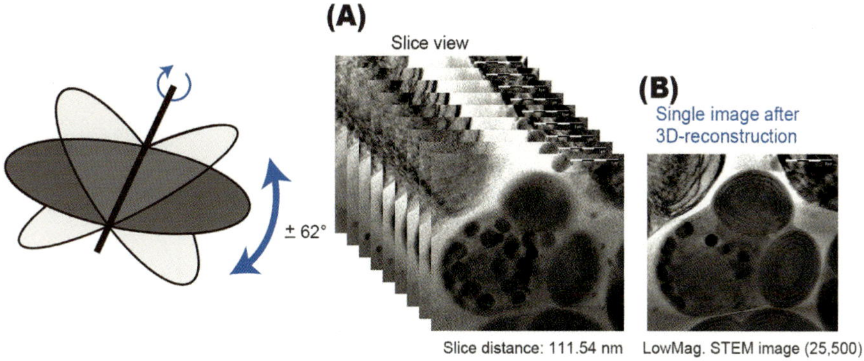


**Fig. 9.13** Cryo-EM image of *C. merolae* cell
Live *C. merolae* cells (final OD750 = 0.01) were cultured on pre-hydrophilized Holey carbon support film, 200 mesh molybdenum EM specimen grids (Quantifoil R0.6/1.0) at 42° – 5% $CO_2$ for 42 h under constant light conditions followed by more 6 h under dark conditions. 40 nm GNPs (Gold nanoparticles) were attached outside the EM specimen grid to act as corrective markers for the 3D reconstruction. EM grid with *C. merolae* cells was fixed by dropping in liquid ethane using Leica EM GP instead of the usual chemical fixation. Physical fixation can prevent the formation of crystals and improve the visibility of the cells. For observing the EM image (STEM) in cell, we used the FEI Titan Krios™, transmission electron microscope with HAADF detector. The Titan Krios™ allows excellent visualization of the intricate mechanisms of individual proteins and/or molecular machines and localization of materials within the three-dimensional architecture of the cell. For tomography, cells were imaged over an angular range from −62° to 62° at 2° × cosθ tilt increments automatically. (**a**) Sequential slice view images (Low Mag. image × 25,500) every 111.54 nm. (**b**) Single x-y slice (Low Mag. STEM image × 25,500) image was reconstructed using inspect 3D analysis software
Other parameters are as follows: image size, 1024 × 1024 pixel; image pixel size, 5.07 nm; exposure time, 25 s/image. Bar is 1 micron

of specimen and resolution, etc. Generally, the resolution of TEM is higher than that of SEM but has a limit to the thickness of the observing biological sample. In this section, we introduced three types of 3D structural models using three types of the advanced electron microscope. At first, in order to obtain 3D structural information of *C. merolae* cell at a whole cell level, FIB-SEM was selected. Then, in order to obtain more detailed information inside of the organelle in *C. merolae* cell, UHVEM tomography was selected. Finally, in order to obtain no distortion structure of *C. merolae* without dehydration, cryo-tomography was selected.

However, disadvantages of electron microscopy observation are dynamic identification of target molecules and complexes. All the gene sequences of *C. merolae* have already been identified (Matsuzaki et al. 2004). Therefore, we think that it is important to effectively utilize recombinants and link them to structural modeling while monitoring important events. For example, the methods also using the same field between cryo-EM observation and cryo-LM observation (Fig. 9.14), which

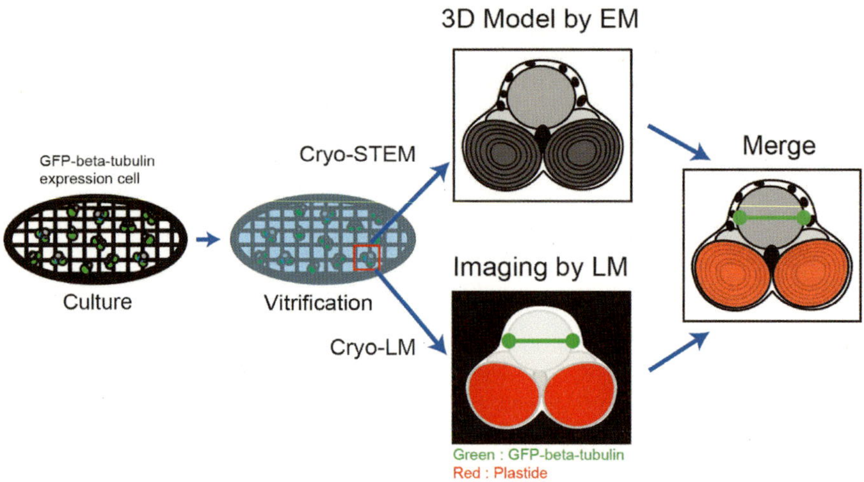


**Fig. 9.14** Correlated target material images of cryo-LM with the images of STEM tomography Illustration of the correlative EM and LM images in the same field is shown using GFP-beta-tubulin-expressed *C. merolae* cell as an example. *C. merolae* cells were grown on the EM grid and cultured under stable condition. (Option: And then, the cells' GFP-tubulin expression condition was checked during cell cycle at room temperature using LM.) As in Fig. 9.13, after physical fixation, GFP microtubules and plastid due to autofluorescence are observed by cryo-LM with a special cryo-chamber system, while on the other hand, STEM tomography observation of the same cell by cryo-EM is carried out, and a 3D structural model is obtained from serial tilted 2D STEM images after 3D reconstruction. By superimposing the two sets of information obtained, it is possible to obtain the high-resolution 3D structural model of the cell while confirming the cell division timing

can identify target molecules, complexes, and organelles by fluorescent protein fused to the target recombinant proteins (Hampton et al. 2017), must become one of the useful methods for obtaining a fine 3D structural model of *C. merolae* cell while identifying the target molecules in the future in the same way as has been done with FIB-SEM and reconstruction. Since further technical developments in milling, a vitrified biological sample by cyo-FIB-SEM and then observing cryo-TEM tomography is progressing (Asano et al. review 2016), the 3D models of a *C. merolae* cell may be able to obtain in the future using tomographic observation without staining the sample, not at the whole cell level.

**Acknowledgments** We thank K. Ohta, Ph.D. (Kurume University) and T. Nishida, Ph.D. (Osaka University) for FIB-SEM operation/discussion and UHVEM operation, respectively.

## References


Asano S, Engel BD, Baumeister W (2016) Situ Cryo-electron tomography: a post-reductionist approach to structural biology. J Mol Biol 428:332–343. https://doi.org/10.1016/j.jmb.2015.09.030. Review.

Dahl R, Staehelin LA (1989) High-pressure freezing for the preservation of biological structure: theory and practice. J Electron Microsc Tech 13:165–174

Denk W, Horstmann H (2004) Serial block-face scanning electron microscopy to reconstruct three-dimensional tissue nanostructure. PLoS Biol 2:e329

Dubochet J, Booy FP, Freeman R, Jones RF, Walter CA (1981) Low temperature electron microscopy. Annu Rev Biophys Bioeng 10:133–149. https://doi.org/10.1146/annurev.bb.10.060181.001025

Dunn DN, Hull R (1999) Reconstruction of three-dimensional chemistry and geometry using focused ion beam microscopy. Appl Phys Lett 75:3414–3416

Fijita H (1986) Ultra-high voltage electron microscopy: past, present and future. EM Tech 3:243–304

Giannuzzi LA, Prenitzer BI, Drown–MacDonald JL, Shofner TL, Brown SR, Irwin RB, Stevie FA (1999) Electron microscopy sample preparation for the biological and physical sciences using focused ion beams. J Process Anal Chem 4:162–167

Hampton CM, Strauss JD, Ke Z, Dillard RS, Hammonds JE, Alonas E, Desai TM, Marin M, Storms RE, Leon F, Melikyan GB, Santangelo PJ, Spearman PW, Wright ER (2017) Correlated fluorescence microscopy and cryo-electron tomography of virus-infected or transfected mammalian cells. Nat Protoc 12:150–167

Hayworth KJ, Morgan JL, Schalek R, Berger DR, Hildebrand DGC, Lichtman JW (2014) Imaging ATUM ultrathin section libraries with WaferMapper: a multi-scale approach to EM reconstruction of neural circuits. Front Neural Circuits 27:68. https://doi.org/10.3389/fncir.2014.00068. eCollection

Ichimura K, Miyazaki N, Sadayama S, Murata K, Koike M, Nakamura K, Ohta K, Sakai T (2015) Three-dimensional architecture of podocytes revealed by block-face scanning electron microscopy. Sci Rep 5:8993. https://doi.org/10.1038/srep08993

Imoto Y, Fujiwara T, Yoshida Y, Kuroiwa H, Maruyama S, Kuroiwa T (2010) Division of cell nuclei, mitochondria, plastids, and microbodies mediated by mitotic spindle poles in the primitive red alga *Cyanidioschyzon merolae*. Protoplasma 241:63–74. https://doi.org/10.1007/s00709-010-0107-y

Itoh R, Takano H, Ohta N, Miyagishima S, Kuroiwa H, Kuroiwa T (1999) Two ftsH-family genes encoded in the nuclear and chloroplast genomes of the primitive red alga *Cyanidioschyzon merolae*. Plant Mol Biol 41:321–337

Iwane AH, Ohta K (2015) Three-dimensional microstructural visualization of mitosis using Focused Ion Beam-Scanning Electron Microscope (FIB-SEM) and 3Mv Ultra-High Voltage Electron Microscope (UHVEM) tomography with nanoscale resolution at whole cell level. Biophys J 108. 618p

Iwane AH, Ohta K (2016) 3D microstructural visualization of the simplest of eukaryotic cell (Cyanidioschyzon Merolae) during mitosis process using several new microscopic techniques. Biophys J 110. 155p

Klotz A, Georg J, Bucinska L, Watanabe S, Reimann V, Januszewski W, Sobotka R, Jendrossek D, Hess WR, Forchhammer K (2016) Awakening of a dormant Cyanobacterium from nitrogen Chlorosis reveals a genetically determined program. Cell 26:2862–2872

Knott G, Marchman H, Wall D, Lich B (2008) Serial section scanning electron microscopy of adult brain tissue using focused ion beam milling. J Neurosci 28:2959–2964

Kuijper M, van Hoften G, Janssen B, Geurink R, De Carlo S, Vos M, van Duinen G, van Haeringen B, Storms M (2015) FEI's direct electron detector developments: embarking on a revolution in cryo-TEM. J Struct Biol 192:179–187

Kunimoto K, Yamazaki Y, Nishida T, Shinohara K, Ishikawa H, Hasegawa T, Okanoue T, Hamada H, Noda T, Tamura A, Tsukita S, Tsukita S (2012) Coordinated ciliary beating requires Odf2-Mediated polarization of basal bodies via basal feet. Cell 148:189–200. https://doi.org/10.1016/j.cell.2011.10.052

Kuroiwa T, Nishida K, Yoshida Y, Fujiwara T, Mori T, Kuroiwa H, Misumi O (2006) Structure, function and evolution of the mitochondrial division apparatus. Biochem Biophys Acta 1763(5–6):510–521

Leunissen JLM, Yi H (2009) Self-pressurized rapid freezing (SPRF): a novel cryofixation method for specimen preparation in electron microscopy. J Microsc 235:25–35

Li M, Ghosh S, Rouns TN, Weiland H, Richimond O, Hunt W (1998) Serial sectioning method in the construction of 3-D microstructures for particle-reinforced MMCs. Mater Charact 41:81–95

Matsuzaki M, Misumi O, Shin-I T, Maruyama S, Takahara M, Miyagishima SY, Mori T, Nishida K, Yagisawa F, Nishida K, Yoshida Y, Nishimura Y, Nakao S, Kobayashi T, Momoyama Y, Higashiyama T, Minoda A, Sano M, Nomoto H, Oishi K, Hayashi H, Ohta F, Nishizaka S, Haga S, Miura S, Morishita T, Kabeya Y, Terasawa K, Suzuki Y, Ishii Y, Asakawa S, Takano H, Ohta N, Kuroiwa H, Tanaka K, Shimizu N, Sugano S, Sato N, Nozaki H, Ogasawara N, Kohara Y, Kuroiwa T (2004) Genome sequence of the ultrasmall unicellular red alga *Cyanidioschyzon merolae* 10D. Nature 428:653–757

Maunsbach AB (1966) The influence of different fixatives and fixation methods on the ultrastructure of rat kidney proximal tubule cells. I. Comparison of different perfusion fixation methods and of glutaraldehyde, formaldehyde and osmium tetroxide fixatives. J Ultrastruct Res 15:242–282

Medalia O, Weber I, Frangakis AS, Nicastro D, Gerisch G, Baumeister W (2002) Macromolecular architecture in eukaryotic cells visualized by cryoelectron tomography. Science 298:1209–1213

Miyagishima S, Itoh R, Toda K, Takahashi H, Kuroiwa H, Kuroiwa T (1998) Orderly formation of the double ring structures for plastid and mitochondrial division in the unicellular red alga *Cyanidioschyzon merolae*. Planta 206:551–560

Miyagishima S, Itoh R, Toda K, Kuroiwa H, Nishimura M, Kuroiwa T (1999) Microbody proliferation and segregation cycle in the single-microbody alga *Cyanidioschyzon merolae*. Planta 208:326–336

Miyagishima S, Kuroiwa H, Kuroiwa T (2001) The timing and manner of disassembly of the apparatuses for chloroplast and mitochondrial division in the red alga *Cyanidioschyzon merolae*. Planta 212:517–528

Miyagishima S, Nishida K, Mori T, Matsuzaki M, Higashiyama T, Kuroiwa H, Kuroiwa T (2003) A plant-specific dynamin-related protein forms a ring at the chloroplast division site. Plant Cell 15:655–665

Moor H, Riehle U (1968) Snap freezing under high pressure: a new fixation technique for freeze-etching. Proc 4th Eur Reg Conf Electron Microsc Rome 2:33–34

Nishida K, Takahara M, Miyagishima S, Kuroiwa H, Matsuzaki M, Kuroiwa T (2003) Dynamic recruitment of dynamin for final mitochondrial severance in a primitive red alga. PNAS 100:2146–2151. https://doi.org/10.1073/pnas.0436886100

Ohta K, Sadayama S, Togo A, Higashi R, Tanoue R, Nakamura K (2012) Beam deceleration for block-face scanning electron microscopy of embedded biological tissue. Micron 43:612–620

Peddie CJ, Collinson LM (2014) Exploring the third dimension: volume electron microscopy comes of age. Micron 61:9–19

Pinali C, Kitmitto A (2014) Serial block face scanning electron microscopy for the study of cardiac muscle ultrastructure at nanoscale resolutions. J Mol Cell Cardiol 76:1–11. https://doi.org/10.1016/j.yjmcc

Schalek R, Kasthuri N, Hayworth K, Berger D, Tapia JC, Morgan JL, Turaga SC, Fagerholm E, Seung HS, Lichtman JW (2011) Development of high-throughput, high-resolution 3D reconstruction of large-volume biological tissue using automated tape collection ultramicrotomy and scanning electron microscopy. Microsc Microanal 17:966–967. https://doi.org/10.1017/S1431927611005708

Studer D, Michel M, Muller M (1989) High pressure freezing comes of age. Scanning Microsc Suppl 3:253–268

Takahashi T, Nishida T, Saito C, Yasuda H, Nozaki H (2015) Ultra-high voltage electron microscopy of primitive algae illuminates 3D ultrastructures of the first photosynthetic eukaryote. Sci Rep 5:14735. https://doi.org/10.1038/srep14735

Takaoka A, Hasegawa T, Yoshida K, Mori H (2008) Microscopic tomography with ultra-HVEM and applications. Ultramicroscopy 108:230–238

Titze B, Genoud C (2016) Volume scanning electron microscopy for imaging biological ultrastructure. Biol Cell 108:307–323. https://doi.org/10.1111/boc.201600024

Toda K, Takano H, Miyagishima S, Kuroiwa H, Kuroiwa T (1998) Characterization of a chloroplast isoform of serine acetyltransferase from the thermo-acidophilic red alga *Cyanidioschyzon merolae*. Biochim Biophys Acta 1403:72–84

Vanhecke D, Graber W, Studer D (2008) Close-to-native ultrastructural preservation by high pressure freezing. Methods Cell Biol 88:151–164

Wanner AA, Kirschmann MA, Genoud C (2015) Challenges of microtome-based serial block-face scanning electron microscopy in neuroscience. J Microsc 259:137–142. https://doi.org/10.1111/jmi.12244

Willingham MC, Rutherford AV (1984) The use of osmium -thiocarbohydrazide-osmium (OTO) and ferrocyanide-reduced osmium methods to enhance membrane contrast and preservation in cultured cells. J Hisrochemi Sciety 32:455–460

Yoshida Y, Kuroiwa H, Misumi O, Yoshida M, Ohnuma M, Fujiwara T, Yagisawa F, Hirooka S, Imoto Y, Matsushita K, Kawano S, Kuroiwa T (2010) Chloroplasts divide by contraction of a bundle of nanofilaments consisting of polyglucan. Science 329:949–953. doi: 10.1126/science.1190791

# Chapter 10
# Coherent X-ray Diffraction Imaging of *Cyanidioschyzon merolae*

**Yuki Sekiguchi, Amane Kobayashi, Yuki Takayama, Mao Oide, Asahi Fukuda, Takahiro Yamamoto, Koji Okajima, Tomotaka Oroguchi, Takeshi Hirakawa, Yayoi Inui, Sachihiro Matsunaga, Masaki Yamamoto, and Masayoshi Nakasako**

**Abstract** Coherent X-ray diffraction imaging (CXDI) is a lensless imaging technique for visualizing the structures of noncrystalline particles at a resolution of several tens of nanometers. The targets are particles with dimensions in the sub-micrometer to micrometer range. We carried out CXDI experiments at 66 K to investigate the internal structures of a whole frozen-hydrated cell and a chloroplast of *Cyanidioschyzon merolae* and a cyanobacteria cell by using coherent X-ray light sources, such as synchrotron and X-ray free-electron laser facilities. Owing to the short wavelength of the X-rays used (0.225 nm), the absorption and multiple scattering of X-rays inside the specimens were negligible. Diffraction patterns from each specimen particle adsorbed onto a thin membrane were collected at resolutions better than 50 $\mu m^{-1}$ in reciprocal space. Therefore, structures of specimen particles can be illustrated at a resolution higher than 200 nm in real space. The most probable electron density map was retrieved from each diffraction pattern. The

Y. Sekiguchi • A. Kobayashi • M. Oide • A. Fukuda • T. Yamamoto • K. Okajima
T. Oroguchi • M. Nakasako (✉)
Department of Physics, Faculty of Science and Technology, Keio University, 3-14-1 Hiyoshi, Kohoku-ku, Yokohama 223-8522, Japan

RIKEN SPring-8 Center, 1-1-1 Kouto, Sayo, Hyogo 679-5148, Japan
e-mail: nakasako@phys.keio.ac.jp

Y. Takayama
Department of Physics, Faculty of Science and Technology, Keio University, 3-14-1 Hiyoshi, Kohoku-ku, Yokohama 223-8522, Japan

RIKEN SPring-8 Center, 1-1-1 Kouto, Sayo, Hyogo 679-5148, Japan

Graduate School of Material Science, University of Hyogo, 3-2-1 Kouto, Kamigori-cho, Ako-gun 679-1297, Japan

T. Hirakawa • Y. Inui • S. Matsunaga
Department of Applied Biological Science, Faculty of Science and Technology, Tokyo University of Science, 2641 Yamazaki, Noda, Chiba 278-8510, Japan

M. Yamamoto
RIKEN SPring-8 Center, 1-1-1 Kouto, Sayo, Hyogo 679-5148, Japan



T. Kuroiwa et al. (eds.), *Cyanidioschyzon merolae*,
https://doi.org/10.1007/978-981-10-6101-1_10

internal structures of the specimens are described here, particularly the structural correlation of the chloroplast of *C. merolae* and the cyanobacteria cell. Based on the experimental results, we discuss the feasibility of CXDI in the structural analyses of biological cells and cellular organelles.

**Keywords** Coherent X-ray diffraction imaging • Whole-cell imaging • X-ray free-electron laser • Synchrotron radiation • Chloroplast • Cyanobacterias

## 10.1 Introduction

One of the ultimate goals of cell biology is a gapless description of the spatiotemporal events from external physical and chemical stimuli at the molecular level propagating to higher hierarchy levels in living cells. Fluorescence and super-resolved light microscopy (LM) has made significant contributions to the visualization of the spatiotemporal events in cells at a spatial resolution of approximately 100–200 nm and at a temporal resolution of several tens of milliseconds (Miyawaki 2013; Sengupta et al. 2012). Electron microscopy (EM) has been used to study cellular structures at a spatial resolution of much better than 200 nm. However, whole-cell and cellular components have dimensions larger than 500 nm and are opaque for electrons. EM requires sectioned and chemically fixed specimens (Gan and Jensen 2011) or freeze-fractured specimens (Heuser 2011).

In recent years, the coherent X-ray diffraction imaging (CXDI) (Miao et al. 2012) has been applied to structural studies of cellular components. Since its first demonstration (Miao et al. 1999), CXDI has been developed and used for structural analyses of noncrystalline particles with sizes ranging from sub-micrometer to micrometer at resolutions of several tens of nanometers (Miao et al. 2008). In CXDI experiments, an isolated noncrystalline particle is irradiated by X-ray beam with high spatial coherence, and the Fraunhofer diffraction from the particle is then recorded at a resolution exceeding the limit of LM by an area detector with a large number of pixels. When using X-rays with short wavelengths, the penetration power allows us to record diffraction patterns from thick particles with negligible multiple scattering and fatal absorption, owing to the weak electromagnetic interactions with the atoms composing the particles. From the diffraction pattern alone, the electron density map of the specimen particle projected along the direction of the incident X-ray is, in principle, reconstructed by using iterative phase-retrieval (PR) algorithms (Fienup 1982). Therefore, CXDI provides opportunities to visualize the whole structure of thick specimens, such as biological cells and organelles at resolutions better than that of LM without staining or chemical labeling. In addition, specimens are prepared without the resin fixation, staining by reagents including heavy atoms and sectioning of specimens, which are necessary in EM observation.

CXDI experiments conducted using synchrotron X-rays have demonstrated the potential to study the structures of noncrystalline biological particles, the size of which are several micrometers (Miao et al. 2003a; Song et al. 2008; Nishino et al. 2009; Jiang et al. 2010). However, those experiments were conducted on dehydrated and chemically fixed specimens in vacuum at ambient temperatures. Dehydration and chemical fixation make the structures of the specimens nonnative. To visualize the functional structures of fully hydrated biological specimens under significantly reduced radiation damage, techniques and apparatus for CXDI experiments at cryogenic temperatures have been developed, including the specimen preparation procedures (Huang et al. 2009; Lima et al. 2009; Takayama and Nakasako 2012; Nakasako et al. 2013). When specimens are kept near the temperature of liquid nitrogen, radiation damage is dramatically reduced owing to the immobilization of radicals (Ravelli and Garman 2006; Bamme et al. 2010). Therefore, cryogenic CXDI could potentially be an effective technique for imaging frozen-hydrated biological specimens, such as cells and organelles, although time-resolved measurement would remain impossible.

In the application of CXDI to biological specimens, unicellular organisms with minimal cellular components are suitable for investigating cellular events, such as global reorganization of cellular organelles and genomes, commonly occurring in the cell cycles of eukaryotes. One of our targets in CXDI is the unicellular red alga, *Cyanidioschyzon merolae* (*C. merolae*), classified as the origin of eukaryotes. *C. merolae* is a very small organism with a diameter of 1.5–2.0 μm and lives in highly acidic environments (Matsuzaki et al. 2004). The cells contain the lowest number of genes among free-living eukaryotes and a minimal set of organelles. They include a cell nucleus, a mitochondrion, a chloroplast, a microbody, endoplasmic reticulum, a single Golgi apparatus and lysosomes (Yagisawa et al. 2007, 2009). The cell cycle can be synchronized with the light/dark cycle of the environment. Therefore, *C. merolae* cells offer opportunities to study the cell cycle in eukaryotes. For instance, the structures of the cells are investigated using time-lapse video images under phase-contrast LM and the temporal sequence of EM images of the cell division process (Miyagishima et al. 1999). Aspects of the cell division of *C. merolae*, such as the cell cycle-dependent coordination of cell nuclei, and the division of mitochondria and chloroplasts by different types of ring-shaped macromolecular assemblies, have been insightfully investigated by using fluorescent LM and EM for stained and sectioned specimens (Imoto et al. 2011).

Here we present projection structures of frozen-hydrated *C. merolae* and its chloroplast reconstructed at resolutions better than 150 nm from diffraction patterns recorded in CXDI experiments at cryogenic temperatures. *C. merolae* cells are suitable for structural analysis by CXDI with respect to the size and the total scattering cross section. In addition, the structure of the chloroplast is compared with that of cyanobacteria from the viewpoint of the endosymbiotic hypothesis regarding the evolution of eukaryotes. Based on the presented results, we discuss the merit of cryogenic CXDI and its combinational use with fluorescence LM for structural studies of cells and cellular organelles.

## 10.2 Theoretical Background of CXDI

### 10.2.1 Diffraction Intensity

In this section, we briefly introduce the basic theory of CXDI, particularly the phase-retrieval method to obtain the electron density distribution of a noncrystalline particle from a diffraction pattern (Fienup 1982; Miao et al. 1998). In CXDI experiments, we observe X-ray diffraction from a single noncrystalline particle with an electron density $\rho(\vec{r})$. When the particle is irradiated by a coherent X-ray with flux density, $I_0$, and wavelength, $\lambda$ (Fig. 10.1a), under the Fraunhofer approximation, the diffraction intensity, $I(\boldsymbol{S})$, at a scattering vector, $\boldsymbol{S}$, with a scattering angle of $2\theta$ ($|\vec{S}| = 2\sin\theta/\lambda$), is described by the following equations:

$$I(\boldsymbol{S}) = I_0\, \boldsymbol{r}_e^2 \frac{\lambda^2}{\sigma A} F(\boldsymbol{S})^* F(\boldsymbol{S}) \frac{1}{\boldsymbol{R}^2}, \tag{10.1}$$

$$F(\boldsymbol{S}) = \int \rho(\boldsymbol{r}) \exp(2\pi i\, \boldsymbol{S} \cdot \boldsymbol{r})\, \mathrm{d}^3 r = |F(\boldsymbol{S})| \exp[i\alpha(\boldsymbol{S})], \tag{10.2}$$

where $r_e$ is the classical electron radius ($2.8179 \times 10^{-15}$ m). $\sigma$ is the oversampling (OS) ratio of the diffraction pattern as defined in the subsequent section. $A$ is the projected dimension and ranges from sub-micrometers to several micrometers in most CXDI experiments for biological specimens. $R$ is the distance from the particle to the observation point in the direction of the scattering vector. $F(\boldsymbol{S})$ is designated as the structure factor, a Fourier transform of $\rho(\boldsymbol{r})$ (Eq. 10.2), and is a complex number expressed by the structure amplitude, $|F(\boldsymbol{S})|$, and the phase, $\alpha(\boldsymbol{S})$. Since $r_e$ is very small, the incident flux density, $I_0$, is necessary to be sufficiently large to collect diffraction patterns from a single noncrystalline particle with good signal-to-noise ratios. Owing to the fact that the lengths of the propagation vectors are the same between the incident and diffracted X-rays, the observable intensity distribution is limited to that intersecting a sphere with a radius of $1/\lambda$ (the Ewald sphere in Fig. 10.1a).

The diffraction pattern from a particle is composed of a number of speckle peaks as a result of complicated interference of diffracted X-rays from the specimen particle (Fig. 10.1a). The sizes of the speckle peaks are comparable to the reciprocal of the particle size.

### 10.2.2 Retrieval of Projection Electron Density

In the small-angle diffraction region, the Ewald sphere is approximated as a plane. Then, the structure factor is expressed as the Fourier transform of the projection of $\rho(\vec{r})$ along the direction of the incident X-ray beam as:

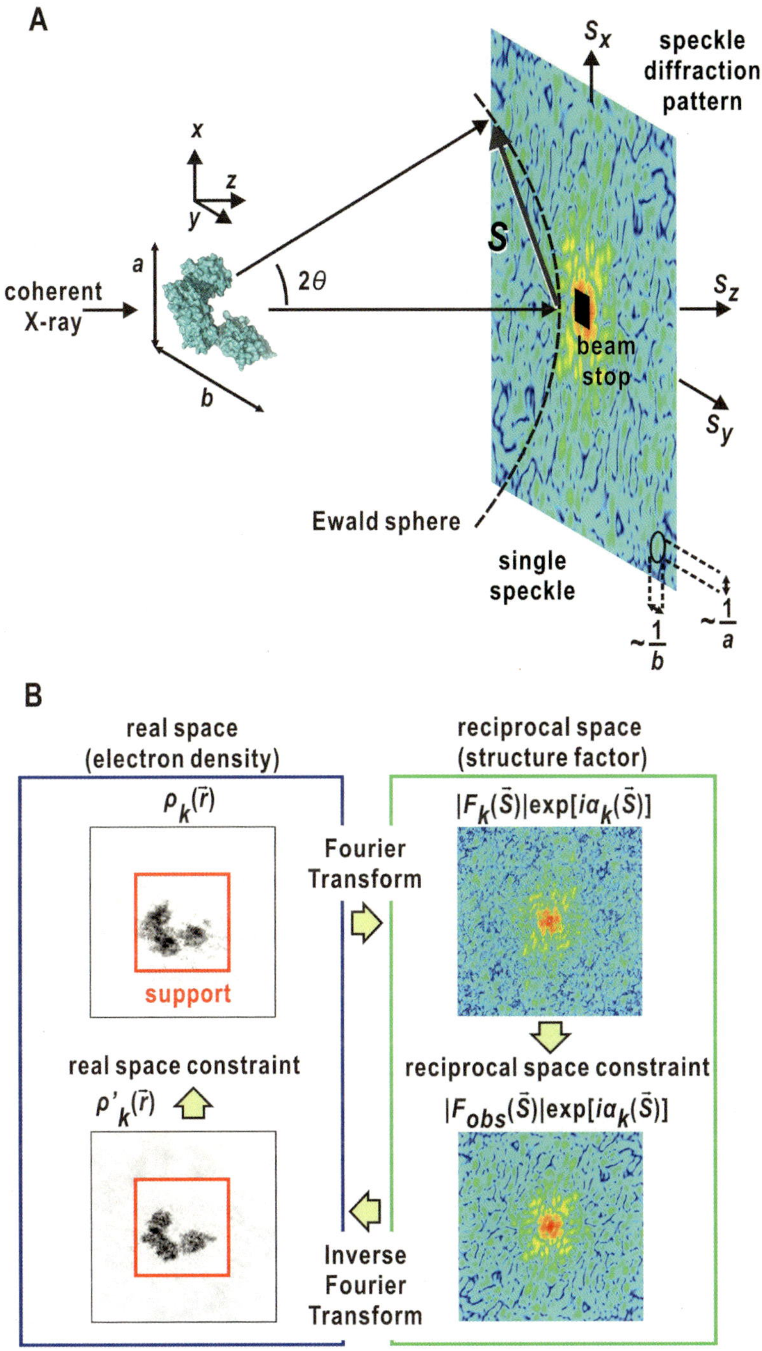


**Fig. 10.1** (**a**) A schematic illustration of CXDI experiment. (**b**) The iterative PR algorithm to reconstruct an electron density map from a diffraction pattern. The theory and the procedure are described in the text

$$\left|F_{\text{OBS}}\left(S_x, S_y, S_z = 0\right)\right| = \left|\sum_{x=0}^{N_x-1} \sum_{y=0}^{N_y-1} \rho_{\text{P}}(x, y) \exp\left[2\pi i \left(S_x \frac{x}{N_x} + S_y \frac{y}{N_y}\right)\right] \Delta x \Delta y\right|, \tag{10.3}$$

where $F_{\text{OBS}}(S_x, S_y, S_z = 0)$ is the structure factor of the projected electron density, $\rho_{\text{P}}(x, y)$, which is expressed as $N_x \times N_y$ pixels with an area element of $\Delta x \Delta y$. When the structure amplitude is recorded using a sufficient number of pixels larger than $N_x \times N_y$, the algebraic equation yielding the projection electron density map can be solved in principle. However, owing to the absence of atoms displaying anomalous scattering from many biological specimens, the centrosymmetry (Friedel's symmetry) appearing in diffraction patterns reduces the net information contained in a set of the diffraction amplitude to $(N_x \times N_y)/2$. Subsequently, the observed diffraction amplitude data is insufficient to solve the algebraic equations. For instance, the diffraction pattern is sampled by $\sigma_x N_x$ along the direction of $S_x$ and by $\sigma_y N_y$ along the direction of $S_y$. When the number of equations, $\sigma_x \times \sigma_y$, is larger than two, $\rho_{\text{P}}(x, y)$ is reconstructed (Miao et al. 2003b):

$$\sigma = \sigma_x \times \sigma_y \geq 2. \tag{10.4}$$

The parameter, $\sigma$, is designated as the OS ratio in Eq. 10.1. Therefore, in CXDI experiments, diffraction patterns are collected so that the OS ratio is sufficiently larger than two.

The projection electron density map, $\rho_{\text{P}}(x, y)$, can be retrieved by using the iterative PR algorithm (Fienup 1982) (Fig. 10.1b). The iterative calculation cycle starts from a map with a random electron density. After the Fourier transform of the map, the structure amplitude of the calculated structure factor is replaced with the observed value. Subsequently, the electron density map is calculated by the inverse Fourier transform of the structure factor with the observed amplitude. The constraints in real space modify the electron density map for the next calculation cycle.

For instance, the constraint used in the hybrid-input-output (HIO) algorithm (Fienup 1982) modifies the electron density map, $\rho_{k+1}\left(\vec{r}\right)$, for the $(k + 1)$-th cycle as:

$$\rho_{k+1}\left(\vec{r}\right) = \begin{cases} \rho'_k\left(\vec{r}\right) & \vec{r} \in \text{Support and } \rho'_k\left(\vec{r}\right) \geq 0 \\ \rho_k\left(\vec{r}\right) - \beta \rho'_k\left(\vec{r}\right) & \text{otherwise} \end{cases}, \tag{10.5}$$

where $\rho_k\left(\vec{r}\right)$ is the map at the start of the $k$-th cycle, and $\rho'_k\left(\vec{r}\right)$ is the inverse Fourier transform of the structure factor, which is composed of the observed diffraction amplitude and the phase calculated from $\rho_k\left(\vec{r}\right)$. A control parameter, $\beta$, reduces the electron densities outside the support.

## 10.3 CXDI Experiment

### *10.3.1 Specimen Preparation for CXDI Experiments*

To reduce scattering and absorption of diffracted X-rays by air, specimens are placed in a vacuum chamber in CXDI experiments. To avoid drying in vacuum, biological specimens are maintained in a frozen-hydrated state (Dubochet et al. 1982; Takayama and Nakasako 2012; Nakasako et al. 2013; Kobayashi et al. 2016a, b). Frozen-hydrated *C. merolae* cells, chloroplasts of *C. merolae*, and cyanobacteria (*Prochlorococcus* strain NIES-2087) cells are prepared by the procedure illustrated in Fig. 10.2a without labeling by fluorescent dyes, fixation by chemical reagents, and sectioning.

Particles are set on thin silicon nitride membranes (thickness 100 nm) attached to support frames made of silicon. The relative humidity around specimen particles is kept at approximately 99% by using our custom-made humidity-controlling device (Takayama and Nakasako 2012; Nakasako et al. 2013; Takayama et al. 2015; Kobayashi et al. 2016b) (Fig. 10.2a). Membrane and support frames tolerate the rapid change in temperature during flash-cooling by liquid ethane. The diffraction patterns from the membranes are negligible in comparison to those from specimens in the small-angle diffraction region. For instance, the diffraction intensity from a silicon nitride membrane is less than 5 X-ray photons per pixel in the small-angle regions, while the diffraction intensity from a specimen particle exceeds $10^5$ X-ray photons per pixel even in the presence of an aluminum attenuator. Since silicon nitride membranes are highly hydrophobic, the surfaces are covered with poly-L-lysine molecules with an approximate molecular weight of 300 k after coating with carbon layers to a thickness of approximately 15 nm. This treatment dramatically improves the affinity of silicon nitride membranes to biological specimens.

For synchrotron CXDI experiments, a single particle is placed at the center of a membrane with an area of $3000 \times 3000\ \mu m^2$ by using a microinjector (Takayama and Nakasako 2012). The size of the membrane allows us to rotate the specimen particle in a wide angle against the direction of the incident X-ray beam for tomography CXDI experiments. For XFEL-CXDI experiments, a number of specimen particles are randomly dispersed onto a membrane at a number density of 7–10 particles per $10 \times 10\ \mu m^2$ area (Takayama et al. 2015; Oroguchi et al. 2015; Kobayashi et al. 2016b). Custom-made silicon frames, with $3 \times 3$ arrays of silicon nitride membrane windows of $1 \times 1\ \mu m^2$ each, are used to reduce the time expended exchanging specimen disks in the vacuum chamber of a diffraction apparatus (Kobayashi et al. 2016b).

Immediately after removing the buffer solutions as much as possible using blotting, specimen disks are rapidly cooled by plunging into liquid ethane, the temperature of which was kept at approximately 80 K. The cooled specimens are stored in liquid nitrogen. The cryogenic specimen preparation is advantageous in order to prepare a large number of frozen-hydrated cells and cellular organelles harvested at the desired period of the cell cycle.

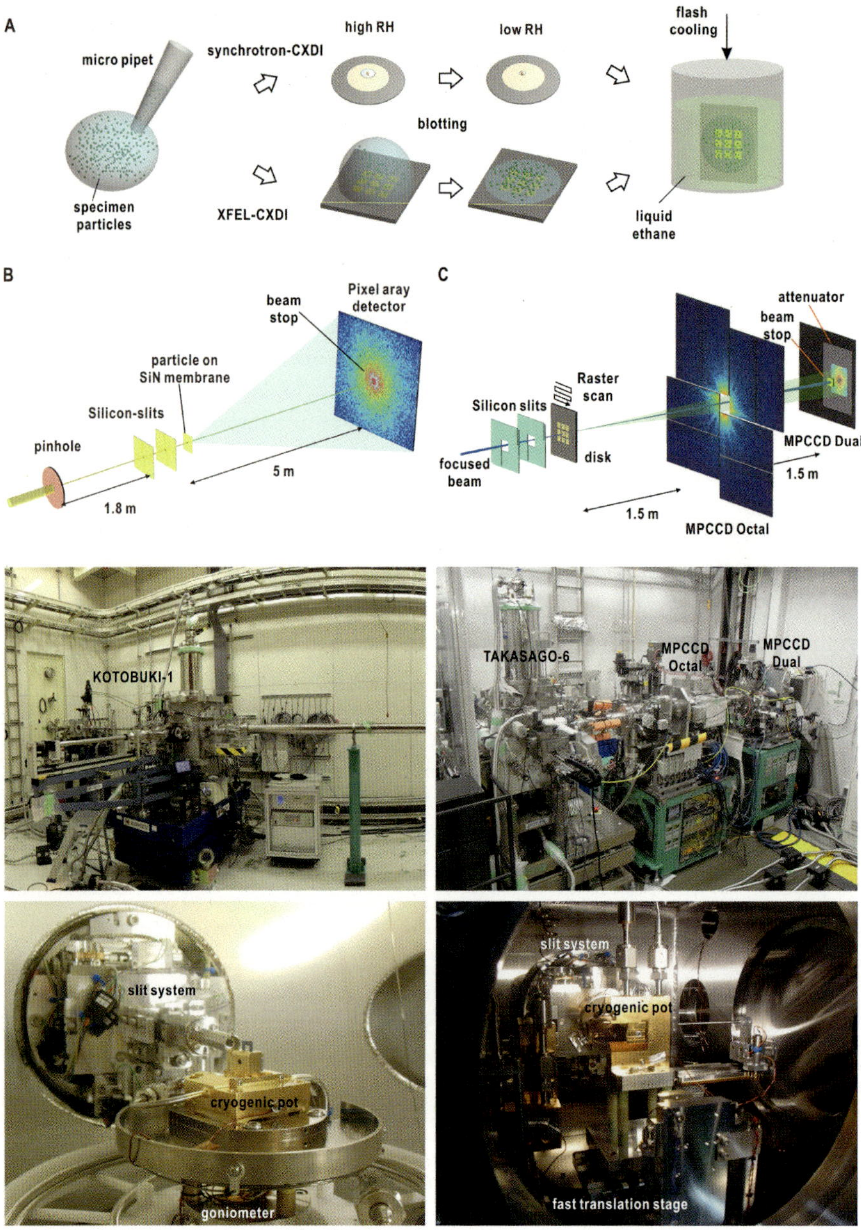


**Fig. 10.2** (**a**) Specimen preparation for cryogenic CXDI experiments under a controlled relative humidity. (**b**) Synchrotron-CXDI experiments at BL29XUL (Tamasaku et al. 2001) of SPring-8 using the KOTOBUKI-1 diffraction apparatus. A schematic illustration of the diffraction experiments (*upper panel*), photographs of the diffraction apparatus (*center*), and the cryogenic specimen stage inside the vacuum chamber of the apparatus (*lower*). (**c**) XFEL-CXDI experiments at SACLA using the TAKASAGO-6 diffraction apparatus. A schematic illustration of the diffraction experiments (*upper panel*), photographs of the diffraction apparatus (*center*), and the cryogenic specimen stage inside the vacuum chamber of the apparatus (*lower*)

### 10.3.2 CXDI Experiments Using Synchrotron X-Rays

In CXDI experiments, specimen particles must be completely bathed in an X-ray beam with a high spatial coherence. In CXDI experiments using synchrotron X-ray (Fig. 10.2b), the coherent X-ray beam is produced by Fraunhofer diffraction from a circular aperture that is placed approximately 2 m upstream of the specimen position. The size and intensity of the produced X-ray beam depends on the aperture size, as predicted by the Fresnel-Kirchhoff integral (Born and Wolf). When using an aperture with the diameter of 40 μm, the X-ray beam supplied to the sample position has a flux density of $10^9$–$10^{10}$ X-ray photons per $mm^2$ per sec, resulting in a diameter for the coherent area of approximately 10 μm.

In the synchrotron CXDI experiments, a custom-made diffraction apparatus, named KOTOBUKI-1, was used (Fig. 10.2b). The apparatus is composed of a vacuum chamber equipped with a goniometer stage, a load-lock chamber, and an alignment table mounting the vacuum chamber and the load-lock chamber. The goniometer stage is itself composed of three translation axes and one rotation axis. The load-lock chamber has a transfer rod for delivering cooled specimens to the goniometer stage. In addition, a slit system is used for reducing parasitic and background scattering from the upstream optics and the pinhole. A cryogenic pot is fixed to the goniometer stage to maintain the temperature of the mounted specimens at approximately 77 K through thermal contact. Liquid nitrogen is supplied to the cryogenic pot, which is then evacuated by a vacuum pump placed outside the vacuum chamber. Using this setup, the temperature of the cryogenic pot is maintained at 66 K by the evaporation cooling effect. The consumption of liquid nitrogen is 6 l per 12 h.

The diffraction apparatus is placed so that the specimen position coincides with the center of the Airy disk diffraction pattern from the pinhole. A pixel array detector located 5 m downstream from the specimen records diffraction patterns.

### 10.3.3 CXDI Experiment Using XFEL Pulses

In recent years, X-ray pulses with very strong intensities and complete spatial coherence are available at X-ray free-electron laser (XFEL) facilities. The weak point of XFEL-CXDI experiments is that single-shot XFEL pulses destroy specimen particles at the atomic level (Chapman et al. 2006b; Nakasako et al. 2013). However, the pulse duration of tens of femtoseconds fortunately allows us to observe the diffraction pattern of a specimen particle before the destruction. Therefore, diffraction patterns can be collected from particles almost radiation damage-free, as demonstrated in crystal structure analysis for proteins using XFEL (Hirata et al. 2014).

X-ray pulses are supplied with a repetition rate of 30 Hz; a large number of diffraction patterns can be collected in a short period of time by supplying fresh specimen particles into the irradiation area (Nakasako et al. 2013; Kimura et al. 2014). Owing to the destruction of both the specimen particle and the silicon nitride membrane by a single XFEL pulse, each membrane window of $1 \times 1$ mm$^2$ is raster-scanned for XFEL pulses at an interval of 25 μm between adjacent irradiation spots. Since diffraction patterns are collected stochastically by scanning specimen disk with randomly dispersed particles, the probability for complete coincidence of the specimen positions with the centers of the X-ray pulses is low. Only when specimen particles are located within the cross section of an X-ray pulse, diffraction patterns with intensities worth analyzing can be collected. The number of high-quality diffraction patterns partly depends on the number density of the dispersed specimen particles.

Cryogenic XFEL-CXDI experiments were performed by using our custom-made diffraction apparatus named TAKASAGO-6 (Kobayashi et al. 2016a) at BL3 (Tono et al. 2013) of SPring-8 Angstrom Compact free-electron LAser (SACLA), an XFEL facility in Japan (Ishikawa et al. 2012). The TAKASAGO-6 diffraction apparatus was developed for XFEL-CXDI experiments on frozen-hydrated noncrystalline biological particles at cryogenic temperatures. The apparatus equipped with a high-speed translation stage carrying a cryogenic pot can raster-scan specimen disks at speeds higher than 25 μm per 33 ms. Therefore, we can use XFEL pulses provided at a repetition rate of 30 Hz for diffraction data collection. A load-lock chamber and miscellaneous devices assist the easy transfer of cooled specimens from a storage filled by liquid nitrogen to the cryogenic pot inside the vacuum chamber. Subsequently, the apparatus allows us to collect more than 2,600,000 diffraction patterns within 75 h (Kobayashi et al. 2016a).

XFEL pulses with a wavelength of 0.225 nm were focused to give an intensity of $10^{10}$–$10^{11}$ photons per $2 \times 2$ μm$^2$ (FWHM) area per 10 f. pulse (Yumoto et al. 2013). The specimen position in the apparatus was adjusted within the focal area. A pair of silicon slits located 5 mm upstream from the focal spot is used to reduce background scattering from the upstream optics to a negligibly low level. Two multi-port CCD (MPCCD) detectors (Kameshima et al. 2014) in a tandem arrangement simultaneously record the diffraction patterns. The MPCCD-Octal detector, composed of eight CCD sensor panels, was placed approximately 1.5 m downstream from the specimen position and covers a resolution range of 6–190 nm. Each sensor panel is composed of $512 \times 1024$ pixels of $50 \times 50$ μm$^2$. The MPCCD-Dual detector, placed 3.0 m downstream from the specimen, recorded the diffraction patterns of the small-angle region in the resolution range of 150–500 nm. To attenuate strong diffraction patterns, a set of aluminum foils with thicknesses of 15–100 μm were placed in front of the MPCCD-Dual detector. A direct beam stopper with a cross section of $2 \times 2$ mm$^2$ was also placed in front of the MPCCD-Dual detector to completely absorb the incident X-ray beam.

### 10.3.4 Structure Analysis

In synchrotron CXDI experiments, diffraction patterns recorded by the pixel array detector are converted to diffraction intensity data after the subtraction of background scattering from the silicon nitride membrane. In XFEL-CXDI experiments, immediately after completing each scan, diffraction patterns recorded by the two MPCCD detectors were processed automatically by the program suite G-SITENNO (Sekiguchi et al. 2014a, b). The suite subtracts the dark current of the MPCCD detectors, extracts diffraction patterns with desired signal-to-noise ratio and resolution, determines the beam center positions in the two detectors, evaluates the degrees of centrosymmetry of patterns, and merges the patterns from the two detectors.

Due to the beam stop and the saturation of the detector pixels, diffraction patterns are missed in the very small-angle region, where the structural information on the shape and size of noncrystalline particles is included. In addition, diffraction patterns are modified by Poisson noise during X-ray detection, particularly for high diffraction angles. These two make the PR calculations for experimental data difficult and require larger OS ratios than those expected from the theoretical consideration for ideal diffraction data (Miao et al. 2003b; Kodama and Nakasako 2011; Oroguchi and Nakasako 2013; Kobayashi et al. 2014).

PR calculations for selected diffraction pattern were carried out according to the ASURA protocol, which retrieves projection electron density maps through two steps (Sekiguchi et al. 2016). In the first step, to estimate the support area, approximately 1000 independent PR calculations were conducted using the HIO algorithm for the diffraction pattern, up to a resolution of 17.5 $\mu m^{-1}$ (corresponding to a resolution of 57 nm). Each electron density map is composed of $J$ pixels and is assigned as a point in a $J$-dimensional hyperspace. By applying a multivariate analysis for PR maps, the most probable support shape of the noncrystalline particle is determined by averaging a selected set of maps. Next, 1000 independent PR calculations by the oversampling smoothness (OSS) algorithm (Rodriguez et al. 2013) for the determined support shape. By the multivariate analysis of 1000 OSS-retrieved maps, the most probable electron density map within the support shape is calculated by averaging the maps composing a cluster of similar maps. The effective resolution of the most probable map was then estimated by the phase-retrieval transfer function (PRTF) (Chapman et al. 2006a) applied to the set of maps. The effective resolution estimated by the PRTF tended to be lower than the resolution of the diffraction pattern used for the PR calculations, because of the small fluctuations of electron densities among the set of maps.

## 10.4 Structures

### *10.4.1* C. merolae *Cells*

The *C. merolae* cells were cultured in the Allen's medium under a 12 hour-illumination/ 12 hour-dark cycle in an incubator. The temperature inside the incubator was controlled at 303 K. In the preparation of the frozen-hydrated *C. merolae* cells, the size of the cells placed on silicon nitride membranes were roughly measured by a digital microscope, which was equipped to the humidity-controlling chamber.

A pinhole of 40 μm in diameter was used for coherent illumination of *C. merolae* cells, the size of the schyzon of which is 2–4 μm in the cell cycle. Diffraction patterns of frozen-hydrated cells in the G1 and M phase were collected by using the KOTOBUKI-1 diffraction apparatus at BL29XU of SPring-8 (Tamasaku et al. 2001). By carefully tuning the beamline optics, the incident intensity reached a flux density of $1.9 \times 10^{10}$ X-ray photons per $mm^2$ per sec, which was sufficient to record diffraction patterns of a *C. merolae* cell in the M phase beyond a resolution of 10 $\mu m^{-1}$ with an exposure time of 60 s.

The diffraction pattern from the G1-phase cell was composed of speckled peaks up to a resolution of approximately 10 $\mu m^{-1}$ (corresponding to a spacing of 100 nm in real space) (Fig. 10.3a). The PR calculations successfully retrieved electron density maps interpreted as the G1-phase cell (Fig. 10.3b). By applying the multi-variate analysis for a number of PR maps, the projection map was obtained at an effective resolution of approximately 137 nm. The electron density map of the cell appeared as an ellipsoid shape with approximate dimensions of $2 \times 1$ μm. The shape was consistent with those of the G1 cells observed in fluorescence LM (Fig. 10.3c). The map had a prominent density peak occupying the area of more than 50% of the cell. By referring to the fluorescence LM images of G1 cells, the prominent peak could be assigned as a chloroplast.

In a trial tomography experiment, diffraction patterns of *C. merolae* in the M phase were collected at an angular interval of 1.5 ° from a single M-phase cell within the Henderson's limit of an irradiation dose (Howells et al. 2009) (Fig. 10.3a). A typical diffraction pattern is shown in Fig. 10.3a. Since the cell in the M phase has almost duplicated the cellular components, the total cross section for the X-ray is approximately three or four times stronger than that of the G1 phase. Therefore, the M-phase cells are better for the demonstration of the two-dimensional and three-dimensional structure analyses of the whole cell beyond a 100 nm resolution. The projection electron density map was retrieved at an effective resolution of 114 nm and displayed a globular shape with the approximate diameter of 2.5 μm rather than the ellipsoid shape observed for the G1 cell. Two prominent density peaks were assignable as two chloroplasts and probably one nucleus.

The apparatus developed for tomography experiments and the pixel array detector allow us to collect a series of diffraction patterns up to a resolution of 30 $\mu m^{-1}$

from a single cell in the M phase in different orientations to the direction of incident X-ray beam. By improving the protocols for data collection and data analysis software, the three-dimensional structure of a whole cell at a resolution better than 50 nm would be possible in the near future.

### *10.4.2 Chloroplast of* C. merolae

Whole-cell imaging of *C. merolae* is difficult in XFEL-CXDI experiments at SACLA, due to the currently available focused beam size smaller than the cell size. Among the organelles in *C. merolae*, chloroplasts have average sizes comparable to or slightly larger than the focused beam size. Since the number of chloroplasts with sizes completely encompassed by the focused beam is small, diffraction patterns from chloroplasts with sizes of approximately 1 μm were extracted from a large number of diffraction patterns.

The intensity, the sizes and shapes of speckle peaks, and the maximum resolution of each diffraction pattern depend on the sizes of the chloroplasts and the relative positions of the chloroplasts against the positions of X-ray pulses. From a number of collected diffraction patterns, those with speckle sizes of approximately $5 \times 5$ pixels and displaying a resolution of 20 $\mu m^{-1}$ (corresponding to a resolution of 50 nm in real space) are selected for PR calculations. In one of the representative diffraction patterns, X-rays were diffracted up to a resolution of 29 nm with a significant signal-to-noise ratio (Takayama et al. 2015).

One of the diffraction patterns displayed a good Friedel's centrosymmetry up to a resolution of 16 $\mu m^{-1}$ and a signal-to-noise ratio greater than 2–3 up to a resolution of 30 $\mu m^{-1}$. The diffraction pattern displayed a concentric arrangement of speckle peaks with an approximate interval of 2.5 $\mu m^{-1}$, suggesting that the chloroplast particle has a globular shape. In addition, the interval of the concentric rings reflects partly the presence of high-electron density regions inside the chloroplast separated by approximately 400 nm. The size of the chloroplast projected along the direction of the incident X-ray was estimated to be approximately 1 μm from the reciprocal of the averaged size of speckle peaks. When the shape of the chloroplast is globular, the Ewald sphere can be approximated as a plane up to a resolution of 11 nm (Oroguchi and Nakasako 2013).

A projection electron density map was obtained at an effective resolution of 41 nm. The electron density map appeared as a C-shape with an approximate size of $600 \times 800$ nm. In the C-shaped structure, high-electron density peaks separated by 400 nm would be a predominant cause for the concentric ring patterns. The average electron density of the central low-density region is comparable to that of vitreous ice embedding in the chloroplast ($0.311 \pm 0.01$ e/A$^3$).

Electron density maps retrieved from small-angle diffraction patterns provide little information on the types of materials observed, i.e., the chemical composition.

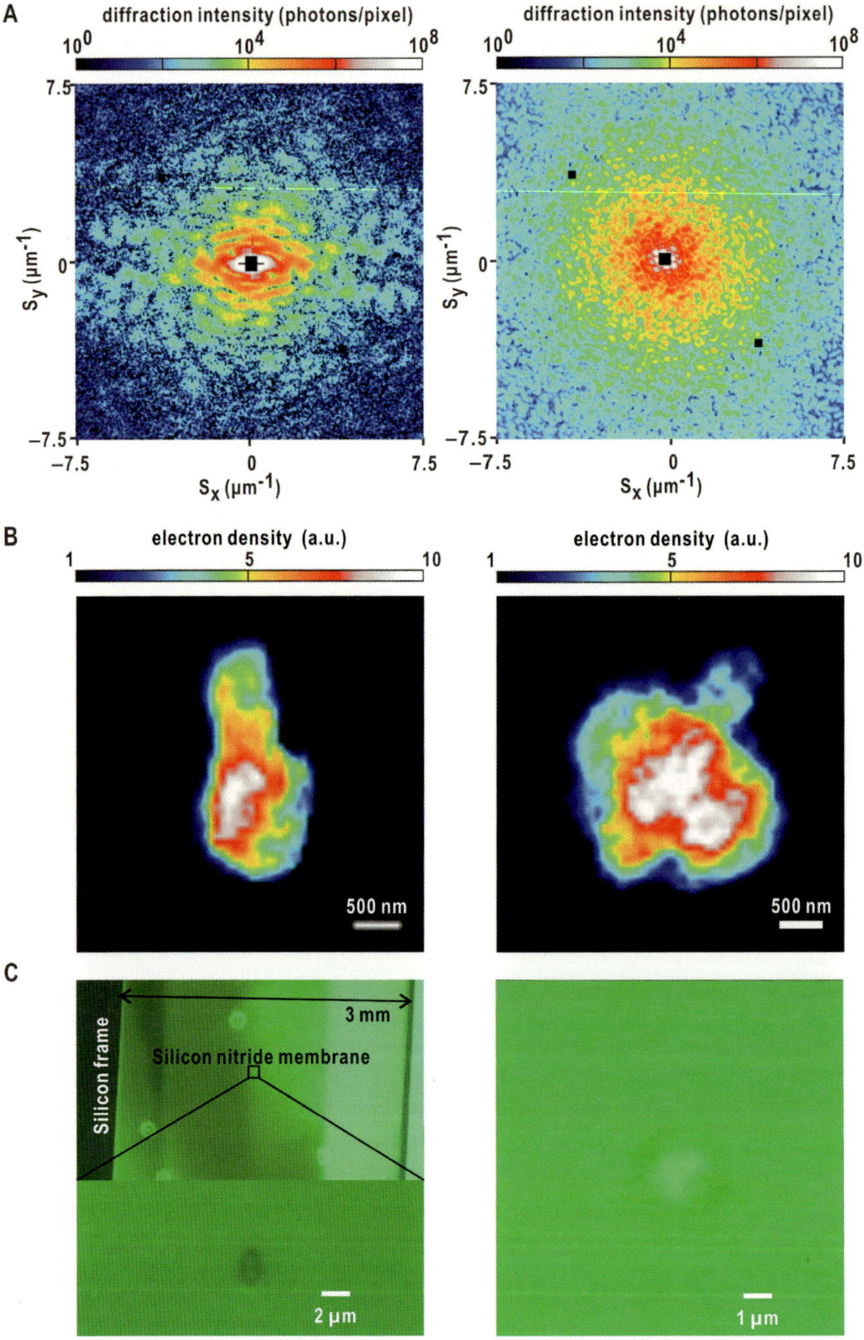


**Fig. 10.3** (**a**) Diffraction patterns from a flash-cooled *C. merolae* cell in the G1 (*left panel*) and M (*right*) phase of the cell cycle. The box-shaped black area is a beam stopper placed in front of the pixel array detector. (**b**) Projection electron density maps of the G1 (*left*) and M (*right*) panels

Therefore, other imaging techniques are necessary to speculate what kind of materials give the electron density distribution that has been acquired. Phase contrast LM, which visualizes the differences of refractive indices inside specimens, gave images of chloroplasts with a small dark region at their centers. When chloroplasts were labeled by dye molecules selectively bound to proteins, the region emitting strong fluorescence arranged in a C-shaped. Since thylakoid membranes, in which membrane proteins, lipid bi-layers, and chlorophylls are abundant, are the major component of chloroplasts, the C-shaped region was assignable as a thylakoid membrane (Kouranov and Schnell 1996; Su et al. 2010). In addition, the region with intense autofluorescence from chlorophylls overlapped with the high-fluorescence intensity region from the dyes attached to the proteins (Takayama et al. 2015).

In chloroplasts, thylakoid membranes contain proteins in far greater quantities than the stroma and the envelope of chloroplasts. Therefore, the C-shape of the electron density map is interpretable as an assembly of thylakoid membranes that is composed of proteins adsorbing chlorophyll molecules. The average electron densities of protein molecules are in the range of 0.41–0.43 e/$A^3$. Both the shapes of high-electron density regions (Fig. 10.4b) and the high-fluorescence intensity region from protein-bound dyes (Fig. 10.4c) resemble the internal structures of chloroplasts observed by EM for sectioned chloroplast treated by heavy-atom staining and resin fixation (Yagisawa et al. 2009; Yoshida et al. 2010).

### *10.4.3 Structure of Cyanobacteria*

A small eukaryote *C. merolae* is evolutionally thought to be an origin of eukaryotes. According to the endosymbiotic theory of the evolution of eukaryotes (Gray and Archibald 2012), major cellular organelles, such as chloroplasts and mitochondria, originated as symbioses between separate single-celled organisms. Cyanobacterium, a photosynthetic organism, is a candidate among organisms incorporating chloroplasts to form eukaryotic cell. In this regard, it is of interest whether or not the structure of a cyanobacterial cell is similar to that of the chloroplast of *C. merolae*, which has a C-shaped distribution of thylakoid membranes. In electron microscopy observations of a section of chemically stained cyanobacterium, lamellar structures of thylakoid membranes distribute in the marginal region rather than in the central part (van de Meene et al. 2006). By comparing the structures of the chloroplast of

**Fig. 10.3** (continued) retrieved from the diffraction patterns in panel (**a**). (**c**) Photograph of silicon nitride membranes (3 × 1 $mm^2$) under a high humidity atmosphere before flash-cooling. Each single cell was placed in the area indicated by a box. Magnified LM images are *C. merolae* cells in the G1 phase (*left*) and M (*right*) phase

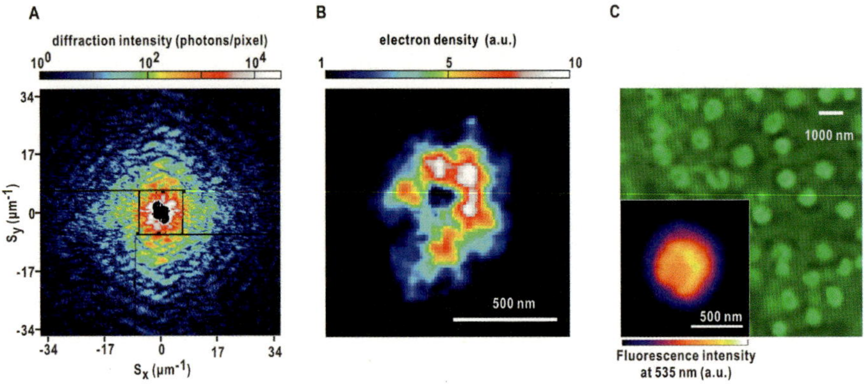


**Fig. 10.4** (**a**) A diffraction pattern from a flash-cooled chloroplast of *C. merolae*. The *black* area in the center is a beam stopper placed in front of the MPCCD-Dual detector and pixels saturated by intense X-rays. (**b**) A projection electron density map retrieved from the diffraction pattern in panel (**a**). (**c**) A phase-contrast image of chloroplasts on a silicon nitride membrane under a high humidity atmosphere before flash-cooling. The *inset* is a fluorescence LM image of a chloroplast stained by dye-adhering proteins (The panels are reused from Takayama et al. (2015) with permission from The Oxford University Press)

*C. merolae* and a whole cyanobacterium cell, the structural relation among evolutionally related organelles would become clear.

We conducted XFEL-CXDI experiments for *Prochlorococcus marinus* (Chisholm et al. 1988). In the diffraction patterns collected through raster scans, the maximum resolution was approximately 20 $\mu m^{-1}$, corresponding to 50 nm in real space (Fig. 10.5a). Typical diffraction patterns of cyanobacterial cells are composed of speckle peaks with sizes of approximately 1.3 $\mu m^{-1}$, which is comparable to the reciprocal sizes of single cyanobacterial cell of approximately 780 nm, which was estimated from dynamic light-scattering measurements. The PR electron density map displayed globular shapes with a diameter of 700–800 nm (Fig. 10.5b) and was consistent with the shape viewed by LM (Fig. 10.5c). High-electron density regions appeared as a C-shape with an approximate diameter of 500 nm. In referring to the tomography study of a frozen-hydrated cyanobacterial cell by using cryogenic transmission electron microscopy (Ting et al. 2007), the C-shaped regions can be assigned as a stack of thylakoid membranes.

Regarding cyanobacteria cells, the three-dimensional electron density maps could be reconstructed from a number of cells. In the maps, the common structures among cells are emphasized and the structural differences between cells that are averaged to be invisible. Even in the averaged maps, the C-shaped structures of thylakoid membranes are usually observed. When comparing the projection electron density maps of a cyanobacteria cell and a chloroplast of *C. merolae,* there are similarities in the distribution of the high-density regions and the dimensions.

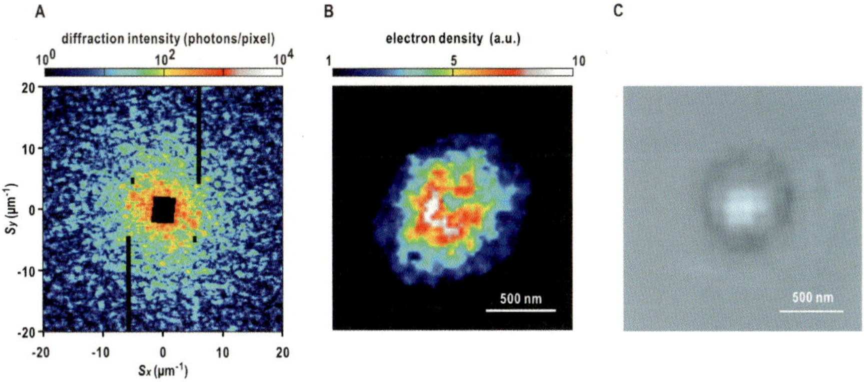


**Fig. 10.5** (**a**) A diffraction pattern from a flash-cooled cyanobacteria cell (*P. marinus* strain NIES-2087 from the National Institute for Environmental Studies, Japan). The box-shaped black area is a beam stopper placed in front of the MPCCD-Dual detector. (**b**) A projection electron density map retrieved from the diffraction pattern in panel (**a**). (**c**) An LM image of a flash-cooled cyanobacteria cell on a silicon nitride membrane under high humidity atmosphere (Panels are reused from Kobayashi et al. (2016b) [doi: 10.1107/S1600577516007736] after modification with permission from The International Union of Crystallography)

### 10.4.4 Feasibility of CXDI in Cell Biology

Here we present three examples to demonstrate the potential of the cryogenic CXDI technique to visualize the structures of noncrystalline biological specimens with dimensions as large as approximately 1–2 μm at effective resolutions better than the limit of LM. Three advantages of the CXDI technique can be postulated from observations of the application presented here. Firstly, because specimens are supplied to experiments in the frozen-hydrated state, specimens of cells and cellular organelles are prepared at a desired state in the cell cycle. Secondly, the resolution of diffraction patterns exceeds the limit of fluorescence microscopy. Finally, because of the penetration power, X-rays with short wavelengths are advantageous to visualize the internal structure of thick objects that are opaque in EM. In addition, cryogenic synchrotron-CXDI experiments using synchrotron X-ray enable us to perform the tomographic structure analyses at a spatial resolution of ~10 nm under the reduction of radiation damage (Howells et al. 2009).

Considering the currently available incident X-ray intensity and focused size of XFEL pulses at SACLA, cellular organelles and small cells are suitable targets of XFEL-CXDI. We demonstrated that, for synthesized metal particles of less than 1 μm, XFEL-CXDI experiments have the potential to analyze the size distribution and structures from thousands of diffraction patterns (Sekiguchi et al. 2014b). For biological specimens, several hundreds to thousands of diffraction patterns from specimen particles can be collected within 1 h. Subsequently, a statistically sufficient size distribution of specimen particles can be constructed as demonstrated for cyanobacteria (Kobayashi et al. 2016b).

Regarding understanding the spatiotemporal events occurring in cell cycles, CXDI would be an imaging tool whole cells and organelles together with sophisticated imaging techniques, such as fluorescence LM, EM, and X-ray microscopy (McDermott et al. 2009). In addition, X-ray ptychography (Rodenburg et al. 2007) can provide projection maps over a large object, such as tissues with far larger sizes than the limitation of X-ray beam sizes. The complementary use of CXD and other sophisticated and established imaging techniques would bridge the resolution gap between cell biology and structural biology at molecular level.

**Acknowledgment** The authors would like to thank Dr. Kensuke Tono, Dr. Takashi Kameshima, Dr. Yasumasa Joti, and the members of the engineering team of SACLA for their great help in XFEL-XCDI experiments. This study was supported by a grant for XFEL key technology and the X-ray Free-Electron Laser Priority Strategy Program from the Ministry of Education, Culture, Sports, Science, and Technology (MEXT) to M.N., M.Y., and S.M.; Grant-in-Aid for Scientific Research on Innovative Areas from MEXT [Nos. jp15076210, jp23120525, jp25120725 to M.N., and Nos. jp24113723, jp26104535 to T.O. jp25120726 to S.M., and jp25114514 to S.M.], Grant-in-Aid for Scientific Research (A) from the Japan Society for the Promotion of Science (JSPS) [Nos. 17654048 and jp16H02218 to M.N.], Grant-in-Aid for Scientific Research (B) from JSPS [jp11558086 to M.N. and jp26291067 to S.M.]; Grant-in-Aid for Young Scientists (B) from JSPS [jp26800227 To T.O.], Grant-in-Aid for Challenging Exploratory Research from JSPS [jp17654048 and jp24654140 To M.N.], Grant-in-Aid for Research Activity Start-up from JSPS [jp25891033 to Y.T.], and Grant-in-Aid for JSPS Fellows (No. jp15J01707 to Y. S., No. jp15J01831 to A.K.); RIKEN Special Postdoctoral Researchers Program to Y.T. We collected diffraction data in cryogenic XFEL-CXDI experiments performed at SACLA (Nos. 2013A8043, 2013B8049, 2014A8033, 2014B8052, 2015A8051, and 2016A8048). The phase-retrieval calculations and multivariate analyses were performed using the mini-K supercomputer system at the SACLA facility.

## References

Bamme BE, Jakana J, Schmid MF, Chiu W (2010) Radiation damage effects at four specimen temperatures from 4 to 100 K. J Struct Biol 169:331–341

Chapman HN, Barty A, Marchesini S, Noy A, Hau-Riege SP, Cui C, Howells MR, Rosen R, He H, Spence JCH, Weierstall U, Beetz T, Jacobsen C, Shapiro D (2006a) High-resolution ab initio three-dimensional x-ray diffraction microscopy. J Opt Soc Am A 23:1179–1200

Chapman HN, Barty A, Bogan MJ, Boutet S, Frank M, Hau-Riege SP, Marchesini S, Woods BW, Bajt S, Benner WH, London RA, Plönjes E, Kuhlmann M, Treusch R, Düsterer S, Tschentscher T, Schnider JR, Spiller E, Möller T, Bostedt C, Hoener M, Shapiro DA, Hodgson KO, Van Der Spoel D, Burmeister F, Bergh M, Caleman C, Huldt G, Seibert MM, Maia FRNC, Lee RW, Szöke A, Timneanu N, Hajdu J (2006b) Femtosecond diffractive imaging with a soft-X-ray free-electron laser. Nat Phys 2:839–843

Chisholm SW, Olson RJ, Zettler ER, Georicke R, Waterbury JB, Welschmeyer NA (1988) A novel free-living prochlorophyte abundant in the oceanic euphotic zone. Nature 334:340–343

Dubochet J, Chang J-J, Freeman R, Lepault J, McDowell AW (1982) Frozen aqueous suspensions. Ultramicroscopy 10:55–61

Fienup JR (1982) Phase retrieval algorithms: a comparison. Appl Opt 21:2758–2769

Gan L, Jensen GJ (2011) Electron tomography of cells. Q Rev Biophys 45:27–56

Gray MW, Archibald JM (2012) Origins of mitochondria and plastids. In: Bock R, Knoop V (eds) Genomics of chloroplasts and mitochondria. Springer, Dordrecht, pp 1–30

Heuser JE (2011) The origins and evolution of freeze-etch electron microscopy. J Electron Microsc 60:S3–S29

Hirata K, Shinzawa-Itoh K, Yano N, Takemura S, Kato K, Hatanaka M, Muramoto K, Kawahara T, Tsukihara T, Yamashita E, Tono K, Ueno G, Hikima T, Murakami H, Inubushi Y, Yabashi M, Ishikawa T, Yamamoto M, Ogura T, Sugimoto H, Shen J-R, Yoshikawa S, Ago H (2014) Determination of damage-free crystal structure of an X-ray sensitive protein using an XFEL. Nat Methods 11:734–736

Howells MR, Beetz T, Chapman HN, Cui C, Holton JM, Jacobsen CJ, Kirz J, Lima E, Marchesini S, Miao H, Sayre D, Shapiro DA, Spence JCH, Starodub D (2009) An assessment of the resolution limitation due to radiation-damage in X-ray diffraction microscopy. J Electron Spectrosc Relat Phenom 170:4–12

Huang X, Nelson J, Kirz J, Lima E, Marchesini S, Miao H, Neiman AM, Shapiro D, Steinbrener J, Stewart A, Turner JJ, Jacobsen C (2009) Soft xray diffraction microscopy of a frozen hydrated yeast cell. Phys Rev Lett 103:198101

Imoto Y, Yoshida Y, Yagisawa F, Kuroiwa H, Kuroiwa T (2011) The cell cycle, including the mitotic cycle and organelle division cycles, as revealed by cytological observations. J Electron Microsc 60:S117–S136

Ishikawa T, Aoyagi H, Asaka T, Asano Y, Azumi N, Bizen T, Ego H, Fukami K, Fukui T, Furukawa Y, Goto S, Hanaki H, Hara T, Hasegawa T, Hatsui T, Higashiya A, Hirono T, Hosoda N, Ishii M, Inagaki T, Inubushi Y, Itoga T, Joti Y, Kago M, Kameshima T, Kimura H, Kirihara Y, Kiyomichi A, Kobayashi T, Kondo C, Kudo T, Maesaka H, Maréchal XM, Masuda T, Matsubara S, Matsumoto T, Matsushita T, Matsui S, Nagasono M, Nariyama N, Ohashi H, Ohata T, Ohshima T, Ono S, Otake Y, Saji C, Sakurai T, Sato T, Sawada K, Seike T, Shirasawa K, Sugimoto T, Suzuki S, Takahashi S, Takebe H, Takeshita K, Tamasaku K, Tanaka H, Tanaka R, Tanaka T, Togashi T, Togawa K, Tokuhisa A, Tomizawa H, Tono K, Wu S, Yabashi M, Yamaga M, Yamashita A, Yanagida K, Zhang C, Shintake T, Kitamura H, Kumagai N (2012) A compact X-ray free-electron laser emitting in the sub-ångström region. Nat Photonics 6:540–544

Jiang H, Song C, Chen C-C, Xu R, Raines KS, Fahimian BP, C-H L, Lee T-K, Nakashima A, Urano J, Ishikawa T, Tamanoi F, Miao J (2010) Quantitative 3D imaging of whole, unstained cells by using X-ray diffraction microscopy. Proc Natl Acad Sci U S A 107:11234–11239

Kameshima T, Ono S, Kudo T, Ozaki K, Kirihara Y, Kobayashi K, Inubushi Y, Yabashi M, Horigome T, Holland A, Holland K, Burt D, Murao H, Hatsui T (2014) Development of an X-ray pixel detector with multi-port charge-coupled device for X-ray free-electron laser experiments. Rev Sci Instrum 85:033110

Kimura T, Joti Y, Shibuya A, Song C, Kim S, Tono K, Yabashi M, Tamakoshi M, Moriya T, Oshima T, Ishikawa T, Bessho Y, Nishino Y (2014) Imaging live cell in micro-liquid enclosure by X-ray laser diffraction. Nat Commun 5:3052

Kobayashi A, Sekiguchi Y, Takayama Y, Oroguchi T, Nakasako M (2014) Dark-field phase retrieval under the constraint of the Friedel symmetry in coherent X-ray diffraction imaging. Opt Express 22:27892–27909

Kobayashi A, Sekiguchi Y, Takayama Y, Oroguchi T, Shirahama K, Torizuka Y, Manoda M, Nakasako M, Yamamoto M (2016a) TAKASAGO-6 apparatus for cryogenic coherent X-ray diffraction imaging of biological non-crystalline particles using X-ray free electron laser at SACLA. Rev Sci Instrum 87:053109

Kobayashi A, Sekiguchi Y, Oroguchi T, Okajima K, Fukuda A, Oide M, Yamamoto M, Nakasako M (2016b) Specimen preparation for cryogenic coherent X-ray diffraction imaging of biological cells and organelles using X-ray free-electron laser at SACLA. JSynchrotron Radiat 23:975–989

Kodama W, Nakasako M (2011) Application of a real-space three-dimensional image reconstruction method in the structural analysis of noncrystalline biological macromolecules enveloped by water in coherent x-ray diffraction microscopy. Phys Rev E 84:021902

Kouranov A, Schnell DJ (1996) Protein translocation at the envelope and thylakoid membranes of chloroplasts. J Biol Chem 271:31009–31012

Lima E, Wiegart L, Pernot P, Howells M, Timmins J, Zontone F, Madsen A (2009) Cryogenic x-ray diffraction microscopy for biological samples. Phys Rev Lett 103:198102

Matsuzaki M, Misumi O, Shin-i T, Maruyama S, Takahara M, Miyagishima S, Mori T, Nishida K, Yagisawa F, Nishida K, Yoshida Y, Nishimura Y, Nakao S, Kobayashi T, Momoyama Y, Higashiyama T, Minoda A, Sano M, Nomoto H, Oishi K, Hayashi H, Ohta F, Nishizaka S, Haga S, Miura S, Morishita T, Kabeya Y, Terasawa K, Suzuki Y, Ishii Y, Asakawa S, Takano H, Ohta N, Kuroiwa H, Tanaka K, Shimizu N, Sugano S, Sato N, Nozaki H, Ogasawara N, Kohara Y, Kuroiwa T (2004) Genome sequence of the ultra-small unicellular red alga *Cyanidioschyzon merolae* 10D. Nature 428:653–657

McDermott G, Le Gros MA, Knoechel CG, Uchida M, Larabell CA (2009) Soft X-ray tomography and cryogenic light microscopy: the cool combination in cellular imaging. Trends Cell Biol 19:587–595

Miao J, Sayre D, Chapman HN (1998) Phase retrieval from the magnitude of the Fourier transforms of nonperiodic objects. J Opt Soc Am A 15:1662–1669

Miao J, Charalambous P, Kirz J, Sayre D (1999) Extending the methodology of X-ray crystallography to allow imaging of micrometre-sized non-crystalline specimens. Nature 400:342–344

Miao J, Hodgson KO, Ishikawa T, Larabell CA, LeGros MA, Nishino Y (2003a) Imaging whole *Escherichia coli* bacteria by using single-particle x-ray diffraction. Proc Natl Acad Sci U S A 100:110–112

Miao J, Ishikawa T, Anderson EH, Hodgson KO (2003b) Phase retrieval of diffraction patterns from noncrystalline samples using the oversampling method. Phys Rev B 67:174104

Miao J, Ishikawa T, Shen Q, Earnest T (2008) Extending X-ray crystallography to allow the imaging of nanocrystalline materials, cells, and single protein complexes. Annu Rev Phys Chem 59:387–410

Miao J, Sandberg RL, Song C (2012) Coherent X-ray diffraction imaging. IEEE J Sel Top Quant Electron 18:399–410

Miyagishima S, Itoh R, Aita S, Kuroiwa H, Kuroiwa T (1999) Isolation of dividing chloroplasts with intact plastid-dividing rings from a synchronous culture of the unicellular red alga *Cyanidioschyzon merolae*. Planta 209:371–375

Miyawaki A (2013) Fluorescence imaging in the last two decades. Microscopy 62:63–68

Nakasako M, Takayama Y, Oroguchi T, Sekiguchi Y, Kobayashi A, Shirahama K et al (2013) KOTOBUKI-1 apparatus for cryogenic coherent X-ray diffraction imaging. Rev Sci Instrum 84:093705

Nishino Y, Takahashi Y, Imamoto N, Ishikawa T, Maeshima K (2009) Three-dimensional visualization of a human chromosome using coherent x-ray diffraction. Phys Rev Lett 102:018101

Oroguchi T, Nakasako M (2013) Three-dimensional structure determination protocol for noncrystalline biomolecules using x-ray free-electron laser diffraction imaging. Phys Rev E 87:022712

Oroguchi T, Sekiguchi Y, Kobayashi A, Masaki Y, Fukuda A, Hashimoto S, Nakasako M, Ichikawa Y, Kurumizaka H, Shimizu M, Inui Y, Matsunaga S, Kato T, Namba K, Yamaguchi K, Kuwata K, Kameda H, Fukui N, Kawata Y, Kameshima T, Takayama Y, Yonekura K, Yamamoto M (2015) Cryogenic coherent X-ray diffraction imaging for biological non-crystalline particles using the KOTOBUKI-1 diffraction apparatus at SACLA. J Phys B 48:184003

Ravelli RBG, Garman EF (2006) Radiation damage in macromolecular cryocrystallography. Curr Opin Struct Biol 16:624–629

Rodenburg JM, Hurst AC, Cullis AG, Dobson BR, Pfeiffer F, Bunk O, David C, Jefimovs K, Johnson I (2007) Hard X-ray lensless imaging of extended objects. Phys Rev Lett 98:034801

Rodriguez J, Xu R, Chen C-C, Zou Y, Miao J (2013) Oversampling smoothness: an effective algorithm for phase retrieval of noisy diffraction intensities. J Appl Crystallogr 46:312–318

Seibert MM, Ekeberg T, Maia FRNC, Svenda M, Andreasson J, Jönsson O, Odić D, Iwan B, Rocker A, Westphal D, Hantke M, DePonte DP, Barty A, Schulz J, Gumprecht L, Coppola N, Aquila A, Liang M, White TA, Martin A, Caleman C, Stern S, Abergel C, Seltzer V, Claverie J-M, Bostedt C, Bozek JD, Boutet S, Miahnahri AA, Messerschmidt M, Krzywinski J, Williams G, Hodgson KO, Bogan MJ, Hampton CY, Sierra RG, Starodub D, Andersson I, Bajt S, Barthelmess M, Spence JCH, Fromme P, Weierstall U, Kirian R, Hunter M, Doak RB,

Marchesini S, Hau-Riege SP, Frank M, Shoeman RL, Lomb L, Epp SW, Hartmann R, Rolles D, Rudenko A, Schmidt C, Foucar L, Kimmel N, Holl P, Rudek B, Erk B, Hömke A, Reich C, Pietschner D, Weidenspointner G, Strüder L, Hauser G, Gorke H, Ullrich J, Schlichting I, Herrmann S, Schaller G, Schopper F, Soltau H, Kühnel K-U, Andritschke R, Schröter C-D, Krasniqi F, Bott M, Schorb S, Rupp D, Adolph M, Gorkhover T, Hirsemann H, Potdevin G, Graafsma H, Nilsson B, Chapman HN, Hajdu J (2011) Nature 470:78–81

Sekiguchi Y, Yamamoto M, Oroguchi T, Takayama Y, Suzuki S, Nakasako M (2014a) IDATEN and G-SITENNO: GUI assisted software for coherent X-ray diffraction imaging experiments and data analyses at SACLA. J Synchrotron Rad 21:1378–1383

Sekiguchi Y, Oroguchi T, Takayama Y, Nakasako M (2014b) Data processing software suite *SITENNO* for coherent X-ray diffraction imaging using the X-ray free-electron laser SACLA. J Synchrotron Radiat 21:600–612

Sekiguchi Y, Oroguchi T, Nakasako M (2016) Classification and assessment of retrieved electron density maps in coherent X-ray diffraction imaging using multivariate statistics. J Synchrotron Radiat 23:312–323

Sengupta P, Van Engelenburg S, Lippincott-Schwartz J (2012) Visualizing cell structure and function with point-localization superresolution imaging. Dev Cell 23:1092–1102

Song C, Jiang H, Mancuso A, Amirbekian B, Peng L, Sun R, Shah SS, Zhou ZH, Ishikawa T, Miao J (2008) Quantitative imaging of single, unstained viruses with coherent X rays. Phys Rev Lett 101:158101

Su H-N, Xie B-B, Zhang X-Y, Zhou B-C, Zhang Y-Z (2010) The supramolecular architecture, function, and regulation of thylakoid membranes in red algae: an overview. Photosynth Res 106:73–87

Takayama Y, Nakasako M (2012) Humidity-controlled preparation of frozen-hydrated biological samples for cryogenic coherent x-ray diffraction microscopy. Rev Sci Instrum 83:054301

Takayama Y, Inui Y, Sekiguchi Y, Kobayashi A, Oroguchi T, Yamamoto M, Matsunaga S, Nakasako M (2015) Cryogenic coherent X-ray imaging for chloroplast from Schyzon. Plant Cell Physiol 56:1272–1286

Tamasaku K, Tanaka Y, Yabashi Y, Yamazaki H, Kawamura N, Suzuki M, Ishikawa T (2001) SPring-8 RIKEN beamline III for coherent X-ray optics. Nucl Instrum Meth A 467-468:686–689

Ting CS, Hsieh C, Sundararaman S, Mannella C, Marko M (2007) Cryo-electron tomography reveals the comparative three-dimensional architecture of Prochlorococcus, a globally important marine cyanobacterium. J Bacteriol 189:4485–4493

Tono K, Togashi T, Inubushi Y, Sato T, Katayama T, Ogawa K, Ohashi H, Kimura H, Takahashi S, Takeshita K, Tomizawa H, Goto S, Ishikawa T, Yabashi M (2013) Beamline, experimental stations and photon beam diagnostics for the hard x-ray free electron laser of SACLA. New J Phys 15:083035

van de Meene AML, Hohmann-Marriott MF, Vermaas WFJ, Roberson RW (2006) The three-dimensional structure of the cyanobacterium *Synechocystis* sp. PCC 6803. Arch Microbiol 184:259–270

Yagisawa F, Nishida K, Kuroiwa H, Nagata T, Kuroiwa T (2007) Identification and mitotic partitioning strategies of vacuoles in the unicellular red alga *Cyanidioschyzon merolae*. Planta 226:1017–1029

Yagisawa F, Nishida K, Yoshida M, Ohnuma M, Shimada T, Fujiwara T, Yoshida Y, Misumi O, Kuroiwa H, Kuroiwa T (2009) Identification of novel proteins in isolated polyphosphate vacuoles in the primitive red alga *Cyanidioschyzon merolae*. Plant J 60:882–893

Yoshida Y, Kuroiwa H, Misumi O, Yoshida M, Ohnuma M, Fujiwara T, Yagisawa F, Hirooka S, Imoto Y, Matsushita K, Kawano S, Kuroiwa T (2010) Chloroplasts divide by contraction of a bundle of nanofilaments consisting of polyglucan. Science 329:949–953

Yumoto H, Mimura H, Koyama T, Kimura T, Matsuyama S, Tono K, Togashi T, Inubushi Y, Sato T, Tanaka T, Kimura T, Yokoyama H, Kim J, Sano Y, Hachisu Y, Yabashi M, Ohashi H, Ohmori H, Ishikawa T, Yamauchi K (2013) Focusing of X-ray free-electron laser pulses with reflective optics. Nat Photonics 7:43–47

# Part V
# Proliferation of Eukaryotic Cells

**Takayuki Fujiwara, Fumi Yagisawa, and Tsuneyoshi Kuroiwa**

Eukaryotic cells possess double membrane-bound organelles that have their own genomes (cell nuclei, mitochondria and chloroplasts); and single membrane-bound organelles (endoplasmic reticulum, Golgi bodies, vacuoles/lysosomes, and peroxisomes). Successful duplication and segregation of these organelles are required for faithful reproduction of eukaryotic cells. However, there has been very limited information on the proliferation of membranous organelles other than cell nuclei because most animal and plant cells contain a large number of these organelles which divide somewhat randomly. In contrast, the unicellular red alga *Cyanidioschyzon merolae* offers the following advantages for investigating proliferation of cell nuclei and organelles: (1) It is extremely simple cellular architecture comprising a single cell nucleus, mitochondrion, chloroplast, and peroxisome; a simple ER structure; a single layer Golgi body; and several vacuoles. (2) The organelle division cycle and cell cycle progression can be synchronized by a 12-h light and 12-h dark cycle. Part V introduces the current knowledge regarding cell proliferation, including the division cycles of membranous organelles.

Chapter 11 summarizes the timing of division and inheritance of the organelles during the cell cycle in *C. merolae*. Chapter 12 introduces underlying mechanisms for temporal segregation of cell cycle progression from photosynthesis and its advantage for safe multiplication of eukaryotic cells. Chapter 13 introduces the coordinative mechanism of DNA replication among double membrane-bound organelles during the cell cycle. Chapter 14 reviews the evolutionary origin of double-membrane-bounded organelles and provides information on the molecular components constituting the cellular nuclear architectures as well as dividing machineries (rings) for mitochondria and chloroplasts. Chapter 15 introduces the dynamics of single-membrane-bound organelles during the cell cycle and their underlying molecular mechanisms in *C. merolae*. Chapter 16 reviews mitotic kinases and cytoskeletal and motor proteins that are involved in cell and organelle proliferation in *C. merolae*.

# Chapter 11
# Cell Cycle and Organelle Division Cycle in *Cyanidioschyzon merolae*

**Takayuki Fujiwara**

**Abstract** A cell replicates by duplicating cellular contents and then dividing into two daughter cells. This is known as the cell cycle, which is characterized by the dynamics of the cell nucleus. However, eukaryotic cells contain organelles other than the nucleus, such as mitochondria, chloroplasts, peroxisomes, endoplasmic reticulum (ER), Golgi bodies, and vacuoles. To understand reproduction of cells, the division and inheritance of organelles as well as the mitotic cycle should be investigated. However, model animal cells and plant cells contain multiples of these organelles which exhibit complex shapes. Therefore, it has been difficult to elucidate their division and inheritance. In contrast, the red alga *Cyanidioschyzon merolae* possesses a minimum set of organelles. Because of its simple cellular architecture, the dynamics and mechanisms underlying its cell division and inheritance have been investigated. In addition, cell cycle stages of *C. merolae* have been determined using an S phase-specific marker proliferating cell nuclear antigen (PCNA) and an M phase-specific marker H3S10ph. In this chapter, I introduce the timing of division and inheritance of organelles during the cell cycle in *C. merolae* and application of the synchronous culture to transcriptome analysis for identifying novel genes related to the division and inheritance of organelles.

**Keywords** PCNA • H3S10ph • Cell cycle • Organelle division • Organelle inheritance • Transcriptome

T. Fujiwara (✉)
Department of Cell Genetics, National Institute of Genetics, 1111 Yata, Mishima, Shi-zuoka 411-8540, Japan

Department of Genetics, Graduate University for Advanced Studies (SOKENDAI), 1111 Yata, Mishima, Shizuoka 411-8540, Japan

Japan Science and Technology Agency, CREST, 4-1-8 Honcho, Kawaguchi, Saitama 332-0012, Japan
e-mail: tkfujiwara@nig.ac.jp



T. Kuroiwa et al. (eds.), *Cyanidioschyzon merolae*,
https://doi.org/10.1007/978-981-10-6101-1_11

## 11.1 Introduction

Cells duplicate their contents, which are then propagated to daughter cells by cell division. This is described as the cell cycle, which is generally understood as a series of events, including cell growth (G1), DNA replication of the cell nucleus (DNA synthesis, S phase), cell growth (G2), chromosome segregation (mitosis, M phase), and cytokinesis. The function of each phase has been characterized in many kinds of cells, such as mammalian cultured cells and yeasts as model cells (Watson et al. 1987; Alberts et al. 2007). It is believed that the functional aspects of a specific stage of the cell cycle are primarily associated with the dynamics of the cell nucleus. However, successful cell production requires division and inheritance of cellular contents, including DNA-containing double membrane-bound organelles (mitochondria and chloroplast) and single membrane-bound organelles [peroxisomes, endoplasmic reticulum (ER), Golgi bodies, and vacuoles], as well as the cell nucleus. Each organelle, except for the nucleus, exists in multiple numbers and exhibits complex shapes in mammalian cells, plant cells, and yeast. For example, the shapes of mitochondria change through fission and fusion (Chan 2012). This feature obscures the organelle dynamics involved in division and inheritance.

In contrast, the unicellular red algae *Cyanidioschyzon merolae* exhibits an extremely simple cellular architecture that consists of a single cell nucleus, mitochondrion, chloroplast, and peroxisome; a simple ER structure; a single layer of Golgi body; and several vacuoles (Kuroiwa et al. 1998; Misumi et al. 2005). The progression of the cell cycle can be synchronized by a 12-h/12-h light/dark cycle (Suzuki et al. 1994). Thus, the simple cellular architecture and synchronous culture offer advantages for examining the spatiotemporal behaviors of organelles during the cell cycle and have largely contributed to the examination of mitochondrial division (Nishida et al. 2003, 2007; Yoshida et al. 2009), chloroplast division (Miyagishima et al. 2003; Yoshida et al. 2010), and peroxisome division and inheritance (Miyagishima et al. 1999; Imoto et al. 2013), as well as the inheritance of the ER (Yagisawa et al. 2012), Golgi body (Yagisawa et al. 2013), and vacuoles (Yagisawa et al. 2007; Fujiwara et al. 2010). The division and inheritance of each organelle take place once in a specific stage of the cell cycle.

Thus, studies on the cell cycle of *C. merolae* can include the behaviors of all organelles as well as the cell nucleus. The detail of the mechanisms and dynamics of division and inheritance of theses organelles will be introduced in Chaps. 14, 15, and 16. Here I introduce the temporal relationship between these organelles and cell cycle progression. In addition, the application of the synchronous culture to transcriptome analysis for identifying novel genes related to the division and inheritance of organelles.

## 11.2 Synchronous Culture of *C. merolae*

The cell cycle of *C. merolae* can be synchronized by a 12-h/12-h light/dark cycle. The first and second synchronized cell division occur around 18 and 40 h after the onset of the first light period during two rounds of a light and dark cycle, respectively (Fig. 11.1a). The representative micrographs of cells are arranged in order of the predicted cell cycle (Fig. 11.1b). 4′,6-Diamidino-2-phenylindole (DAPI)-

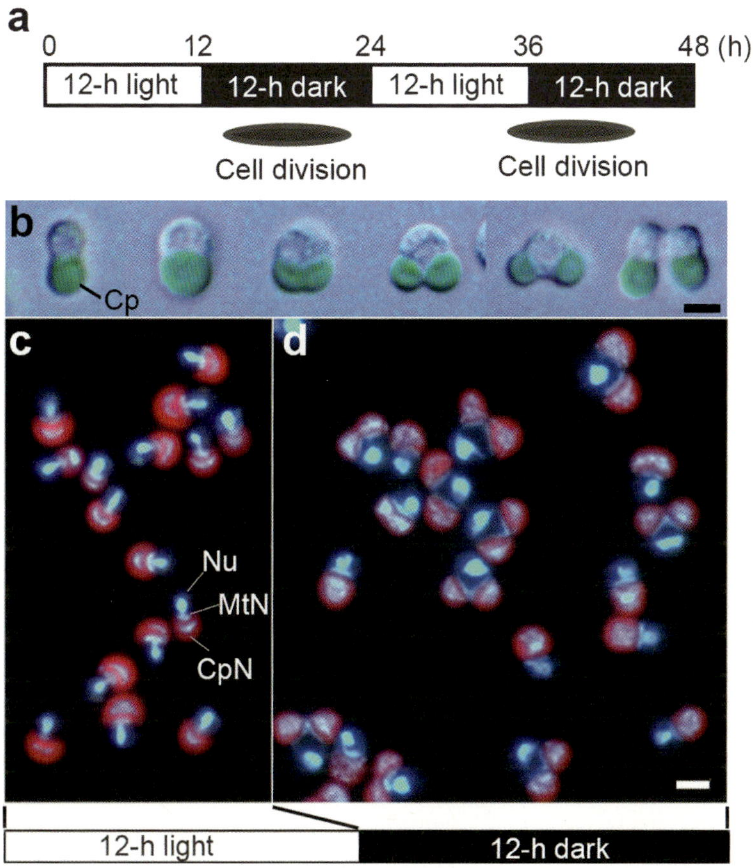


**Fig. 11.1** Synchronous culture of *C. merolae*. (**a**) A diagram showing the timing of cell division in synchronous culture. The numbers indicate time after the onset of a first light period. (**b**) Microscopic images showing the course of cell division of *C. merolae*. Cp, chloroplast. (**c** and **d**) Fluorescence microscopy images showing representative G1-phase cells (**b**) and M phase cells (**c**). The DNA of cell nuclei, mitochondria, and chloroplast was stained with DAPI and emits fluorescence. The chloroplasts emit autofluorescence (the red autofluorescence from chloroplasts is shown in the online version). Mitochondrial nucleoids reflect the position and shape of mitochondria. Nu, cell nucleus; MtN, mitochondrial nucleoid; Cp, chloroplast. Scale bars = 2 μm in (**b**) and (**d**) (Reproduced from Fujiwara et al. 2009)

staining and fluorescent microscopy show typical G1 phase cells during the light period and M phase cells during the dark period, respectively (Figs. 11.1c and d). Mitochondrial nucleoid represents the position and shape of mitochondria in *C. merolae*. The cell nucleus, the flat mitochondrion, and the spherical chloroplast are observed in the G1 phase cells (Fig. 11.1c), whereas dividing cell nucleus, mitochondrion, and chloroplast are observed in the M phase cells (Fig. 11.1d).

## 11.3 Determination of the Cell Cycle Stages of *C. merolae*

Cell cycle stage-specific markers have been developed to objectively show the stages of the cell cycle (G1, S, G2, and M phases). Proliferating cell nuclear antigen (PCNA/CMS101C), a processivity factor for DNA polymerase δ, and the phospho-epitope of histone H3 (CMN165C and CMN176C) at serine 10 (H3S10ph) are used as S phase-specific markers (Celis and Celis 1985) and an M phase-specific marker (Hendzel et al. 1997), respectively. Immunofluorescence microscopy shows that both signals of PCNA and H3S10ph are detectable within the cell nucleus, but in distinct patterns (Fig. 11.2a). The signal of PCNA is observed in cells with a single spherical chloroplast (S phase), which then disappears before constriction of a chloroplast dividing plane (G2). In contrast, the signals of H3S10ph are detectable in cells with divided chloroplasts, which persist until cytokinesis is completely concluded (M phase). The indices of the S and M phase cells indicate that DNA replication of the cell nucleus (S phase) occurs 2–3 h prior to mitosis (Fig. 11.2b). The S and M phases are temporally proximate during the cell cycle of *C. merolae* under the culture conditions examined.

In addition, the stages of the M phase are further determined by the behavior of the centromere-specific histone H3 variant [known as centromere protein A (CENPA)/CME099C; Maruyama et al. 2007] (Fig. 11.2c). The signals of CENPA are detected as a speckle-like structure within the nucleus at the beginning of the M phase (prophase), following which they are converted to two discrete structures (metaphase). Each structure is partitioned into daughter nuclei (anaphase/telophase), and the signals remain for a certain time after cytokinesis in separated daughter cells (telophase/G1 phase), following which they disappear (early G1). The structure of the metaphase nucleus is converted from an elliptical shape in the prophase to a characteristic rod-like shape (Fig. 11.2d). These results indicate that the bulk of *C. merolae* chromosomes undergo condensation during mitosis, although individual chromosomes are indiscernible from each other. Concurrent with metaphase transition, the nuclear envelope dissolves at least partially, and the nuclear pore complexes disperse into the cytoplasm. These results suggest that the cell nucleus undergoes semi-open mitosis (Fujiwara et al. 2013).

By using the synchronous culture of *C. merolae*, the division and inheritance of organelles including cell nuclei, mitochondria, chloroplasts, peroxisome ER, and Golgi bodies have been examined (Misumi et al. 2005).

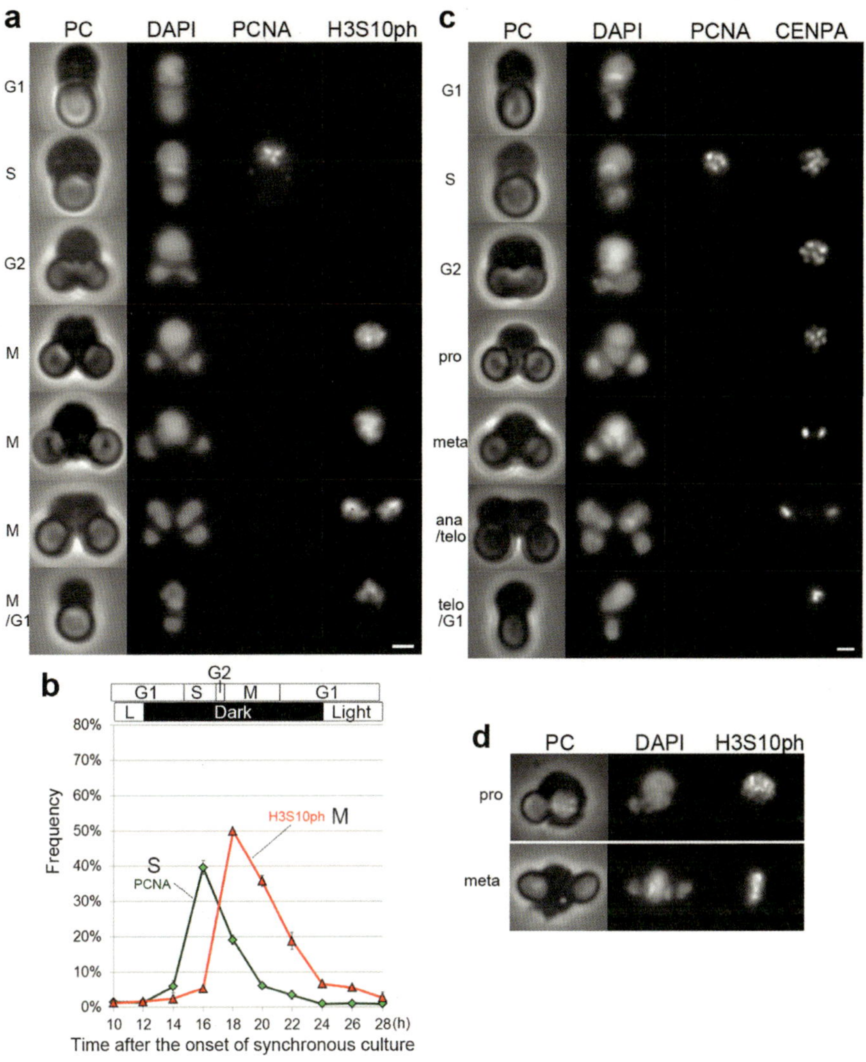


**Fig. 11.2** Determination of the cell cycle stages in *C. merolae*. (**a**) Immunofluorescence labeling of *C. merolae* cells with antibodies against PCNA and H3S10ph. Phase contrast (PC) and DAPI-stained images are also shown. The images are arranged according to the predicted stages of the cell cycle. (**b**) Time course of frequencies of cells positive for PCNA (S phase) and H3S10ph (M phase) in a synchronous culture [shown in (**a**)]. Error bars represent the SD. L indicates a light period. (**c**) Immunofluorescence labeling with antibodies against CENH3. (**d**) Immunofluorescence labeling of *C. merolae* cells with an antibody against H3S10ph. The prophase and metaphase cells are viewed from a different angle. Bar = 1 μm in (**a**), (**c**), and (**d**) (Reproduced from Fujiwara et al. 2013)

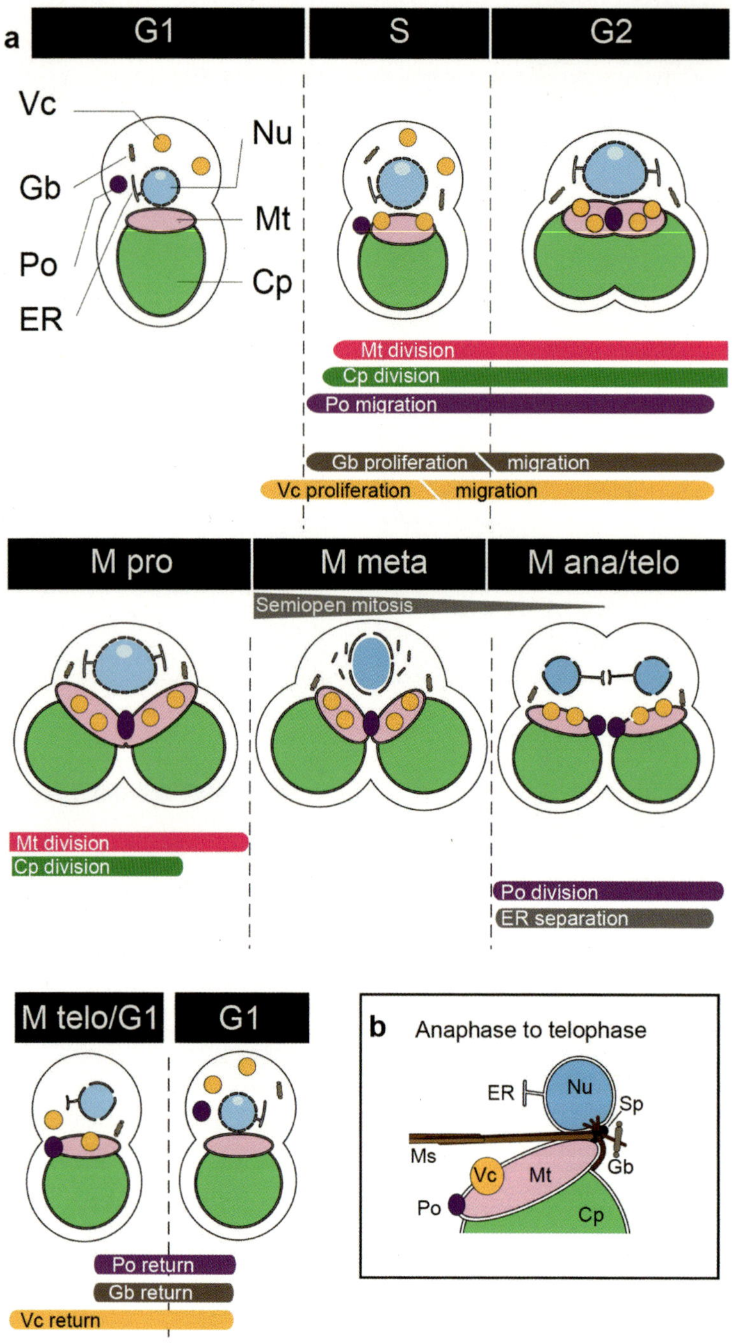


**Fig. 11.3** Schematic diagrams showing the timing of organelle division and inheritance during the cell cycle of *C. merolae* (**a**) and the daughter organelle complex that is inherited to daughter cells by mitotic spindles (**b**). The spindle pole and mitotic spindles associate with the daughter cell

## 11.4 Timing of the Division and Inheritance of Organelles During Cell Cycle Stages

As introduced in Chaps. 14, 15, and 16, the division and inheritance of the mitochondrion, chloroplast, peroxisome, ER, Golgi body, vacuole, as well as the cell nucleus occur in a cell cycle-dependent manner. Here, I summarize the temporal relationship between cell shape, stages of the cell cycle, and timing of organelle division and inheritance as follows (Fig. 11.3a). (1) Cell cycle stages: G1 phase (PCNA-negative and H3S10ph-negative cells), S phase (PCNA-positive cells), G2 phase (PCNA-negative and H3S10ph-negative cells, wider and dividing chloroplasts), and M phase (PCNA-negative and H3S10ph-positive cells). Stages during M phase: prophase (speckle-like CENPA signals), metaphase (two clusters of CENPA signals), anaphase/telophase (two clusters of CENPA signals in partitioning sister chromosomes before cytokinesis), and telophase/G1 phase (a cluster of CENPA signals in separated daughter cells) (Fig. 11.2). (2) Mitochondrial division occurs from the S phase through to the prophase. (3) Chloroplast division occurs from the S phase through to the prophase. (4) Peroxisome migrates from the vicinity of the cell nucleus to the surface of the mitochondrion or chloroplast in the S phase and moves to a mitochondrial dividing site in the G2 phase. It then divides after mitochondrial division in the anaphase, and daughter peroxisomes are inherited into daughter cells while associating with daughter mitochondria. (5) Peripheral ER is divided in the anaphase. (6) Golgi bodies proliferate in the S phase, following which they are transferred to the spindle poles near both ends of the mitochondrion in G2. Daughter Golgi bodies are inherited into daughter cells while associating with spindle poles (Fig. 11.3b). (7) Vacuoles proliferate in the G1/S phase and migrate to the mitochondrial surface from the S phase to the G2 phase. Daughter vacuoles are inherited into daughter cells while associating with daughter mitochondria. Thus, daughter organelles construct the complexes as shown in Fig. 11.3a (G2 and M phases) and are then inherited to daughter cells. The complexes are transferred by mitotic spindles that associate with sister chromosomes and daughter mitochondria (Nishida et al. 2005) (Fig. 11.3b). The microtubules are time specifically organized from the S phase through the M phase in *C. merolae* (the details are introduced in Chap. 16), and the mitotic spindles are involved in not only the segregation of the chromosomes but also the inheritance of the organelle complexes.

**Fig. 11.3** (continued) nucleus, mitochondrion, and Golgi body. The daughter mitochondrion directly associates with the daughter vacuoles, peroxisome, and chloroplast. *Nu* cell nucleus; *Mt.* mitochondrion; *Cp* chloroplast; *Po* peroxisome; *ER* endoplasmic reticulum; *Gb* Golgi body; *Vc* vacuole; *Sp* spindle pole; *Ms.* mitotic spindle. The following reports were referred to show the timing of the division and inheritance of organelle during a cell cycle: mitochondria (Nishida et al. 2007), chloroplasts (Yoshida et al. 2010), peroxisomes (Miyagishima et al. 1999), ER (Yagisawa et al. 2012), Golgi bodies (Yagisawa et al. 2013), vacuoles (Yagisawa et al. 2007), and cell cycle (Imoto et al. 2010, 2011)

## 11.5 Gene Expression Profiles for Identifying Novel Genes Related to the Division and Inheritance of Organelles

As described above, the division and inheritance of organelles occur in a cell cycle-dependent manner. Thus, transcriptome analysis of *C. merolae* synchronous culture has become a powerful tool for understanding the behavior of organelles and identifying novel genes related to the division and inheritance of organelles. The analysis of two rounds of the light and dark cycle by using a DNA microarray has shown that 158 genes in total are induced during the S phase or G2-M phase (Fujiwara et al. 2009). Genes numbering 95 and 63 are included into the S phase genes and in the G2-M phase, respectively. In brief, the S phase genes include *cyclin1* (CML219C G1/S cyclin, Miyagishima et al. 2014) and *PCNA*/CMS101C (DNA replication in the cell nucleus). The G2-M phase genes include *cyclinA*/CMI203C (potential S to G2/M cyclin), *cyclinB*/CML227C (M phase cyclin), $\alpha$- and $\beta$-tubulin ($\alpha-$/CMT504C, $\beta-$/CMN263C, mitotic spindle), and condensin subunits (*SMC2*/CMG189C, *SMC4*/CME029C, chromosome organization, and segregation). In addition, the S phase genes also contain *FtsZ1*/CMS361C (mitochondrial division) and *FtsZ2*/CMS004C (chloroplast division). The G2-M–phase genes contain *CmDnm2*/*DRP5B*/CMN262C (chloroplast division). The microarray data contains genes related to the division of organelles. Thus, the data likely includes novel genes related to the division of inheritance of organelles. In actual fact, it has contributed to identifying three novel genes: *ZED* (mitochondrial division, Yoshida et al. 2009), *PDR1* (chloroplast division, Yoshida et al. 2010), and *VIG1* (vacuole inheritance, Fujiwara et al. 2010). In the future, microarray data will offer further novel findings.

**Acknowledgments** Our study was partly supported was by JSPS KAKENHI (no. JP15K18588 to T.F.; no. 25251039 to S.M.) and by the Core Research for Evolutional Science and Technology Program of the Japan Science and Technology Agency (S.M.).

## References

Alberts B, Johnson A et al (2007) Molecular biology of the cell. Garland science, 5th edn. Taylor & Francis Group, New York

Celis JE, Celis A (1985) Cell cycle-dependent variations in the distribution of the nuclear protein cyclin proliferating cell nuclear antigen in cultured cells: subdivision of S phase. Proc Natl Acad Sci U S A 82:3262–3266

Chan DC (2012) Fusion and fission: interlinked processes critical for mitochondrial health. Annu Rev Genet 46:265–287

Fujiwara T, Kuroiwa H et al (2010) The coiled-coil protein VIG1 is essential for tethering vacuoles to mitochondria during vacuole inheritance of *Cyanidioschyzon merolae*. Plant Cell 22:772–781

Fujiwara T, Misumi O et al (2009) Periodic gene expression patterns during the highly synchronized cell nucleus and organelle division cycles in the unicellular red alga *Cyanidioschyzon merolae*. DNA Res 16:59–72

Fujiwara T, Tanaka K et al (2013) Spatiotemporal dynamics of condensins I and II: evolutionary insights from the primitive red alga *Cyanidioschyzon merolae*. Mol Biol Cell 24:2515–2527

Hendzel MJ, Wei Y et al (1997) Mitosis-specific phosphorylation of histone H3 initiates primarily within pericentromeric heterochromatin during G2 and spreads in an ordered fashion coincident with mitotic chromosome condensation. Chromosoma 106:348–360

Imoto Y, Fujiwara T et al (2010) Division of cell nuclei, mitochondria, plastids, and microbodies mediated by mitotic spindle poles in the primitive red alga *Cyanidioschyzon merolae*. Protoplasma 241:63–74

Imoto Y, Yoshida Y et al (2011) The cell cycle, including the mitotic cycle and organelle division cycles, as revealed by cytological observations. J Electron Microsc 60:S117–S136

Imoto Y, Kuroiwa H et al (2013) Single-membrane-bounded peroxisome division revealed by isolation of dynamin-based machinery. Proc Natl Acad Sci U S A 110:9583–9588

Kuroiwa T, Kuroiwa H et al (1998) The division apparatus of plastids and mitochondria. Int Rev Cytol 181:1–41

Maruyama S, Kuroiwa H et al (2007) Centromere dynamics in the primitive red alga *Cyanidioschyzon merolae*. Plant J 49:1122–1129

Misumi O, Matsuzaki M et al (2005) *Cyanidioschyzon merolae* genome. A tool for facilitating comparable studies on organelle biogenesis in photosynthetic eukaryotes. Plant Physiol 137:567–585

Miyagishima SY, Itoh R et al (1999) Microbody proliferation and segregation cycle in the single-microbody alga *Cyanidioschyzon merolae*. Planta 208:326–336

Miyagishima SY, Nishida K et al (2003) A plant-specific dynamin-related protein forms a ring at the chloroplast division site. Plant Cell 15:655–665

Miyagishima SY, Fujiwara T et al (2014) Translation-independent circadian control of the cell cycle in a unicellular photosynthetic eukaryote. Nat Commun 5:3807

Nishida K, Takahara M et al (2003) Dynamic recruitment of dynamin for final mitochondrial severance in a primitive red alga. Proc Natl Acad Sci U S A 100:2146–2151

Nishida K, Yagisawa F et al (2005) Cell cycle-regulated, microtubule-independent organelle division in *Cyanidioschyzon merolae*. Mol Biol Cell 16:2493–2502

Nishida K, Yagisawa F et al (2007) WD40 protein Mda1 is purified with Dnm1 and forms a dividing ring for mitochondria before Dnm1 in *Cyanidioschyzon merolae*. Proc Natl Acad Sci U S A 104:4736–4741

Suzuki K, Ehara T et al (1994) Behavior of mitochondria, chloroplasts and their nuclei during the mitotic cycle in the ultramicroalga *Cyanidioschyzon merolae*. Eur J Cell Biol 63:280–288

Watson JD, Hopkins NH et al (1987) Molecular biology of the gene. The Benjamin/Cummings Publishing Company, Inc., Menlo Park

Yagisawa F, Nishida K et al (2007) Identification and mitotic partitioning strategies of vacuoles in the unicellular red alga *Cyanidioschyzon merolae*. Planta 226:1017–1022

Yagisawa F, Fujiwara T et al (2012) Mitotic inheritance of endoplasmic reticulum in the primitive red alga *Cyanidioschyzon merolae*. Protoplasma 249:1129–1135

Yagisawa F, Fujiwara T et al (2013) Golgi inheritance in the primitive red alga, *Cyanidioschyzon merolae*. Protoplasma 250:943–948

Yoshida Y, Kuroiwa H et al (2009) The bacterial ZapA-like protein ZED is required for mitochondrial division. Curr Biol 19:1491–1497

Yoshida Y, Kuroiwa H et al (2010) Chloroplasts divide by contraction of a bundle of nanofilaments consisting of polyglucan. Science 329:949–953

# Chapter 12
# Regulation of Cell Cycle Progression by Circadian Rhythms in *Cyanidioschyzon merolae*

**Shinya Miyagishima**

**Abstract** Cell division in several lineages of eukaryotic algae occurs predominantly during the night. Cell cycle progression is shown to be regulated by circadian rhythms. However, the advantages and their underlying mechanisms conferred by this restriction of cell division to night are poorly understood. By using the unicellular red alga *Cyanidioschyzon merolae*, we recently showed that the retinoblastoma (RB)-E2F-DP pathway inhibits G1/S transition during the daytime. In *C. merolae*, E2F is phosphorylated in a time-dependent manner, peaking during the evening, which in turn permits the phosphorylation of RB only when the cell has grown to a certain size threshold. In addition, it is suggested that temporal segregation of photosynthesis during the daytime, which produces reactive oxygen species (ROS), and DNA replication and mitosis during the night is important for eukaryotic algae. Because the temporal segregation of respiratory activity and cell cycle progression has been observed in yeasts and mammalian cells, the temporal restriction of cell cycle progression is probably important for facilitating the safe multiplication of eukaryotic cells.

**Keywords** Cell cycle • Circadian rhythms • *Cyanidioschyzon merolae* • G1/S transition • Retinoblastoma • Synchronous culture

## 12.1 Introduction

The Earth's 24-h rotation generates the day and night environmental cycle. Many organisms, including bacteria and unicellular and multicellular eukaryotes, have evolved endogenous oscillators of an approximately 24-h cycle called circadian

---

S. Miyagishima (✉)
Department of Cell Genetics, National Institute of Genetics, 1111 Yata, Mishima, Shizuoka 411-8540, Japan
e-mail: smiyagis@nig.ac.jp



T. Kuroiwa et al. (eds.), *Cyanidioschyzon merolae*,
https://doi.org/10.1007/978-981-10-6101-1_12

clocks. Circadian clocks produce circadian rhythms that are daily oscillations in biological processes that function to maximize the fitness of the organism in response to environmental changes produced by external day and night cycles. For example, in land plants, circadian rhythms regulate photosynthesis, stomatal opening, and flowering time by regulation of respective gene expressions (McClung 2006; Dodd et al. 2005).

To be defined as a circadian rhythm, a rhythmic phenomenon should meet the following three criteria: (1) the persistence of a ~24-h rhythm under a constant condition, such as continuous light or a dark condition; (2) entrainability of the rhythm by environmental changes, for example, by illumination or temperature shift; and (3) temperature compensation of the rhythm (the ~24-h rhythm is kept relatively constant within a physiological range of temperatures) (McClung 2006; Johnson 2010).

It is well established that one of the core mechanisms that produces an internal ~24-h rhythm is the transcription and translation feedback loop (TTFL) of transcription factors. For example, in the land plant *Arabidopsis thaliana*, the main feedback loop is composed of the transcription factors CCA1, LHY, and TOC1. TOC1 induces the expression of CCA1 and LHY, and messenger RNA (mRNA) levels of *CCA1* and *LHY* peak around morning. CCA1 and LHY repress transcription of *TOC1* and *TOC1* mRNA peaks around evening. However, recent studies have shown that this simplified loop is, in fact, composed by multiple interlocked TTFL involving several other proteins. In addition, the transcription factors constituting TTFLs are largely distinct in different eukaryotic lineages. Thus, it is thought that the TTFLs have emerged in different eukaryotic lineages independently during the course of evolution (McClung 2006).

One of the biological processes that is known to be regulated by circadian rhythms in various lineages of eukaryotes, including mammals, is cell division (Mori and Johnson 2000; Johnson 2010). Regulation of G1/S and G2/M by circadian rhythms has been observed in mammalian cells. In the regenerating liver hepatocytes, expression of *wee1* (a suppressor of G2/M transition) is regulated by TTFL, thereby facilitating the restriction of G2/M transition to the morning when the WEE1 protein level reaches its lowest point (Matsuo et al. 2003). In fibroblasts, the multifunctional nuclear protein NONO binds PER (a component of TTFL), and NONO-PER binds to the p16-Ink4A (a suppressor of G1/S transition) promoter to direct its circadian transcription (Kowalska et al. 2013). Although the mechanism of circadian control of cell division has remained largely unknown in eukaryotic algae, recent studies, particularly in *C. merolae*, have started to reveal the mechanisms and the advantages conferred by this temporal restriction of cell division in eukaryotic algae (Miyagishima et al. 2014).

## 12.2 Circadian Gating of G1/S Transition in *C. merolae*

Several lineages of eukaryotic algae are known to divide predominantly during the night (Tamiya et al. 1953; Goto and Johnson 1995; Mori and Johnson 2000; Moulager et al. 2007, 2010; Johnson 2010) (Fig. 12.1). This characteristic of eukaryotic algae led to the first experimental procedure of cell division synchronization in eukaryotes, which was developed in the green alga *Chlorella ellipsoidea* by subjecting the culture to a 24-h light and dark cycle (Tamiya et al. 1953). Although the mechanism was not uncertain at that point, it was later shown that the gating of cell division is restricted to night by circadian rhythms in green algae (Goto and Johnson 1995; Moulager et al. 2007, 2010), red algae (Miyagishima et al. 2014), and euglenids (Mori and Johnson 2000). A previous transcriptome analysis in the green alga *Ostreococcus tauri* suggested a model in which the G1/S transition is regulated by the light- and time-dependent transcription of G1/S regulators (Moulager et al. 2007, 2010). However, the mechanism of the circadian gating of cell division and the advantage conferred by this temporal restriction of cell division in eukaryotic algae have remained poorly understood.

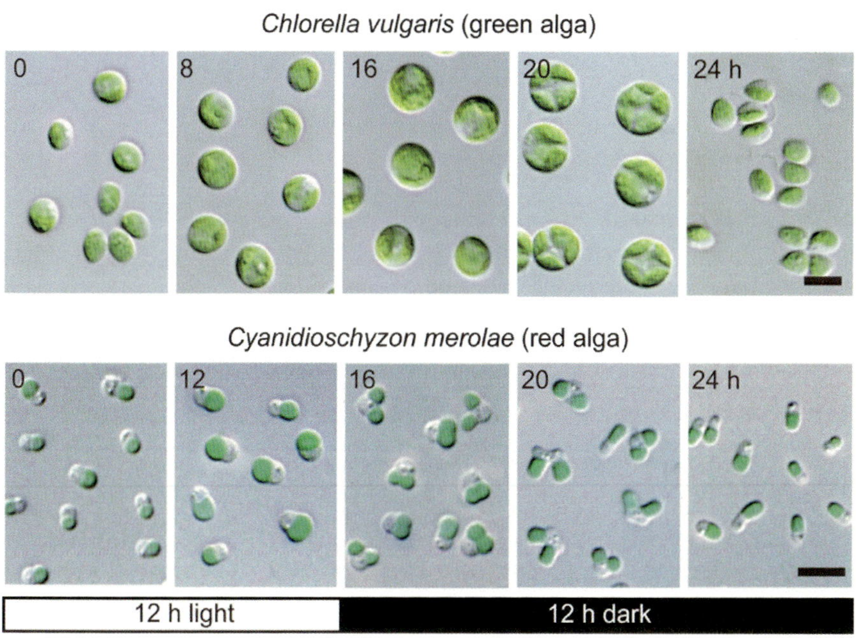


**Fig. 12.1** Cell division in the *green* alga *Chlorella vulgaris* and *red* alga *Cyanidioschyzon merolae* under a 12-h light/12-h dark cycle. The periods of cell division of both algae are restricted to the night. *C. merolae* cells proliferate by binary fission. *C. vulgaris* cells grow in size by manyfold, but never divide during the daytime. During the night, these grown cells undergo two or more rounds of S and M phases without growth to produce 4–32 daughter cells. Scale bars = 5 μm

To address these issues, we recently examined the circadian regulation of cell cycle progression by using a synchronous culture of *C. merolae* (Miyagishima et al. 2014). First, we examined whether the *C. merolae* cell cycle progression is regulated by circadian rhythms and, if so, which point(s) of the cell cycle is regulated by the rhythms. After entrainment by one round of a 12-h light/12-h dark (LD) cycle, cell division occurred predominantly during night under LD or subjective night under a continuous light condition (LL) (Fig. 12.2a). Under LD, the cell number increased by a factor of 1.48 during one round of LD, indicating that 48% of the cells divided (Fig. 12.2a). Under LL, 65% of the cells divided during one round (24 h) of LL. However, 35% did not divide during the first day, and for these cells, the following round of cell division started at the beginning of the next subjective night, even though these cells kept growing continuously under LL (Fig. 12.2a). These results suggest that G1/S transition has to wait until subjective night, even when the cell has grown to the size sufficient for cell division (Miyagishima et al. 2014). Thus, cell division is prohibited during (subjective) daytime. In addition, the study showed that the timing of G1/S transition (E2F phosphorylation as described below) occurred almost at the same timing at 38, 42, and 45 °C in the synchronized culture, although the durations of S and M phases changed depending on the temperature (Miyagishima et al. 2014). Thus, G1/S transition of *C. merolae* meets all three criteria of circadian rhythms, namely, the persistence of an ~24-h rhythm under constant conditions (LL), the entrainability of the rhythm by environmental changes (this was already confirmed by the cell cycle synchronization under LD), and temperature compensation of the rhythm.

G1/S transition in green algae has been shown to be regulated by the E2F–DP–retinoblastoma (RB) pathway (Fang et al. 2006; Fang and Umen 2008; Moulager et al. 2010; Olson et al. 2010) as in animal cells (Trimarchi and Lees 2002). RB represses the G1/S transition by binding to the E2F-DP heterodimer, which in turn binds certain specific cis elements of S phase genes. During the G1 phase and as the result of cell growth, G1 cyclin (such as cyclin D in animal cells) accumulates, and the cyclin-dependent kinase (CDK)-G1 cyclin phosphorylates RB. This phosphorylation inactivates RB, thus activating the transcription of the S phase genes by E2F-DP (Trimarchi and Lees 2002) (Fig. 12.2c).

The *C. merolae* genome (Matsuzaki et al. 2004) encodes single copies of CDKA, homologues of RB called RB-related (RBR), E2F, cyclin 1 (the G1 cyclin), cyclin A (the S phase cyclin), and two copies of DP (Moriyama et al. 2010; Miyagishima et al. 2014). To examine the mechanism restricting G1/S transition to subjective night, we examined the level and status of phosphorylation of these proteins. The results showed that E2F is phosphorylated and then dephosphorylated in a time-dependent manner and that the phosphorylation peaks around evening under LD (Miyagishima et al. 2014) (Fig. 12.2b). In addition, this time-dependent phosphorylation of E2F occurred even under a continuous dark condition (DD) after entrainment by one round of LD in which cells did not grow and all cells remained in the G1 phase (Miyagishima et al. 2014) (Fig. 12.2b). Thus, the circadian E2F phosphorylation occurs independently of cell cycle progression in *C. merolae*. In addition, we observed that RBR is phosphorylated only when cells have grown and enter into the S phase (Miyagishima et al. 2014) (Fig. 12.2b). We then tested the role of E2F phosphorylation in *C. merolae* cell cycle progression by transiently overexpressing a phospho-mutated or phospho-mimicked version of E2F. The phospho-mutated E2F inhibited the G1/S transition, whereas phospho-mimicked E2F accelerated G1/S transition (Miyagishima et al. 2014). Thus, the above results suggest that the peaking of circadian E2F phosphorylation around evening permits RBR to be phosphorylated (inactivated) only when the cell has grown beyond a certain size threshold (Fig. 12.2c).

The above model was tested by constitutive expression of phospho-mimicked *E2F* (E2F-p) and knockout of *RBR* (*rbr* KO). The *rbr* KO, but not E2F-p, exhibited a small cell phenotype [the volume was approximately 1/8 of the wild type (WT)] (Miyagishima et al. 2014). This result is consistent with the role of RBR, which suppresses G1/S transition until the cell has grown to a certain size, as well as with a previous study which showed that a *Chlamydomonas reinhardtii rb* (*mat3*) mutant exhibits a small cell phenotype (Fang et al. 2006). The culture of E2F-p and *rbr* KO under LD and LL showed that cell cycle progression in these strains is uncoupled from circadian rhythms, thus supporting the above model (Miyagishima et al. 2014) (Fig. 12.2c).

To obtain insights into the benefits provided by the inhibition of cell cycle progression during the daytime, we examined the doubling time and expression of oxidative stress marker genes in the E2F-p and the *rbr* KO. The doubling time of E2F-p (55.2 h under LD; 36.4 h under LL; 40 μE $m^{-2}$ $s^{-1}$) and *rbr* KO (63.2 h under LD; 48.7 h under LL) was longer than that of the WT (48.7 h under LD; 31.7 h under LL) (Miyagishima et al. 2014). In addition, the expression of oxidative stress markers such as catalase (a hydrogen peroxide scavenger), RAD51 (involved in DNA repair), and 8-oxoguanine-DNA glycosylase (responsible for the excision of 8-oxoguanine, a mutagenic base by-product by reactive oxygen species, from DNA) was upregulated in E2F-p and *rbr* KO compared with the WT (Miyagishima et al. 2014). Thus, it is suggested that uncoupling of cell cycle progression from circadian rhythms reduces the cell growth and elevates oxidative stress.

## 12.3 Temporal Segregation of Oxygenic Photosynthesis and Respiration and Cell Cycle Progression in Eukaryotes

It is well known that circadian rhythms regulate the peak of photosynthetic activities during subjective daytime (Farre and Weise 2012). Photosynthesis in chloroplasts supports the growth and proliferation of eukaryotic cells by providing reducing power and synthesizing ATP using light energy. However, photosystems produce reactive oxygen species such as superoxide ($O_2^-$), hydrogen peroxide ($H_2O_2$), hydroxyl radical (OH), and singlet oxygen ($^1O_2$) that oxidize biomolecules indiscriminately, leading to deleterious effects (Apel and Hirt 2004). Thus, the

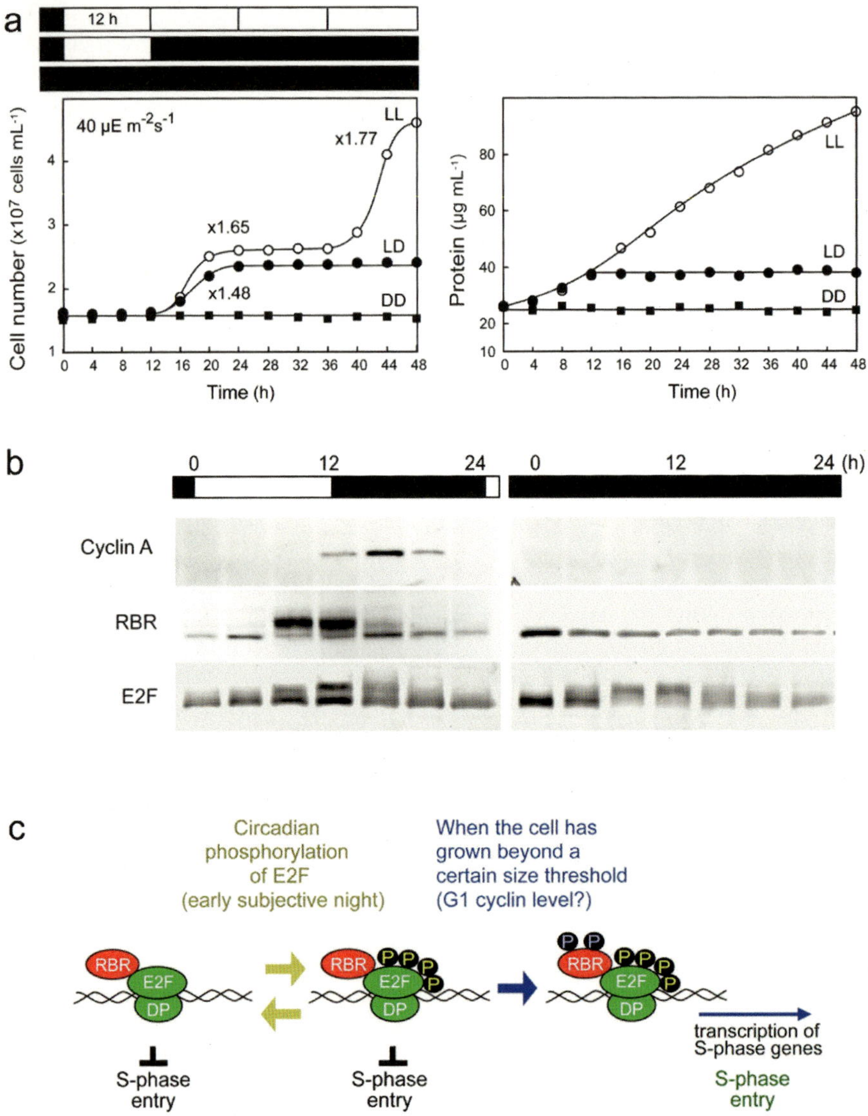


**Fig. 12.2** The circadian regulation of G1/S transition in *Cyanidioschyzon merolae*. (**a**) The changes in the cell number and protein level per culture during synchronous culture. Cells were cultured at 42 °C. Cells were entrained by a 12-h light (40 μE $m^{-2}$ $s^{-1}$) and 12-h dark cycle, following which they were cultured under the same condition (LD), continuous light (LL), or continuous dark (DD). (**b**) Immunoblotting showing circadian E2F phosphorylation under LD or DD. RBR is phosphorylated, and cyclin A is expressed when cells enter into the S phase. The *upper* and *lower* bands of RBR correspond to phosphorylated and nonphosphorylated RBR, respectively. The *upper* and *lower* bands of E2F correspond to phosphorylated or nonphosphorylated E2F, respectively. (**c**) The circadian regulation of the G1/S transition by the

study in *C. merolae* suggests that the temporal segregation of the reactive oxygen species (ROS) producing photosynthesis and cell cycle progression is important for facilitating the safe proliferation of eukaryotic algae. However, in land plants, circadian regulation of cell division is not evident. In seed plants, cell division is restricted to meristematic tissues that possess nonphotosynthetic proplastids, whereas leaf cells with photosynthetic chloroplasts expand without any cell division (but with endoreduplication) (Lopez-Juez and Pyke 2005). Thus, it appears that land plants spatially segregate ROS-producing photosynthesis and cell proliferation.

The temporal segregation of ROS-producing respiration and cell cycle progression has been observed in mammals and yeasts. In mouse skin, metabolic processes and cell cycle progression are regulated by circadian rhythms at transcription levels. Genes involved in oxidative phosphorylation are upregulated during the daytime, whereas cell cycle genes are upregulated during the night. In addition, the cells are most sensitive to ultraviolet (UV)-induced DNA damage during the night when the cells enter into the S phase (Geyfman et al. 2012). These results suggest that the restriction of DNA replication and cell division to the night is important to protect the cells from UV and oxidative stresses (Geyfman et al. 2012). In budding yeast under nutrient-limited conditions, the cells rhythmically alternate the balance between glycolysis and respiration. The glycolysis (reductive phase) and respiration (oxidative phase) are upregulated alternately over a cycle of 4–5 h, and DNA replication is restricted to the reductive phase (Chen et al. 2007). In addition, it was shown that this temporal restriction of the S phase reduces the occurrence of mutations in the genome (Chen et al. 2007). Thus, the temporal segregation of ROS-producing photosynthesis and respiration from cell cycle progression is likely to be critical for eukaryotic cells.

**Acknowledgments** Our study was partly supported by Japan Society for the Promotion of Science Grant-in-Aid for Scientific Research 25251039 (to S.M.) and by the Core Research for Evolutional Science and Technology Program of the Japan Science and Technology Agency (to S.M.).

**Fig. 12.2** (continued) RBR–E2F–DP pathway in *C. merolae*. E2F-DP binds the *cis* element of S phase genes throughout the cell cycle. RBR inhibits the transcriptional activity of E2F-DP. The G1/S transition is gated by the circadian rhythm of E2F phosphorylation, peaking during early subjective night. The E2F phosphorylation grants permission for the phosphorylation of RBR. RBR is phosphorylated only when the cell has grown beyond a certain size threshold. As in the case of animal cells, the G1 cyclin level likely functions to monitor the cell size. RBR phosphorylation inactivates RBR, thereby enabling E2F-DP to transcribe S phase genes (Reproduced from Miyagishima et al. 2014)

## References


Apel K, Hirt H (2004) Reactive oxygen species: metabolism, oxidative stress, and signal transduction. Annu Rev Plant Biol 55:373–399

Chen Z, Odstrcil EA et al (2007) Restriction of DNA replication to the reductive phase of the metabolic cycle protects genome integrity. Science 316:1916–1919

Dodd AN, Salathia N et al (2005) Plant circadian clocks increase photosynthesis, growth, survival, and competitive advantage. Science 309:630–633

Fang SC, Umen JG (2008) A suppressor screen in chlamydomonas identifies novel components of the retinoblastoma tumor suppressor pathway. Genetics 178:1295–1310

Fang SC, de los Reyes C et al (2006) Cell size checkpoint control by the retinoblastoma tumor suppressor pathway. PLoS Genet e167:2

Farre EM, Weise SE (2012) The interactions between the circadian clock and primary metabolism. Curr Opin Plant Biol 15:293–300

Geyfman M, Kumar V et al (2012) Brain and muscle Arnt-like protein-1 (BMAL1) controls circadian cell proliferation and susceptibility to UVB-induced DNA damage in the epidermis. Proc Natl Acad Sci U S A 109:11758–11763

Goto K, Johnson CH (1995) Is the cell division cycle gated by a circadian clock? The case of *Chlamydomonas reinhardtii*. J Cell Biol 129:1061–1069

Johnson CH (2010) Circadian clocks and cell division: what's the pacemaker? Cell Cycle 9:3864–3873

Kowalska E, Ripperger JA et al (2013) NONO couples the circadian clock to the cell cycle. Proc Natl Acad Sci U S A 110:1592–1599

Lopez-Juez E, Pyke KA (2005) Plastids unleashed: their development and their integration in plant development. Int J Dev Biol 49:557–577

Matsuo T, Yamaguchi S et al (2003) Control mechanism of the circadian clock for timing of cell division in vivo. Science 302:255–259

Matsuzaki M, Misumi O et al (2004) Genome sequence of the ultrasmall unicellular red alga *Cyanidioschyzon merolae* 10D. Nature 428:653–657

McClung CR (2006) Plant circadian rhythms. Plant Cell 18:792–803

Miyagishima SY, Fujiwara T et al (2014) Translation-independent circadian control of the cell cycle in a unicellular photosynthetic eukaryote. Nat Commun 5:3807

Mori T, Johnson CH (2000) Circadian control of cell division in unicellular organisms. Prog Cell Cycle Res 4:185–192

Moriyama T, Terasawa K et al (2010) Characterization of cell-cycle-driven and light-driven gene expression in a synchronous culture system in the unicellular rhodophyte *Cyanidioschyzon merolae*. Microbiology 156:1730–1737

Moulager M, Monnier A et al (2007) Light-dependent regulation of cell division in Ostreococcus: evidence for a major transcriptional input. Plant Physiol 144:1360–1369

Moulager M, Corellou F et al (2010) Integration of light signals by the retinoblastoma pathway in the control of S phase entry in the picophytoplanktonic cell Ostreococcus. PLoS Genet 6: e1000957

Olson BJ, Oberholzer M et al (2010) Regulation of the Chlamydomonas cell cycle by a stable, chromatin-associated retinoblastoma tumor suppressor complex. Plant Cell 22:3331–3347

Tamiya H, Iwamura T et al (1953) Correlation between photosynthesis and light-independent metabolism in the growth of Chlorella. Biochim Biophys Acta 12:23–40

Trimarchi JM, Lees JA (2002) Sibling rivalry in the E2F family. Nat Rev Mol Cell Biol 3:11–20

# Chapter 13
# Control of Cell Nuclear DNA Replication by Chloroplast and Mitochondrion

**Yuki Kobayashi, Yu Kanesaki, Mitsumasa Hanaoka, and Kan Tanaka**

**Abstract** α-Proteobacteria and cyanobacteria endosymbiosis has been crucial to the evolution of eukaryotic cells. The descendants of these bacteria gave rise to mitochondria and chloroplasts, and these organelles still retain their own genome proliferation systems. Coordination between the proliferation processes of these organelles and the eukaryotic cell cycle is indispensable for cellular maintenance, and we have studied this using the red alga *Cyanidioschyzon merolae*. During the cell cycle progression of *C. merolae*, organelle DNA replication (ODR) in both of the mitochondrion and the chloroplast occurs prior to nuclear DNA replication (NDR). We found that Mg-protoporphyrin IX (Mg-ProtoIX), a type of tetrapyrrole synthesized in the chloroplast, accumulates with the onset of ODR, thereby inducing NDR. Binding of the F-box protein Fbx3 to Mg-ProtoIX was also shown to be involved in the polyubiquitination of Cyclin 1, which activates cyclin-dependent kinase. Moreover, Mg-ProtoIX–Fbx3 binding inhibits Fbx3-mediated polyubiquitination of Cyclin 1. These results suggest that Fbx3 is a receptor for Mg-ProtoIX in the chloroplast signal to the nucleus and that it appears to function as a checkpoint for the coordination of ODR and NDR. In this chapter, we discuss the ODR and NDR coordination system in the cell cycle.

**Keywords** Cell cycle • DNA replication • Endosymbiosis • Evolution • Organelle • Retrograde signal • Tetrapyrrole • Ubiquitination

---

Y. Kobayashi • K. Tanaka (✉)
Laboratory for Chemistry and Life Science, Institute of Innovative Research, Tokyo Institute of Technology, R1-29, 4259 Nagatsuta, Midori-ku, Yokohama 226-8503, Japan
e-mail: kntanaka@res.titech.ac.jp

Y. Kanesaki
Research Institute of Green Science and Technology, Shizuoka University, 836 Ohya, Suruga-ku, Shizuoka 422-8529, Japan

M. Hanaoka
Graduate School of Horticulture, Chiba University, 1-33 Yayoi, Inage-ku, Chiba, Chiba 263-8522, Japan

 
T. Kuroiwa et al. (eds.), *Cyanidioschyzon merolae*,
https://doi.org/10.1007/978-981-10-6101-1_13

## 13.1 Introduction

Oxygen-respiring α-proteobacteria were involved in the symbiotic origin of eukaryotic cells, resulting in the origination of the respiratory organelles mitochondria. Similarly, the endosymbiosis of cyanobacteria led to the evolution of plant cells with chloroplasts (Leister 2005). Most genes of these endosymbionts have been lost or transferred to the nuclear genome (Gray 1993), but mitochondria and chloroplasts have retained their own unique genomes, and proliferation processes such as genome replication and division are performed distinctly from those of the nuclear system (Palenik 1992). Therefore, mechanisms exist to coordinate the proliferating cycles of nucleus and organelles to maintain the integrity of eukaryotic cells. In typical eukaryotic model cells, mitochondria and chloroplasts are present as multiple or filamentous networks, and their proliferation cycles are not clearly recognized. To better understand the coordinating process, we have studied the unicellular red alga *Cyanidioschyzon merolae* which contains only one nucleus, one mitochondrion, and one chloroplast per cell and clearly coordinated proliferating cycles. Information about the three genomes is now available, and basic experimental techniques such as cell synchronization, DNA introduction into the cell, and gene targeting have been developed (Kuroiwa 1998, 2004; Ohta et al. 1998, 2003; Matsuzaki et al. 2004; Nozaki et al. 2007; Kobayashi et al. 2010; Fujiwara et al. 2013). In this chapter, we summarize what is known about the coordinative mechanism of DNA replication between organelles and the nucleus in the cell cycle.

## 13.2 Nuclear DNA Replication Is Controlled by a Retrograde Signal

*C. merolae* is a photoautotrophic organism for which synchronized cultures can be established by entraining in light/dark cycles. The cell cycle arrests at G1 phase in the dark and shifts to the S phase under illumination (Suzuki et al. 1994). Based on detailed microscopic observations, it was previously proposed that the occurrence of mitochondria and chloroplast DNA replication (organelle DNA replication, ODR) preceded nuclear DNA replication (NDR) in *C. merolae* (Fig. 13.1; Kuroiwa 1998; Sakai et al. 2004). However, no information was available about the ODR/NDR coordinating mechanism.

In eukaryotic cells, activation of a G1 cyclin–cyclin-dependent kinase (CDK) complex is essential for the initiation of NDR (Ekholm and Reed 2000; Zegerman and Diffley 2007). Genome sequence analysis of *C. merolae* has revealed that G1 CDK (CDKA) and G1 cyclin (Cyclin 1) are homologous to *Arabidopsis* CDKA;1 and E-type cyclin, respectively. We examined the function of CDKA in the *C. merolae* cell cycle using a CDK2 inhibitor, which was expected to be specific for CDKA. The inhibition of CDKA by the CDK2 inhibitor was confirmed by the

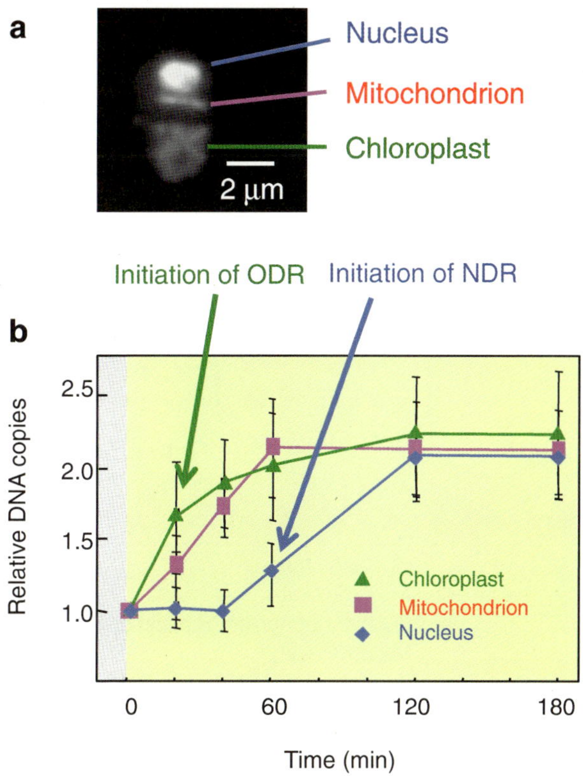


**Fig. 13.1** Observation of DNA replication in organelles and nucleus by DAPI staining. Organellar and nuclear DNA was stained with DAPI and quantitated under a fluorescence microscope. (**a**) Photomicrograph of cells after DAPI staining. (**b**) Organelle DNA replication initiates in advance of nuclear DNA replication under synchronous conditions. Synchronous cultured cells were sampled at indicated times, and the amount of DNA was plotted

direct monitoring of CDKA phosphorylation activity. We found that NDR was completely inhibited by the CDK2 inhibitor, whereas ODR was not affected. This indicated that ODR is regulated independently of CDKA. We next examined the effects of adding aphidicolin, an inhibitor of nuclear DNA polymerase, and nalidixic acid, an inhibitor of organellar DNA gyrase. Aphidicolin inhibited NDR but not ODR or CDKA activity. In contrast, nalidixic acid inhibited not only ODR but also CDKA activation and NDR. These data strongly suggested that ODR occurs independently of CDKA activation, whereas NDR requires ODR and subsequent CDKA activation (Fig. 13.2). We hypothesized that a signal of ongoing or completed ODR from the organelle leads to the activation of CDKA.

## 13.3 Tetrapyrrole Molecules as the Signal Coordinating ODR and NDR

Recent studies showed that the organelle status is relayed to the nucleus by a retrograde signal, resulting in the transcriptional regulation of genes for organellar maintenance. Reactive oxygen species, the intracellular oxidation–reduction

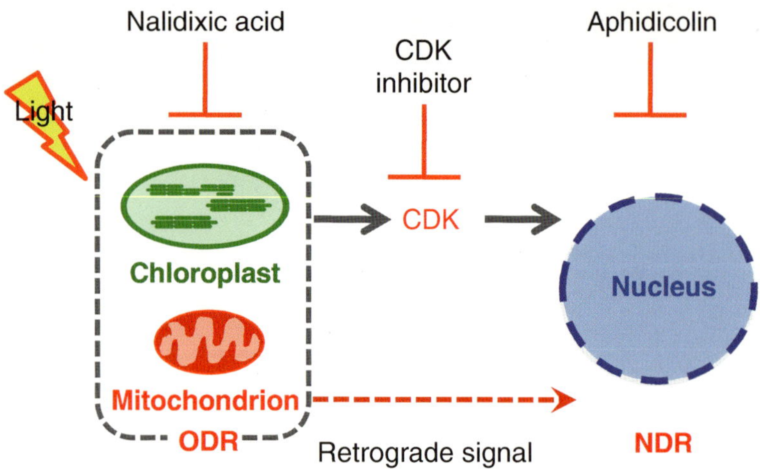


| | ODR | NDR |
|---|---|---|
| CDK inhibitor | + | - |
| Aphidicolin (Inhibitor of nuclear replication) | + | - |
| Nalidixic acid (Inhibitor of organelle replication) | - | - |

**Fig. 13.2** A retrograde signal controls nuclear replication.
Schematic of NDR activation by a retrograde signal. The ODR inhibitor nalidixic acid inhibits both CDKA activation and NDR. The CDK inhibitor inhibits NDR but not ODR. The NDR inhibitor aphidicolin has no effect on ODR or CDKA activation

potential, and tetrapyrroles have all been reported to be typical retrograde signals for nuclear transcriptional regulation (Nott et al. 2006; Mense and Zhang 2006; Gough 2009). We hypothesized that retrograde signaling might play a role in the coordination of ODR and NDR.

Tetrapyrrole molecules are represented by chlorophyll, heme, and their biosynthetic intermediates (Fig. 13.3). In land plants, the chlorophyll precursors ProtoIX, Mg-ProtoIX, and Mg-ProtoIX methylester are known to participate in chloroplast stress response signaling (Nott et al. 2006). Watanabe et al. (2013) showed that homologs of tetrapyrrole synthetic genes are found in the *C. merolae* genome and that tetrapyrrole molecules are synthesized in both mitochondrial and chloroplast compartments (Fig. 13.3). To determine if tetrapyrrole molecules act as the retrograde signal to inform the nucleus of the ODR status, we examined the effect of extrageneously supplied tetrapyrrole molecules to *C. merolae* growth medium. First, *C. merolae* cells were synchronized at the G1 phase by incubating the culture in the dark. Then, 4′,6-diamidino-2-phenylindole (DAPI) staining was used to periodically monitor the effects of tetrapyrrole addition under continuous dark conditions on the ODR and NDR (Kobayashi et al. 2009). While no DNA replication occurred in control cells, the addition of ProtoIX or Mg-ProtoIX induced NDR

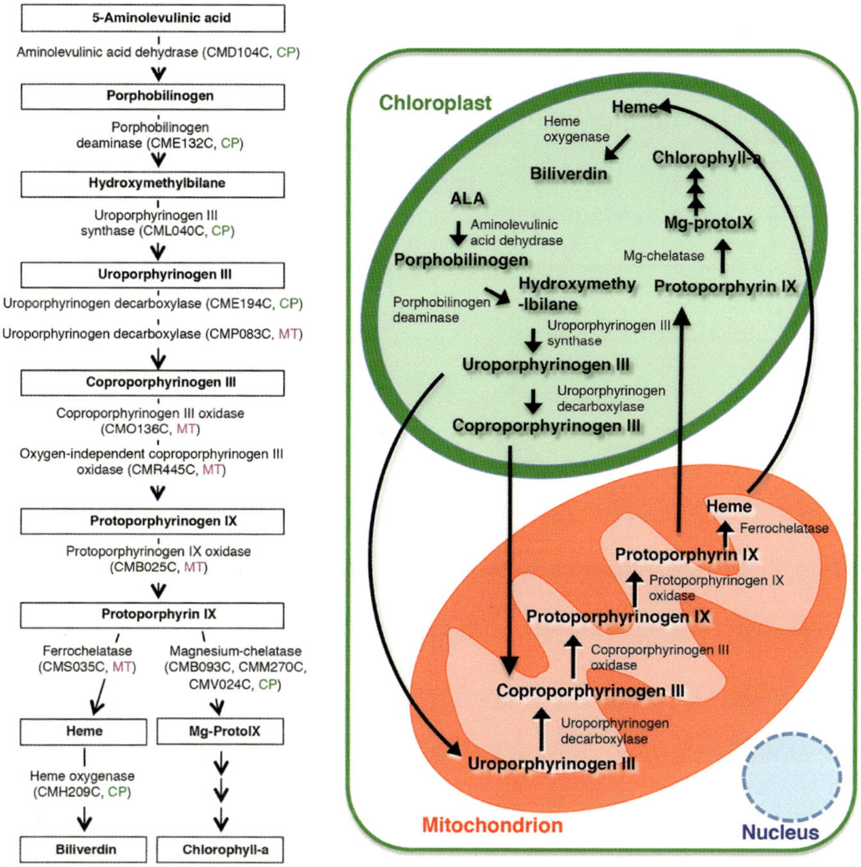


**Fig. 13.3** Schematic of the tetrapyrrole biosynthetic pathway.
Schematic representation of the tetrapyrrole biosynthetic pathway. *Boxes* indicate tetrapyrrole intermediates, and the predicted localization of each enzyme is shown as mitochondrion (MT) or chloroplast (CP) (Watanabe et al. 2013). *Boldface and plane texts* indicate tetrapyrrole intermediates and synthesizing enzymes, respectively

without inducing ODR, whereas other tetrapyrrole molecules including heme had no effect. The monitoring of CDKA activity revealed that the effect of Mg-ProtoIX on NDR was mediated through the activation of CDKA. Furthermore, the addition of an inhibitor of Mg-ProtoIX synthesis inhibited NDR even under light illumination.

Next, we measured the intracellular accumulation of ProtoIX and Mg-ProtoIX at G1 and S phases in the light using the synchronized culture. The transient accumulation of Mg-ProtoIX was observed, which correlated with the occurrence of ODR, whereas ProtoIX was not detected. These observations indicated that Mg-ProtoIX is a signal from the chloroplast that informs of the occurrence of ODR prior to CDKA activation (Fig. 13.4; Kobayashi et al. 2009).

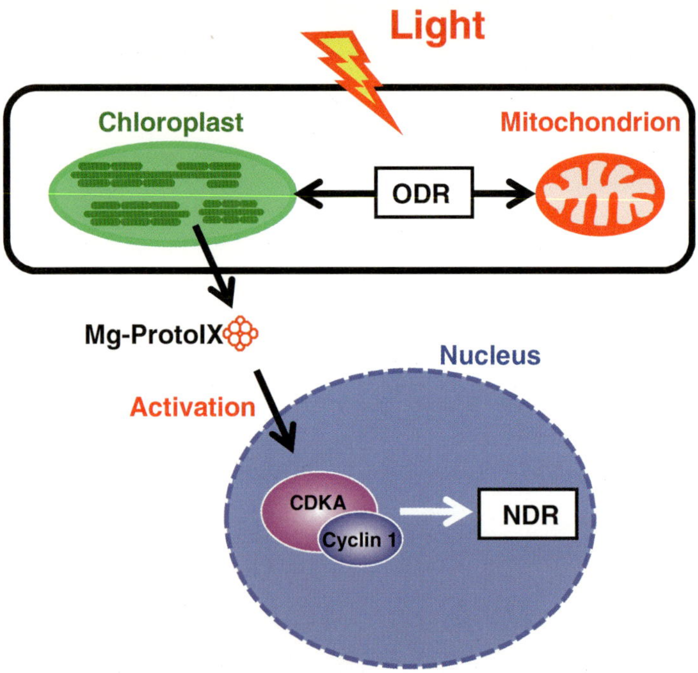


**Fig. 13.4** Model of cooperative mechanism of DNA replication between organelles and nucleus. ODR initiation by light induces release of the signaling molecule Mg-ProtoIX from chloroplasts, which activates CDKA and initiates NDR

## 13.4 Involvement of Ubiquitination in the Coordination of DNA Replication

Specific proteolysis by proteasomes is one of the basic regulatory mechanisms used to control cell cycle progression in eukaryotes (Reed 2006; Jung et al. 2009). Therefore, we examined the effect of proteasome inhibition on ODR and NDR and found that the addition of a proteasome inhibitor to G1-arrested cells in the dark resulted in CDKA activation and induced NDR without activating ODR. This result suggested that Mg-ProtoIX activates CDKA through the inhibition of proteasome-dependent protein degradation. Because the CDKA–Cyclin 1 complex was suggested to induce NDR, we assessed Cyclin 1 protein accumulation during cell cycle progression from G1 to S phase under synchronized cultivation conditions. Cyclin 1 protein was found to accumulate in association with ODR. The addition of a proteasome inhibitor as well as Mg-ProtoIX to G1-arrested cells in the dark induced Cyclin 1 accumulation. However, while the proteasome inhibitor induced the accumulation of ubiquitinated Cyclin 1, this ubiquitination did not occur with the addition of Mg-ProtoIX. These observations suggested that Mg-ProtoIX inhibits Cyclin 1 ubiquitination, thereby preventing its proteasome-dependent degradation.

## 13.5 An F-Box Protein Is a Novel Mg-ProtoIX Receptor

Protein ubiquitination is performed by the concerted activity of ubiquitin-activating enzyme, ubiquitin-conjugating enzyme, and ubiquitin ligase (E3). These three enzymes conjugate ubiquitin to the ε-amino group of the target protein lysine side chain using an isopeptide bond. During this process, E3 ubiquitin ligase plays an important role in recognizing the target protein. Among the different types of ubiquitin ligases, the SCF complex, composed of Skp1, Cullin 1, and F-box proteins, is known to be involved in cell cycle control (Reed 2006; Jung et al. 2009) mainly through target protein recognition (Reed 2006). Because several recent studies reported the involvement of specific ligand-binding F-box proteins in various signaling processes (Dharmasiri et al. 2005; Kepinski and Leyser 2005; Jung et al. 2009; Somers and Fujiwara 2009), we hypothesized that an F-box protein might serve as the receptor for Mg-ProtoIX. To investigate this possibility, we examined the interaction of four *C. merolae* genome-encoded F-box proteins (Fbx1–4) with Cyclin 1 using the *Escherichia coli* recombinant system. Pull-down assays of these recombinant tagged proteins revealed the interaction of Fbx3 and Fbx4 with Cyclin 1. Only Fbx3 was shown to interact with Mg-ProtoIX. Moreover, the interaction of Cyclin 1 with Fbx3, but not with Fbx4, was inhibited by the addition of Mg-ProtoIX but not by that of other tetrapyrrole molecules examined. These results strongly suggested that Fbx3 is the Mg-ProtoIX receptor protein involved in the modulation of Cyclin 1 ubiquitination and degradation in *C. merolae*.

## 13.6 In Vivo Evidence for the Mg-ProtoIX–Fbx3 Signaling Regulation of NDR

To examine whether Mg-ProtoIX–Fbx3 signaling occurs in vivo, we overexpressed glutathione S-transferase (GST)-fused Fbx3 in *C. merolae* to produce strain GF3. Pull-down assays of GF3 proteins confirmed the interaction of Cyclin 1 with GST-Fbx3 and showed that this was inhibited by Mg-ProtoIX which was consistent with our working hypothesis. To examine the control of NDR by Fbx3, we constructed the Fbx3-null strain DS19. After arresting the cells at G1 phase in the dark, Cyclin 1 protein accumulation was shown to be kept at basal levels and induced by light illumination in wild-type cells. In DS19 cells, however, Cyclin 1 levels remained consistently high in the dark, and the addition of proteasome inhibitor did not induce the accumulation of ubiquitinated Cyclin 1. The observation of NDR irrespective of the occurrence of ODR in DS19 cells supported the proposed role for Fbx3 in the negative control of NDR (Kobayashi et al. 2011). Cyclin 1 is polyubiquitinated by SCF-type E3 ubiquitin ligase containing Fbx3 and is then rapidly degraded in dark-arrested G1 phase cells. Light illumination induces ODR by a currently unknown mechanism, which results in the accumulation of

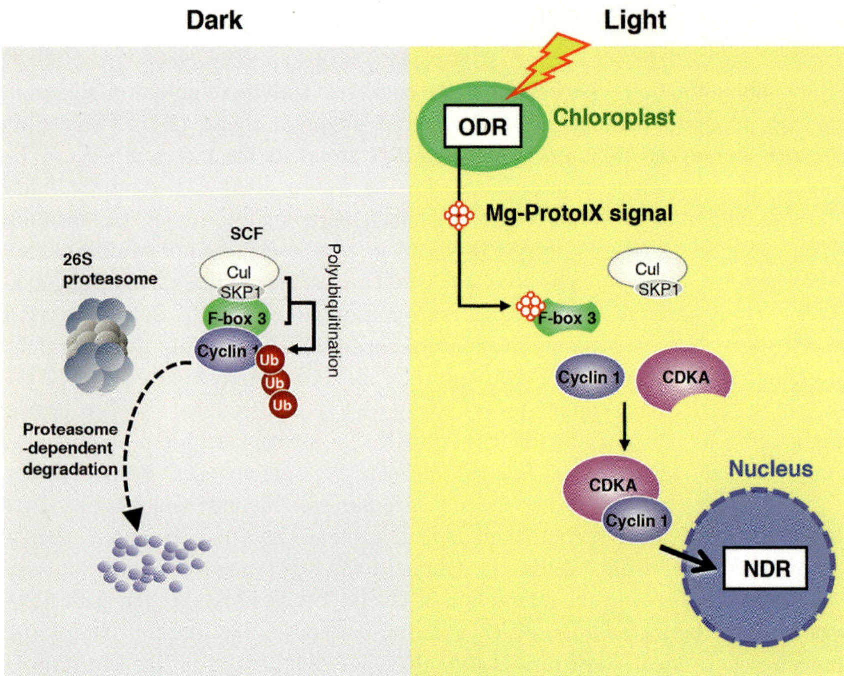


**Fig. 13.5** Model of the molecular mechanism of ODR and NDR DNA replication. Under dark conditions, Cyclin 1 is polyubiquitinated by SCF containing Fbx3 (F-box protein) and then undergoes proteasome degradation. The signaling molecule Mg-ProtoIX is induced by ODR initiation and released from chloroplasts and then binds to cytosolic Fbx3. This inhibits the polyubiquitination of Cyclin 1, allowing CDKA activation and the initiation of NDR

Mg-ProtoIX. Because Mg-ProtoIX is synthesized by Mg-chelatase, which is predicted to be localized in the chloroplast, Mg-ProtoIX is also expected to be synthesized in the chloroplast and released into the cytosol as the signaling molecule. By binding to Fbx3, Mg-ProtoIX inhibits the ubiquitination of Cyclin 1, which induces the activation of CDKA and NDR (Fig. 13.5).

## 13.7 Heme Is Necessary for ODR Initiation

Little is known about the mechanism of controlling ODR. We recently showed that heme, another tetrapyrrole molecular species synthesized in the mitochondrion (Watanabe et al. 2013), is required for the occurrence of ODR (Kobayashi et al. 2016). This was demonstrated during functional analysis of the phytohormone abscisic acid (ABA) in *C. merolae*. ABA plays important roles in the vascular plant response to abiotic stresses such as salinity, drought, low temperatures, stomatal closure, and seed dormancy (Hartung 2010; Hauser et al. 2011). It has also been extracted from unicellular photosynthetic organisms, but its function here is unknown. We found that ABA accumulates in *C. merolae* under salt stress conditions and that the exogenous application of ABA arrested the cell cycle at the G1/S checkpoint. Further analysis showed that ABA arrested the occurrence of ODR by preventing intracellular heme accumulation (Kobayashi et al. 2016). ABA induces the free heme-scavenging protein TSPO that binds intracellular unbound heme, resulting in the depletion of unbound heme. The addition of exogenous heme canceled this ABA inhibitory effect, indicating that intracellular unbound heme accumulation is required for the occurrence of ODR. Thus, heme that is synthesized in the mitochondrion works as another organelle-to-nucleus signal, in addition to Mg-ProtoIX, in the regulation of DNA replication.

## 13.8 Conclusion

Our studies revealed the existence of rigid mechanisms to coordinate DNA replication in the eukaryotic cell cycle. Tetrapyrrole molecules synthesized in the organelles act as signals from the organelle to the nucleus, suggesting that cell cycle checkpoints monitor the organelle status. Prior to the endosymbiotic origin of chloroplasts, host cell growth would have occurred irrespective of environmental light conditions. However, after endosymbiosis, the ancestral chloroplast could only proliferate in the light, so the endosymbiotic association was unlikely to be maintained unless a mechanism evolved to prevent host proliferation in the dark, as proposed in our recent studies. While the presence of a similar mechanism in land plants is uncertain, we observed that ODR inhibition also inhibits NDR and that this NDR inhibition is negated by the inhibition of proteasomes (Kobayashi et al. 2011). These observations suggested that these mechanisms are even conserved in land plants. Future analyses will help clarify the underlying mechanisms of the coordination of ODR and NDR in plant cells.

## References

Dharmasiri N, Dharmasiri S, Estelle M (2005) The F-box protein TIR1 is an auxin receptor. Nature 435(7041):441–445

Ekholm SV, Reed SI (2000) Regulation of G1 cyclin-dependent kinases in the mammalian cell cycle. Curr Opin Cell Biol 12(6):676–684

Fujiwara T, Ohnuma M, Yoshida M et al (2013) Gene targeting in the red alga *Cyanidioschyzon merolae*: single- and multi-copy insertion using authentic and chimeric selection markers. PLoS One 8:e73608

Gough NR (2009) Wait for me. Sci Signal 2(55):ec27

Gray MW (1993) Origin and evolution of organelle genomes. Curr Opin Genet Dev 3(6):884–890

Hartung W (2010) The evolution of abscisic acid (ABA) and ABA function in lower plants, fungi and lichen. Func Plant Biol 37(9):806–812

Hauser F, Waadt R, Schroeder JI (2011) Evolution of abscisic acid synthesis and signaling mechanisms. Curr Biol 21(9):R346–R355

Jung T, Catalgol B, Grune T (2009) The proteasomal system. Mol Asp Med 30(4):191–296

Kepinski S, Leyser O (2005) The Arabidopsis F-box TIR1 is an auxin receptor. Nature 435(7041):446–451

Kobayashi Y, Kanesaki Y, Tanaka A et al (2009) Tetrapyrrole signal as a cell-cycle coordinator from organelle to nuclear DNA replication in plant cells. Proc Natl Acad Sci U S A 106(3):803–807

Kobayashi Y, Ohnuma M, Kuroiwa T et al (2010) The basics of cultivation and molecular genetic analysis of the unicellular red alga *Cyanidioschyzon merolae*. Endocytobiosis Cell Res 20:53–61

Kobayashi Y, Imamura S, Hanaoka S et al (2011) A tetrapyrrole-regulated ubiquitin ligase controls algal nuclear DNA replication. Nat Cell Biol 13(4):483–487

Kobayashi Y, Ando H, Hanaoka M et al (2016) Abscisic acid participates in the control of cell-cycle initiation through heme homeostasis in the unicellular red alga *Cyanidioschyzon merolae*. Plant Cell Physiol 57(5):953–960

Kuroiwa T (1998) The primitive red algae: *Cyanidium caldarium* and *Cyanidioschyzon merolae* as model system for investigating the dividing apparatus of mitochondria and plastids. BioEssays 20(4):344–354

Kuroiwa T (2004) Organelle nuclei in higher plants: structure, composition, function, and evolution. Int Rev Cytol 238:59–117

Leister D (2005) Genomics-based dissection of the cross-talk of chloroplasts with the nucleus and mitochondria in Arabidopsis. Gene 354:110–116

Matsuzaki M, Misumi O, Shin-I T et al (2004) Genome sequence of the ultrasmall unicellular red alga *Cyanidioschyzon merolae* 10D. Nature 428(6983):653–657

Mense SM, Zhang L (2006) Heme: a versatile signaling molecule controlling the activities of diverse regulators ranging from transcription factors to MAP kinase. Cell Res 16(8):681–692

Nott A, Jung H, Koussevitzky S et al (2006) Plastid-to-nucleus retrograde signaling. Annu Rev Plant Biol 57:739–759

Nozaki H, Takano H, Misumi O et al (2007) A 100%-complete sequence reveals unusually simple genomic features in the hot spring red alga *Cyanidioschyzon merolae*. BMC Biol 5:28

Ohta N, Sato N, Kuroiwa T (1998) Structure and organization of the mitochondrial genome of the unicellular red alga *Cyanidioschyzon merolae* deduced from the complete nucleotide sequence. Nucleic Acids Res 26(22):5190–5198

Ohta N, Matsuzaki M, Misumi O et al (2003) Complete sequence and analysis of the plastid genome of the unicellular red alga *Cyanidioschyzon merolae*. DNA Res 10(2):67–77

Palenik B (1992) Polymerase evolution and organism evolution. Curr Opin Genet Dev 2(6):931–936

Reed SI (2006) The ubiquitin proteasome pathway in cell cycle control. Results Probl Cell Differ 42:147–181

Sakai A, Takano H, Kuroiwa T (2004) Organelle nuclei in higher plants: structure, composition, function, and evolution. Int Rev Cytol 238:59–117

Somers DE, Fujiwara S (2009) Thinking outside the F-box: novel ligands for novel receptors. Trends Plant Sci 14(4):206–213

Suzuki K, Ehara T, Osafune T et al (1994) Behavior of mitochondria, chloroplasts and their nuclei during the mitotic cycle in the ultramicroalga *Cyanidioschyzon merolae*. Eur J Cell Biol 63(2):280–288

Watanabe S, Hanaoka M, Ohba Y et al (2013) Mitochondrial localization of ferrochelatase in a red alga *Cyanidioschyzon merolae*. Plant Cell Physiol 54(8):1289–1295

Zegerman P, Diffley JF (2007) Phosphorylation of Sld2 and Sld3 by cyclin-dependent kinases promotes DNA replication in budding yeast. Nature 445(7125):281–285

# Chapter 14
# Double-Membrane-Bounded Organelles: Recent Findings Regarding Division, Inheritance, Structure, and Evolution of the Nucleus, Mitochondria, and Chloroplasts

**Yamato Yoshida, Yuki Sakamoto, Katsuya Iwasaki, Shinichiro Maruyama, and Sachihiro Matsunaga**

**Abstract** The nucleus of *Cyanidioschyzon merolae* has quite a limited set of components. The core and linker histones, H2A.Z, and a centromeric histone variant CenH3 are preserved, but H2A variants including H2A.X, H2A.M, and H2A.W are absent. Some proteins of the nuclear pore complex and nucleoli are also preserved, but almost all homologs of nuclear lamina and nuclear envelope proteins are lacking. The chloroplast of this species is divided by a ring, which is composed of two inner rings, the FtsZ ring and inner plastid-dividing (PD) ring, and two outer rings, the outer PD ring and dynamin ring. Identification of the PD ring components in *C. merolae* enabled us to elucidate the molecular mechanism behind the PD machinery. For example, a glycosyltransferase protein, plastid-dividing ring 1, acts in establishing the PD ring by forming ring-shaped polyglucan filaments. The division of mitochondria is performed by the mitochondrion-dividing (MD), FtsZ, and dynamin rings. Mda1 functions as a mediator for stably linking a dynamin. In addition, an FtsZ-associated protein, ZED, is involved in completing the inner rings of the MD machinery. Analyses of the mechanisms of mitochondrial and chloroplast division in *C. merolae* contribute to our understanding of the endosymbiotic process of double-membrane-bounded organelles.

---

Y. Yoshida
Laboratory for Single Cell Gene Dynamics, Quantitative Biology Center, RIKEN, 6-2-3 Furuedai, Suita, Osaka 565-0874, Japan

Y. Sakamoto
Imaging Frontier Center, Organization for Research Advancement, Tokyo University of Science, 2641 Yamazaki, 2641 Yamazaki, Noda, Chiba 278-8510, Japan

K. Iwasaki • S. Matsunaga (✉)
Department of Applied Biological Science, Faculty of Science and Technology, Tokyo University of Science, 2641 Yamazaki, Noda, Chiba 278-8510, Japan
e-mail: sachi@rs.tus.ac.jp

S. Maruyama
Department of Environmental Life Sciences, Graduate School of Life Sciences, Tohoku University, 6-3, Aoba, Aramaki, Aoba, Sendai, Miyagi 980-8578, Japan



T. Kuroiwa et al. (eds.), *Cyanidioschyzon merolae*,
https://doi.org/10.1007/978-981-10-6101-1_14

**Keywords** Nucleus • Nuclear envelope • Nuclear pore • Histone • Chloroplast • Plastid • PD ring • FtsZ • Dynamin • Mitochondria • MD ring

## 14.1 Introduction

Generally, eukaryotic cells have double-membrane-bounded organelles, including the nucleus, mitochondria, and chloroplasts. A eukaryotic nucleus consists of the nucleoplasm, nuclear envelope (NE), nuclear lamina, and nucleolus (Matsuanga et al. 2013). DNA is tightly packed in chromatin, consisting of a basic structural unit, the nucleosome. The nucleosome contains DNA and a histone octamer composed of H2A, H2B, H3, and H4. A linker DNA, located between nucleosomes, binds to a linker histone, H1. The NE wrapping the nucleoplasm consists of a double membrane formed by a lipid bilayer; the outer NE directly connects to the endoplasmic reticulum. Proteins at the NE and nuclear pores function in transporting substances between the nucleoplasm and the cytoplasm and in regulating nuclear shape and movement. The nuclear lamina is a mesh-like structure inside the inner NE and provides mechanical support to the NE. Finally, the nucleolus is a compartment responsible for the synthesis of ribosomes. These nuclear structures except for the nuclear lamina have been observed by electron microscopy in *Cyanidioschyzon merolae* (Maruyama et al. 2004; Yagisawa et al. 2012), but their molecular components have remained unknown. Here, we present analyses of the components of the nucleus, including of the nucleosome, NE, nuclear pore, and nucleolus, in *C. merolae*.

*C. merolae* contains only one chloroplast (plastid) per cell. Chloroplast DNA is located in the center of this spherical chloroplast, which has a similar localization to that of cyanobacterial nuclei. Owing to their cyanobacterial origin, the proliferation of chloroplasts is performed by binary fission. The single chloroplast in *C. merolae* cells divides by the following process: The chloroplast starts with a spherical shape or acorn-like shape in the interphase cell. Subsequently, it transforms into cup-shaped, trapezoidal and then dumbbell-shaped structures in M phase. Next, the dumbbell-shaped chloroplast is severed into two daughter chloroplasts, which are inherited by the daughter cells during the cytokinesis phase. The structural details of the process of chloroplast division were first unveiled by using electron microscopy in the unicellular red alga *Cyanidium caldarium* RK-1 (Mita et al. 1986). A series of ultrastructural observations of *C. caldarium* cells subjected to rapid freezing fixation revealed a specialized ring structure at the bridge of a dividing chloroplast, which was later named the plastid-dividing (PD) ring. However, the hard and thick cell wall of *C. caldarium* cells obstructed further investigation of the molecular mechanisms of chloroplast division. However, this was overcome by utilizing the unicellular red alga *C. merolae* (Kuroiwa 1998). The following biological features of *C. merolae* cells enabled further study of chloroplast division: (1) *C. caldarium* cells multiply by two sequential endospore division, while *C. merolae* cells multiply by binary fission, namely, the life cycle of *C. merolae* is simpler than that of *C. caldarium*; (2) by not being surrounded by a rigid cell wall, *C. merolae* cells are accessible for various isolation approaches; and (3) the simple and compact structure of the *C. merolae* genome makes it suitable for multi-omics analysis. As a result, it has been revealed that chloroplast division is performed by a ring-shaped supermolecule structure, the plastid-dividing (PD) machinery. Here, we introduce the mechanism of chloroplast division by the PD machinery in *C. merolae*.

Similar to chloroplasts, mitochondria also originated from an endosymbiotic ancestor, such as alpha-proteobacteria; therefore, they have their own genome and gene system and proliferate via division (fission), as is the case in chloroplast proliferation (Gillham et al. 1994; Lane and Martin 2010). Mitochondria govern various metabolic pathways in almost all eukaryotes and are associated with critical biological processes such as energy generation and cell death (Friedman and Nunnari 2014). To date, more than five kinds of proteins and some specialized ring structures for mitochondrial division have been identified throughout primitive alga *C. merolae* to human and higher plants; however, the fundamental conserved mechanism of mitochondrial division has remained unclear (van der Bliek et al. 2013; Roy et al. 2015). One of the obstacles to elucidating this is the complexity of mitochondrial dynamics, which refers to the continuous processes of intracellular movement, morphological transformation, and fusion, in addition to division (Yaffe 1999; Hoppins 2014). Furthermore, mitochondria usually form narrow and long cylindrical shapes in the yeast *Saccharomyces cerevisiae* and human cultured cells, which results in difficulty distinguishing whether the two ends of a long cylindrical mitochondrion have been formed by the fission process or are at a stage just before fusion. In contrast, *C. merolae* contains only one disc-shaped mitochondrion per cell, which matches the chloroplast number, and the *C. merolae* mitochondrion undergoes binary division once during a cell-division cycle but does not undergo fusion (Nishida et al. 2003; Matsuzaki et al. 2004). Thus, in addition to chloroplast division, *C. merolae* is a very useful organism for shedding light on mitochondrial division. Interestingly, some of the mechanisms of mitochondrial division are very similar to those of chloroplast division. The division of mitochondria involves at least three types of ring structures, the mitochondrion-dividing (MD) ring, the FtsZ ring, and the dynamin ring, which assemble a supramolecular complex called "the mitochondrion-dividing (MD) machinery," as homologous structure with the PD machinery for chloroplast division (Kuroiwa et al. 2006). The striking resemblance between mitochondrial and chloroplast division suggests that the division systems of endosymbiotic organelles might have evolved from the original membrane fission mechanism. Furthermore, several higher-organism-specific factors for mitochondrial dynamics have recently been characterized (Roy et al. 2015), indicating development and complication of mitochondrial division mechanism in metazoans and higher plants, and resultantly, the original factors for mitochondrial proliferation are not easy to identify in these higher organisms. On the other hand, it is assumed that evolutionally ancestral alga *C. merolae* cell may make it possible to elucidate fundamental and primitive mechanism of mitochondrial proliferation.

This chapter provides an overview of the cellular and molecular components constituting the double-membrane-bounded organelles in eukaryotic cells, with special emphasis on *C. merolae*. Upon summarizing the current knowledge on these organelles, we further discuss the challenges and promising research fields for future study.

## 14.2 Evolution of Double-Membrane-Bounded Organelles

### *14.2.1 The Origin of the Nucleus*

The evolutionary history of the cell nucleus is central to addressing the origin of eukaryotes. To date, numerous hypotheses (and speculations) have been proposed and some refuted, but the debate on this issue is far from reaching a consensus (Martin et al. 2015). These hypotheses can be represented by two extremes: the "nucleus-first" and "mitochondrion-first" scenarios. The nucleus-first scenario envisions that an amitochondriate but cytoskeleton-possessing "proto-eukaryote" engulfed an alpha-proteobacterium-like endosymbiont, which subsequently gave rise to the current mitochondrion, by phagocytosis. The mitochondrion-first or perhaps "endosymbiosis-first" scenario assumes that the eubacterial ancestor of mitochondria was integrated into, or fused with, a nucleus-lacking cell derived from an archaeal lineage, like the "Asgard" archaea, which was recently identified by meta-genome sequencing (Spang et al. 2015; Zaremba-Niedzwiedzka et al. 2017). It is thought that this then started to establish a unified organismal system surrounded by a plasma membrane, which led to the formation of more sophisticated nucleus-like inner membrane structures. Other hypotheses are intermediate between these proposals or go beyond them: One example of this latter category is the recently reported "inside-out" hypothesis, which explains the origin of eukaryotic cells by proposing a hypothetical ancestral "eocyte"-like archaeon, which developed a membrane protrusion to surround the proto-mitochondrial alpha-proteobacterium and fused at the periphery leading to the formation of plasma membranes and endoplasmic reticulum (Baum and Baum 2014). Although it is difficult to prove any of this empirically, more genome sequence data of poorly sampled lineages like the Asgard archaea—of which the living cells have yet to be isolated but the genome has been shown to contain eukaryotic genes encoding proteins responsible for membrane trafficking and signaling systems (Spang et al. 2015; Zaremba-Niedzwiedzka et al. 2017)—can shed light on how cell nuclei emerged and what the first eukaryote looked like.

### 14.2.2 The Origins of Endosymbiont-Derived Organelles and Their Outer Membranes

Endosymbiotic theory, which is widely acknowledged to have been established by the Russian botanist Konstantin Mereschkowsky and was promulgated by the American biologist Lynn Margulis, postulates that the mitochondrion and chloroplast (plastid) evolved through a series of endosymbioses of free-living alphaproteobacteria and cyanobacteria, respectively (Martin et al. 2015; Archibald 2015). In many textbooks, it is assumed that this theory is correct, and thus explanations with easy-to-understand cartoons are presented about how the organelles evolved. However, as mentioned above, although we are fairly sure that this actually happened, we are far from understanding how it happened.

One of the conundrums regarding the endosymbiotic origin of organelles is where the organelles' outer membranes have come from. In contrast to a classic and pervasive idea that the outer membranes of endosymbiont-derived organelles are remnants of the host's phagosome (food vacuole) membrane, Cavalier-Smith proposed an updated explanation, based on the observations that many bacterial envelopes are made of two membranes, that the outer membranes had been derived from the outer envelope of the endosymbionts (Cavalier-Smith 1987). Although it may be extremely difficult to obtain evidence to refute either of these theories, recent findings of bacterial proteins localized on the outer membranes seem to be consistent with the latter (Hayashi et al. 2009; Lee et al. 2014; Miyagishima et al. 2014; Muñoz-Gómez et al. 2015). Either way, to a greater or lesser degree, the organelle membranes are still evolutionary chimeras, as many of their proteins are encoded by endosymbiont-derived genes, present in the endosymbiont or host nuclear genomes, and host-derived genes, of which the products are targeted and functioning on the membranes, in every eukaryotic lineage (Cavalier-Smith 1987).

### 14.2.3 Nuclear Envelope Dynamics

Every eukaryotic cell divides in order to proliferate. The same applies for endosymbiont-derived organelles but does not for every cell nucleus. The nucleus is the organelle that defines the eukaryotic cell and is characterized by a structure containing genome DNA and proteinaceous machinery for accurate duplication and transcription. Some eukaryotes including multicellular animals and plants perform "open" mitosis where the nuclear envelope (NE) breaks down during mitosis, while complete NE breakdown is not seen in other eukaryotes, including unicellular fungi and algae, the mitosis of which is "semi-open" or "closed." De Souza and Osmani discussed that whether the nuclear envelope breaks down or not depends on the position of the microtubule-organizing center (MTOC): NE breakdown is not essential if the MTOC (e.g., spindle pole body in fungi) localizes on or is embedded in the nuclear envelope, but NE requires breakdown to connect mitotic spindles to chromosomes if the MTOC (e.g., centriole in mammals) is located in the cytosol (De Souza and Osmani 2007).

*C. merolae* performs closed mitosis with partial NE disruption and possesses conspicuous spindle pole bodies, which are associated with nuclear and organelle division (Nishida et al. 2005; Maruyama et al. 2007; Fujiwara et al. 2013; Yoshida et al. 2013). In fact, orthologs of the mitotic kinases of Polo-like kinase and Aurora kinase in *C. merolae* localize at both spindle poles and mitochondrial polar microtubules or mitochondrial division sites (Kato et al. 2011; Yoshida et al. 2013; Okamura et al. 2017). These findings suggest that organelle division is regulated by protein phosphorylation through coordination with nuclear division. Considering that the physical interaction among the nucleus, mitochondrion, and plastid mediated by spindle structures is potentially a powerful and efficient way to integrate and synchronize organelle division and inheritance, this should be a promising research field in *C. merolae* as well as other eukaryotes.

## 14.3 Structure and Components of the Nucleus

### *14.3.1 Nucleosome*

#### 14.3.1.1 Histone H2A Variants

Histones are encoded by multiple genes, which generate several histone variants such as canonical H2A, H2A.X, H2A.M, H2A.W, and H2A.Z in higher plants (Kawashima et al. 2015). *C. merolae* has only three H2A variants belonging to the canonical H2A and H2A.Z group (Fig. 14.1) (Matsuzaki et al. 2004; Misumi et al. 2005). Canonical H2A in other photosynthetic organisms such as *Marchantia polymorpha*, *Physcomitrella patens*, *Selaginella moellendorffii*, and *Arabidopsis thaliana* contains the conserved KYAE/T amino acids in the L1 loop that are responsible for interacting with H2B and the acidic amino acids at the C-terminal (Kawashima et al. 2015). In contrast, *C. merolae* H2As contain the GYAE amino acids in the L1 loop and no acidic motif at the C-terminal, although *C. merolae* H2As belong to the canonical H2A group (Fig. 14.1). The replacement of K by G in the L1 loop may be a characteristic of the ancestral histone H2A. *C. merolae* harbors canonical H2A and H2A.Z but not H2A.X, H2A.M, and H2A.W, while *Ostreococcus tauri* has canonical H2A, H2A.Z, and H2A.X (Fig. 14.1) (Kawashima et al. 2015). These results suggest that a plant ancestor had canonical H2A and H2A.Z, that H2A.X was the first to diverge from these, and then finally that the H2A.M and H2A.W variants appeared in the plant lineage.

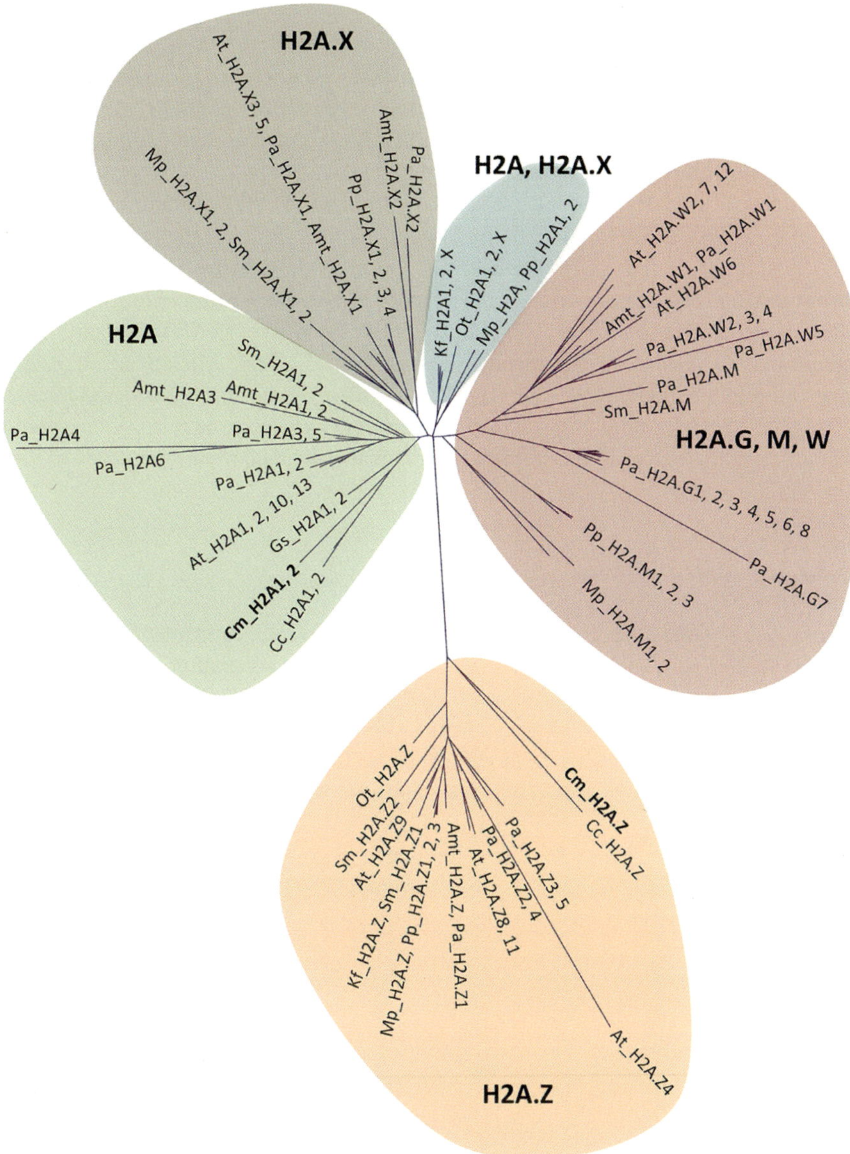


**Fig. 14.1** A phylogenetic tree of H2A variants in photosynthetic organisms. This tree of H2A was analyzed by neighbor-joining methods after its establishment from *Cyanidioschyzon merolae* (Cm), *Galdieria sulphuraria* (Gs), *Chondrus crispus* (Cc), *Ostreococcus tauri* (Ot), *Klebsormidium flaccidum* (Kf), *Marchantia polymorpha* (Mp), *Physcomitrella patens* (Pp), *Selaginella moellendorffii* (Sm), *Picea abies* (Pa), *Amborella trichopoda* (Amt), and *Arabidopsis thaliana* (At). H2A variants in *C. merolae* are shown in *bold*

#### 14.3.1.2 H2A.X

When DNA is damaged by environmental stresses (UV, ionizing radiation, etc.) or endogenous metabolic processes, a DNA double-strand break (DSB) can occur in the nucleus. In the repair process after a DSB, H2A.X in the chromatin around the DSB is phosphorylated by Ataxia Telangiectasia Mutated (ATM) and Ataxia Telangiectasia and Rad3 (ATR) kinases (Burma et al. 2001; Wang et al. 2005; Stiff et al. 2004). DNA repair protein complexes then gather at the phosphorylated H2A.X and repair the DSB (Paull et al. 2000). Although the H2A.X variant is absent in *C. merolae*, homologs of ATRkinase, DNA damage recognition proteins Mitotic recombination (Mre) 11 and Radiation-sensitive (RAD) 50, and DNA recombinase RAD51 are preserved (Matsuzaki et al. 2004; Misumi et al. 2005), suggesting that canonical H2A or H2A.Z at DSB regions is phosphorylated instead of H2A.X. These findings suggest that research on DNA repair in *C. merolae* should provide new insights into the minimal mechanism of DNA repair.

#### 14.3.1.3 Other Histones

*C. merolae* has one homolog each of histone H1, H2B, H3, centromeric histone H3 (CENH3), and H4 (Table 14.1) (Matsuzaki et al. 2004; Misumi et al. 2005). H1 is not a core histone and binds to linker DNA between nucleosomes, which contributes to chromatin compaction, transcriptional regulation, nucleosome spacing, and chromosome spacing (Kalashnikova et al. 2016). *Galdieria sulphuraria* and *O. tauri* do not have H1 variants, although all other photosynthetic organisms including *C. merolae* do (Table 14.1). CENH3, which is a centromere-specific histone H3 variant, is required for chromatin segregation during cell division. Most eukaryotes including *C. merolae* harbor CENH3 (Table 14.1) (Matsuzaki et al. 2004; Misumi et al. 2005).

### *14.3.2 Nuclear Lamina and Envelope Proteins*

The nuclear lamina is mainly composed of lamin, Crowded nuclei (CRWN), NE81, and NUP-1 in animals, plants, *Dictyostelium discoideum*, and *Trypanosoma brucei*, respectively (McKeon et al. 1986; Masuda et al. 1997; Sakamoto and Takagi 2013; Krüger 2012; DuBois et al. 2012). No lamina protein homolog has been found in the *C. merolae* genome, suggesting that the *C. merolae* nucleus does not have the nuclear lamina structure (Table 14.2) (Matsuzaki et al. 2004; Misumi et al. 2005). Various nuclear envelope proteins have been reported in animals and yeast. Emerin, Man1, Lamina-associated polypeptide (LAP) 2, and lamin B receptor (LBR) are inner nuclear envelope (NE) proteins and tether DNA and chromatin. The protein complex including SUN (Sad1p, UNC-84) and KASH (Klarsicht, ANC-1, Syne Homology) forms a linker between the nuclear lamina and the cytoskeleton (Olins

**Table 14.1** Number of genes encoding histone variants in photosynthetic organisms

| | H1 | H2A | H2B | H3 | CenH3 | H4 |
|---|---|---|---|---|---|---|
| *Arabidopsis thaliana* | 3 | 12 | 4 | 7 | 1 | 1 |
| *Oryza sativa* | 4 | 12 | 4 | 3 | 1 | 1 |
| *Amborella trichopoda* | 3 | 8 | 6 | 4 | 2 | 1 |
| *Picea abies* | 26 | 35 | 24 | 16 | – | 18 |
| *Selaginella moellendorffii* | 3 | 7 | 7 | 3 | 1 | 1 |
| *Physcomitrella patens* | 5 | 11 | 9 | 4 | 1 | 3 |
| *Marchantia polymorpha* | 1 | 6 | 5 | 2 | 1 | 1 |
| *Klebsormidium flaccidum* | 1 | 4 | 4 | 2 | 1 | 1 |
| *Ostreococcus tauri* | – | 3 | 1 | 2 | 1 | 1 |
| *Chondrus crispus* | 1 | 3 | 2 | 1 | 1 | 2 |
| *Galdieria sulphuraria* | – | 1 | 3 | 1 | 1 | 1 |
| *Cyanidioschyzon merolae* | 1 | 2 | 1 | 1 | 1 | 1 |

**Table 14.2** Number of genes encoding nuclear envelope and lamina proteins in photosynthetic organisms

| | NUP107 | SUN | WIP | SINE | CRWN |
|---|---|---|---|---|---|
| *Arabidopsis thaliana* | 1 | 5 | 3 | 4 | 4 |
| *Oryza sativa* | 1 | 4 | 2 | 2 | 2 |
| *Selaginella moellendorffii* | 1 | 3 | – | 2 | 2 |
| *Physcomitrella patens* | 1 | 2 | – | 1 | 1 |
| *Klebsormidium flaccidum* | 1 | 1 | – | 1 | 1 |
| *Ostreococcus tauri* | 1 | 1 | – | – | – |
| *Chondrus crispus* | 1 | 1 | – | – | – |
| *Galdieria sulphuraria* | 1 | 2 | – | – | – |
| *Cyanidioschyzon merolae* | 1 | – | – | – | – |

et al. 2010; Berk et al. 2013; Meinke and Schirmer 2015). Plant KASH were classified into two classes, WPP domain-interacting proteins (WIP) and SUN domain–interacting NE protein (SINE) (Zhou et al. 2015) (Fig. 14.3). In *C. merolae*, homologs of these NE proteins are not preserved, except for LBR (Table 14.2) (Matsuzaki et al. 2004). A comparison of NE components between *A. thaliana* and *C. merolae* was performed based on the complete genome sequences (Fig. 14.2), which suggested that the NE of *C. merolae* consists of an extremely simplified set of proteins.

### *14.3.3 Nuclear Pore*

The nuclear pore is a channel in the NE that regulates the transportation of nucleic acids, proteins, and some metabolites between the nucleus and the cytoplasm. This pore is composed of approximately 30 nuclear pore complexes (NPCs), forming the largest protein complex in cells (estimated at 60 MDa and 125 MDa in yeast and

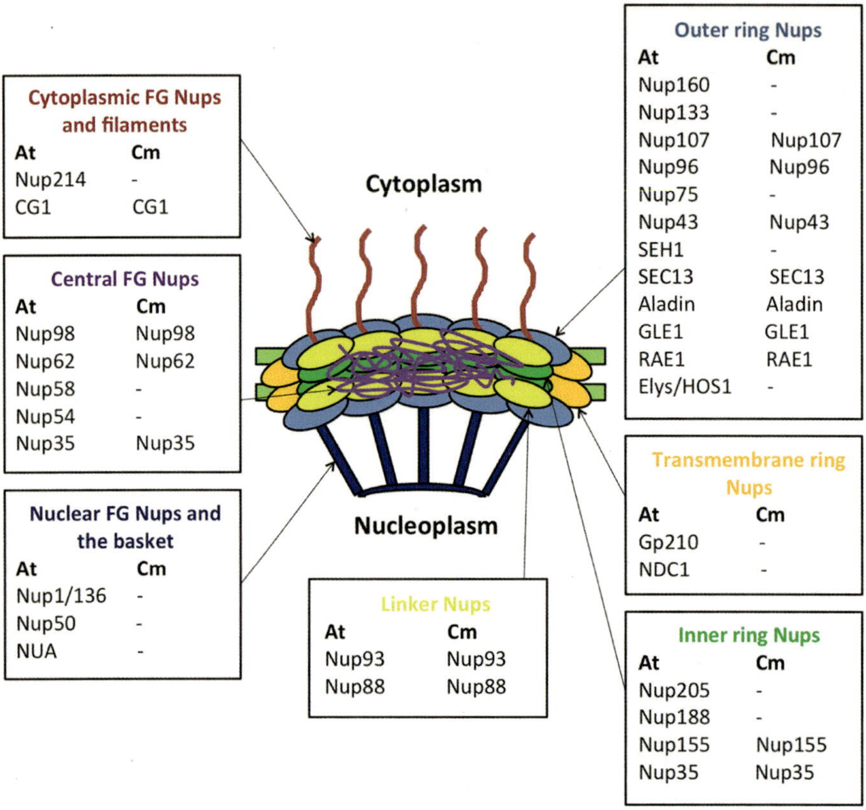


**Fig. 14.2** Comparison of the nuclear structures between *Arabidopsis thaliana* (At) and *Cyanidioschyzon merolae* (Cm)

vertebrates, respectively). According to the architectural map of yeast and *A. thaliana* NPCs, their components are categorized into five classes: transmembrane ring, core scaffold (inner ring, outer ring, and linker), cytoplasmic filaments, nuclear basket, and central phenylalanine-glycine repeat NPCs (Fig. 14.3) (Tamura and Hara-Nishimura 2013). In the case of *C. merolae*, only 19 NPC homologs including Nucleoporin (NUP) 107 have been found, while the proteins of the transmembrane ring and nuclear basket are absent (Fig. 14.3). Gp210, which is one of the transmembrane ring components, plays an important role in NE breakdown (Galy et al. 2008). NE is partially broken down, and NPCs are detached from it at the metaphase in *C. merolae* (Fujiwara et al. 2013). Because *C. merolae* does not have transmembrane ring components, the partial NE breakdown and NPC disassembly are induced by other NPC proteins. The nuclear basket, which is the most internal part of NPCs, works as a docking site for transport complexes from the nucleoplasm. For example, plant nuclear basket proteins, NUA and Nup136, and vertebrate Tpr are involved in mRNA export from the

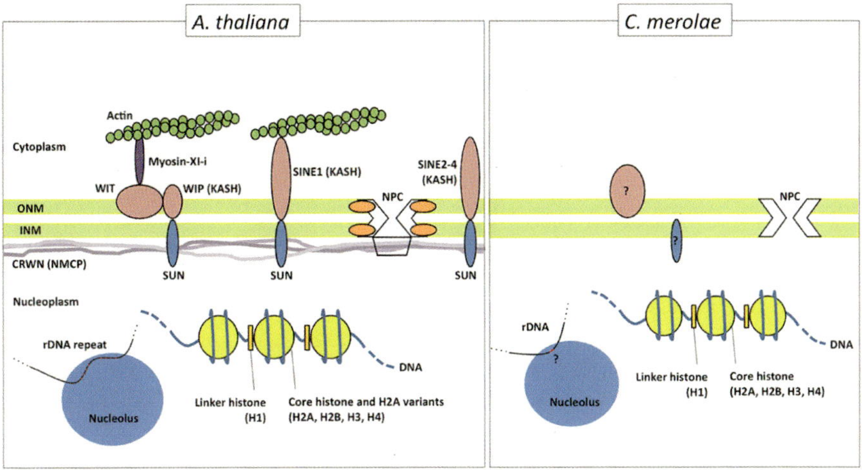


**Fig. 14.3** Comparison of NPCs between *Arabidopsis thaliana* and *Cyanidioschyzon merolae*

nucleoplasm to the cytoplasm (Shibata et al. 2002; Xu et al. 2007; Tamura et al. 2010). Additionally, *Xenopus* Nup153 corresponding to Nup136 in *A. thaliana* can interact with lamin A and B (Smythe et al. 2000; Al-Haboubi et al. 2011). As described above, *C. merolae* does not harbor lamin and CRWN, suggesting its simple structure of the NE and nuclear pore without a nuclear basket.

### 14.3.4 Nucleolus

The nucleolus can be clearly observed in the nucleus of *C. merolae* by fluorescent and electron microscopy (Maruyama et al. 2004). The nucleolus is usually formed by ribosomal DNA (rDNA) and proteins. rDNA units including the 18S, 5.8S, and 28S rRNA genes, which are essential components for translation, form high-order tandem repeats in the genome of most eukaryotes. For example, there are ~700–800 copies of rDNA repeats distributed on chromosomes 2 and 4 in *A. thaliana* (Copenhaver and Pikaard 1996). It has been proposed that these repeats are necessary for the synthesis of sufficient numbers of ribosomes and the formation of the nucleolus. Surprisingly, only three different rDNA units have been detected in the *C. merolae* genome, and each unit encodes only a single copy of the 18S, 5.8S, and 28S rRNA genes (Maruyama et al. 2004). The number of *C. merolae* rDNA units is the smallest among the eukaryotes studied to date. Major nucleolar protein homologs are conserved in *C. merolae*, suggesting that these nucleolar proteins can synthesize a sufficient number of ribosomes without multiple repeats of rDNA genes.

## 14.4 Chloroplast Division

### *14.4.1 Composition of the PD Machinery*

#### 14.4.1.1 The PD Ring

Electron microscopic observation has revealed that the PD ring appears during chloroplast division at the division site in numerous photosynthetic eukaryotes. Two (or three) types of electron-dense specialized ring structures are named as the PD ring: the outer PD ring is the main skeletal structure of the PD machinery and composed of a ring-shaped bundle of nanofilaments 5–7 nm in width on the cytosolic side of the outer envelope membrane (OEM) (Figs. 14.4 and 14.5a, b) (Miyagishima et al. 2001a; Yoshida et al. 2006). The PD ring usually refers to the outer PD ring structure. In addition, the inner PD ring is observed as a 5-nm-thick belt-like ring on the stromal side of the inner envelope membrane (IEM) (Hashimoto 1986), and a middle PD ring has occasionally been observed in the intermembrane space (Fig. 14.4) (Miyagishima et al. 1998, 1999). The outer and inner PD rings have been widely observed in the plant kingdom (Kuroiwa et al. 1998), whereas the middle PD ring has only been identified in *C. merolae* (Miyagishima et al. 1998) and the green alga *Nannochloris bacillaris* (Sumiya et al. 2008). Many angiosperms also have double rings at the constricted isthmi of dividing plastids, such as proplastids, amyloplasts, and chloroplasts. As the PD ring structure cannot be found during the early phases of plastid division in land plants, the number and electron density of the PD rings in these organisms might be too low to detect by electron microscopy. Therefore, in land plants including *Arabidopsis thaliana*, it is more difficult to study the morphological dynamics of the PD rings throughout the division of chloroplasts. Interestingly, phylogenetic study of the PD ring has suggested a clear trend in which the PD rings of the primitive unicellular eukaryotes with smaller genomes are larger than those of the multiplastidic cells of land plants; therefore, *C. merolae* is one of the most suitable organisms for studying chloroplast division with the PD ring. On the basis of these previous studies, it has been concluded that PD rings, especially the outer rings, are universal in the plant kingdom and that they play an important role in plastidokinesis. Although the molecular components of the outer PD ring have not been identified during the more than 25 years since the discovery of the PD rings, our recent study has shed light on this issue (Yoshida et al. 2010). Chemical staining screening enabled additional findings to be obtained because periodic acid (PA)-horseradish peroxidase (HRP) staining indicated that the outer PD ring is likely to contain saccharic components. Based on this finding, we performed a proteomic analysis of the isolated PD machinery fraction and identified a novel glycosyltransferase protein, later designated as plastid-dividing ring 1 (PDR1) (Yoshida et al. 2010). Interestingly, gene expression of PDR1 was specifically detected in the chloroplast division phase, and PDR1 proteins assembled a single-ring structure at the chloroplast division site (Fig. 14.5c). Ultrastructural studies

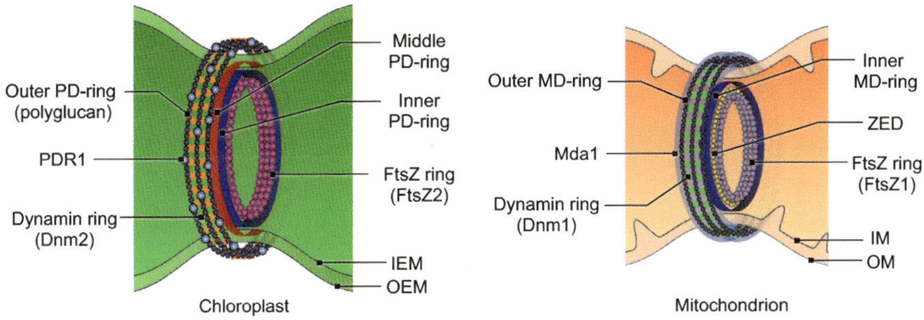


**Fig. 14.4** Schematic representations of the division machinery for chloroplasts (plastids) and mitochondria in *Cyanidioschyzon merolae*. *IEM* inner envelope membrane; *OEM* outer envelope membrane; *IM* inner membrane; *OM* outer membrane

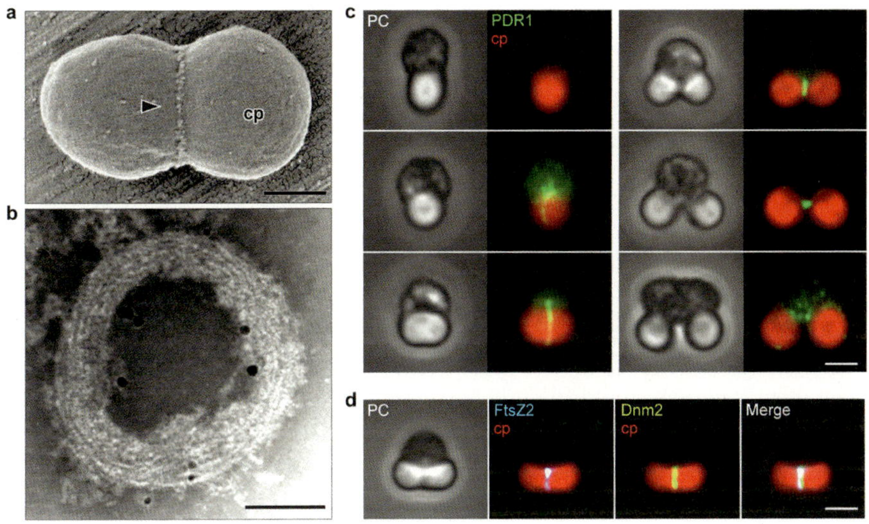


**Fig. 14.5** Chloroplast division in *Cyanidioschyzon merolae*. (**a**) A scanning electron micrograph of a dividing chloroplast. A black arrowhead indicates the PD machinery. cp, chloroplast. Scale bar, 200 nm. (**b**) An immunoelectron micrograph of isolated PD machinery. Immunogold signals indicate protein localizations of Dnm2 (*small black dots*) and FtsZ2 (*large black dots*). Scale bar, 100 nm. (**c**) Distributions of glycosyltransferase protein PDR1 through chloroplast division. Scale bar, 1 μm. (**d**) Immunolocalization of GTPase protein Dnm2 and tubulin-like protein FtsZ2. Scale bar, 1 μm. The chloroplasts (cp) emit *red* autofluorescence (Images were reproduced and modified, with permission, from Yoshida et al. 2006, 2008, 2010)

clearly showed that PDR1 proteins localized on the outer PD ring filaments. Furthermore, analyses of the components of the purified PD ring filaments revealed the existence of glucan molecules as components of the outer PD ring filament. Taking these findings together, although the mechanism of synthesis is still not clear, PDR1 is probably involved in biosynthesis of the polyglucan filaments of the PD ring. Potential orthologs of PDR1 have also been shown to be encoded in land plant genomes.

#### 14.4.1.2 The FtsZ Ring

Chloroplast FtsZ is a homolog of the bacterial fission protein FtsZ (Bi and Lutkenhaus 1991; Osteryoung and Vierling 1995; Osawa et al. 2008; Adams and Errington 2009; Miyagishima and Kabeya 2010; TerBush et al. 2013) and assembles into a ring beneath the inner envelope membrane at the division site (Figs. 14.4 and 14.5d) (Vitha et al. 2001; Mori et al. 2001; Miyagishima et al. 2001b). Both chloroplast and bacterial FtsZ proteins are composed of two functional domains, a GTP-binding domain at the N-terminal and a GTPase-activating domain at the C-terminal, and assemble protofilaments via interaction of the GTPase-activating C-terminal domain of one subunit with the GTP-binding N-terminal domain of the other to produce a completely one-dimensional polymer (Nogales et al. 1998; Löwe and Amos 1998; Oliva et al. 2004). Interestingly, whereas bacteria have one FtsZ gene in their genome, the chloroplast *FtsZ* gene underwent duplication, and these duplicated loci, now present in the nuclear genomes, are widely conserved throughout photosynthetic eukaryotes (Miyagishima et al. 2004; Schmitz et al. 2009; TerBush et al. 2013). Thus, *C. merolae* genomes contain two types of *FtsZ* gene, *FtsZ2–1* (*FtsZ2* in *A. thaliana*) and *FtsZ2–2* (*FtsZ1* in *A. thaliana*), for chloroplast division (Takahara et al. 2000; Matsuzaki et al. 2004; Miyagishima et al. 2004). Phylogenetic studies have classified the chloroplast FtsZs into two groups. One of these groups including FtsZ2–1 is more ancestral and contains a conserved C-terminal core motif similar to that of bacterial FtsZ, which is assumed to interact with other chloroplast division proteins to tether the FtsZ ring to the inner membrane. The other FtsZ group including FtsZ2–2 lacks the C-terminal core motif and probably evolved by duplication of the original *FtsZ* gene or its precursor during the establishment of chloroplasts after the endosymbiotic event. Although the mechanistic details of FtsZ ring assembly and dynamics remain unclear, our recent study of chloroplast FtsZs using the heterologous yeast system revealed that two types of chloroplast FtsZs spontaneously formed heteropolymers and then assembled into a single ring that could generate contractile force in the absence of any other related proteins. Interestingly, the chloroplast FtsZ heteropolymer contained higher kinetic dynamics for protofilament assembly, mobility, and flexibility than that of homopolymers and is structurally similar to microtubules comprised of α-tubulin and β-tubulin, suggesting that the gene duplication of chloroplast *FtsZ* led to innovation in the kinetic functions of the FtsZ ring for chloroplast division and indicates the occurrence of convergent evolution in the abilities of chloroplast FtsZs and eukaryotic tubulins (Yoshida and Mogi 2016; Yoshida et al. 2016). Consistent with molecular genetic studies using *A. thaliana*, it is suggested that the chloroplast FtsZ ring assembly in vivo is further promoted by several regulatory factors. ARC6 positively regulates FtsZ protofilament assembly by interacting with the FtsZ2–1 type of FtsZ molecules (Glynn et al. 2008). PARC6 also interacts with the FtsZ2–1 type of FtsZ molecules and induces remodeling of the FtsZ protofilaments for ring formation (Glynn et al. 2009; Zhang et al. 2015). Moreover, ARC3 (Shimada et al. 2004) and GC1 (Maple et al. 2004) negatively regulate FtsZ ring assembly, and ARC3 coordinately supports the remodeling of the FtsZ protofilament with PARC6 (TerBush and Osteryoung 2012). However, as these FtsZ-associated genes are not conserved in the *C. merolae* genome, another regulatory system with unknown factors might coordinate the assembly of the FtsZ ring in this species.

#### 14.4.1.3 The Dynamin Ring

A nuclear-encoded protein and member of the dynamin superfamily, dynamin 2 (Dnm2), in *C. merolae* (also known as ARC5/DRP5B in *A. thaliana*) is also observed as a ring structure at the chloroplast division site (Miyagishima et al. 2003) (Figs. 14.4 and 14.5d). Originally, the dynamin protein (called classical dynamin) was identified as a 100-kDa mechanochemical GTPase for the scission of clathrin-coated vesicles from the plasma membrane (Shpetner and Vallee 1989; Praefcke and McMahon 2004). The function of classical dynamin was initially determined in the paralyzed *Drosophila* mutant *shibire* (Koenig and Ikeda 1989; Chen et al. 1991; van der Bliek and Meyerowitz 1991). Temperature-sensitive mutants of classical dynamin showed the accumulation of budding synaptic vesicles on the cell membrane in nerve terminals, suggesting that classical dynamin plays a role in membrane remodeling in synaptic vesicles for endocytosis via a mechanochemical function. It is known that dynamin family proteins feature several functional domain subunits as follows: a large GTPase domain (approximately 300 amino acids), a middle domain, a pleckstrin homology (PH) domain, a GTPase effector domain (GED), and a carboxy-terminal proline-rich domain (PRD) (Praefcke and McMahon 2004; Faelber et al. 2011; Ford et al. 2011). However, although their molecular mass is similar to that of classical dynamin, chloroplast dynamins, such as Dnm2, contain only a GTPase domain at the amino-terminal region, which is similar to the position of the GTPase domain of classical dynamin, while the rest of the protein region does not show any clear similarity with known protein functional domains. The functional importance of a dynamin-related protein in chloroplast division was first shown by a study utilizing *C. merolae* cells (Miyagishima et al. 2003). Interestingly, chloroplast dynamin proteins formed a discontinuous ring structure at the chloroplast division site in the early phase of chloroplast division, and dynamin patches appeared to link up and assembled into a single ring in the later phase of division. Similar results were also obtained in *A. thaliana*. Mutants of ARC5 exhibit defects in chloroplast division and GFP-fused ARC5 localized as a discontinuous ring at the chloroplast division site (Gao et al. 2003). Surprisingly, the group of chloroplast division dynamins is phylogenetically related to a group of cytokinesis-relating dynamins (Miyagishima et al. 2008). Considering that the *C. merolae* genome contains only two dynamin-related proteins, namely, Dnm2 for chloroplast division and Dnm1 for mitochondrial division, these findings raise the question of which was the original function of the ancestral dynamin protein.

#### 14.4.1.4 Assembly and Dynamics of the PD Machinery

The significant functions of the dynamic trio of rings, PD, FtsZ, and dynamin, in chloroplast division have been clarified by a series of studies using *C. merolae* cells. Specifically, it has been shown that the chloroplast divides by the action of supramolecular complexes comprising this dynamic trio of rings as the PD machinery across the double membranes of the chloroplast (Fig. 14.4). Although the molecular function of each ring in the PD machinery has not been fully clarified, a series of studies using *C. merolae* cells has begun to uncover the molecular mechanisms by which this machinery functions. Formation of the PD machinery is executed during a specific phase of the cell-division cycle and is regulated by the cell cycle-dependent gene expression of components of the PD machinery. As in bacteria, assembly of the FtsZ ring in the stromal region is the first known event that occurs at the chloroplast division site. Then, the inner PD ring appears at the division site beneath the inner envelope membrane. After formation of the inner rings, the interaction of FtsZ and some membrane proteins might transfer information about the location of the FtsZ ring to the outside of the chloroplast. Based on this information, the glycosyltransferase protein PDR1 probably binds to the outer membrane. PDR1 has sequence similarity to glycogenin, which acts as a priming protein for glycogen synthesis; therefore, it is now hypothesized that PDR1 can elongate the glucan chain to synthesize PD ring filaments from UDP-glucose molecules, analogous to the synthesis of glycogen (Yoshida et al. 2010). Another possibility has also been suggested by a series of electron microscopic observations in *C. caldarium*. Small electron-dense vesicles were visualized along the line at the division site before the assembly of the outer PD ring. Interestingly, the boundary of the vesicles appeared to coincide with one end of a PD ring filament, suggesting that PD ring filaments might be biosynthesized on the surface of these vesicles from their components (Kuroiwa et al. 1998).

In chloroplasts, these inner and outer rings appeared to be linked to each other through nanoscale holes that appeared on the groove of the division site, as revealed by scanning electron microscopy. Recent yeast two-hybrid studies using several chloroplast division proteins in *A. thaliana* showed that the FtsZ ring interacts with the inner membrane proteins ARC6 and PARC6 in the stromal region, and ARC6 further interacts with the outer membrane protein PDV2 in the intermembrane space (Glynn et al. 2008, 2009). However, these genes were not found in the *C. merolae* genome (Miyagishima and Kabeya 2010), indicating that functional analogs of these genes might play a similar role in *C. merolae* and/or primitive photosynthetic eukaryotes. In the final step of the PD machinery assembly, dynamin protein molecules cross-link the outer PD ring filaments. A series of assembly steps was confirmed by a range of knockdown experiments of the PDR1 gene in

*C. merolae*. When *PDR1* gene expression was downregulated, the FtsZ ring assembled, but dynamin proteins were not recruited to the division site. Thus, FtsZ ring formation does not depend on the subsequent formation of the PD and dynamin rings, and dynamin localization relies on the formation of the outer PD ring.

There is still much to reveal about the contractile mechanism of the PD machinery, so many studies on this issue are being undertaken. Sequential electron microscopic (EM) observations of cells in chloroplast division showed that the thickness of the inner PD ring does not change, but its volume decreases at a constant rate during contraction (Miyagishima et al. 2001b). On the other hand, the width and thickness of the outer PD ring monotonically increase during contraction, retaining its density and volume, in *C. merolae* (Miyagishima et al. 2001a, 2001b). Similar EM observations of the transition of the outer PD ring have also been made in the green alga *N. bacillaris* (Sumiya et al. 2008) and in land plants (Kuroiwa et al. 2002). These morphological transitions of the outer PD ring during chloroplast division have prompted the idea that the outer PD ring filaments can slide and squeeze chloroplast membranes. The establishment of a technique for isolating the PD machinery from *C. merolae* cells in an intact form provided a major breakthrough on this issue (Yoshida et al. 2006). Isolated PD machinery formed not only a circular structure but also formed super-twisted and spiral structures featuring both clockwise and anticlockwise spirals. The existence of isolated PD machinery in these twisted states indicates the motive force involved in the contraction of chloroplast membranes. Indeed, it was observed that membrane-free PD machinery autonomously became contracted when chloroplast membranes were dissolved by detergent treatment. In addition, reconstituted chloroplast FtsZ rings also displayed contractile ability. Although the detailed molecular mechanism involved in this is still unclear, the contraction process of the FtsZ ring accompanies a transition in the FtsZ protofilament from a less dynamic to a more dynamic state. Therefore, it is assumed that the dynamic transition of protofilament states induces a decrease in the average protofilament length in the FtsZ ring and thus causes contraction (Yoshida et al. 2016). Thus, some challenging studies were performed to reveal how the PD machinery is transformed morphologically during contraction and which component proteins of the PD machinery play a key role in generating this force.

## 14.5 Division of Mitochondria

### *14.5.1 Composition of the MD Machinery*

#### 14.5.1.1 The MD Ring

In the slime mold *Physarum polycephalum*, an electron-dense specialized ring-like structure was found on the bridge of daughter mitochondria at the final phase of mitochondrial division (Kuroiwa et al. 1977, 1994, 2006). The ring structure was finally designated as the outer mitochondrion-dividing (MD) ring and seemed to constrict the mitochondrial division site, as determined by electron microscopic observations of *C. merolae* cells (Kuroiwa et al. 1995). By ultrastructural studies, it was concluded that the outer MD ring is composed of a bundle of nanofilaments 4–6 nm in width and is assembled on the outer mitochondrial membrane on the cytoplasmic side (Figs. 14.4 and 14.6a b). In addition, a thin belt-shaped ring, the inner MD ring, was also discovered beneath the inner mitochondrial membrane in the matrix region (Fig. 14.4). The cross-sectional area of the outer MD ring is about 50 nm wide in the initial phase of mitochondrial division and gradually increases to

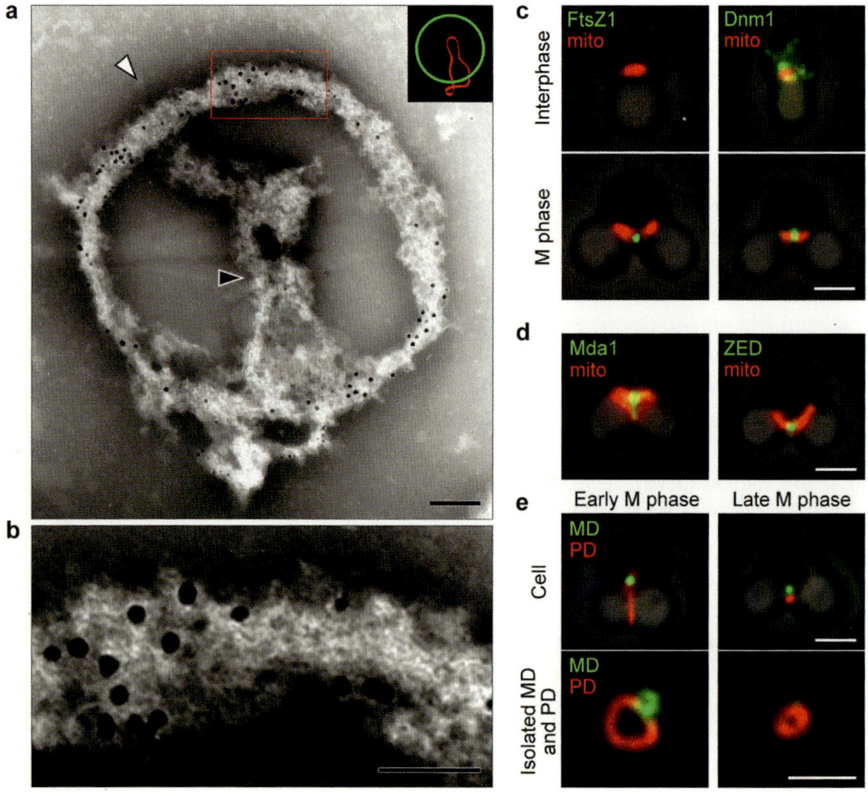


**Fig. 14.6** Mitochondrial division in *Cyanidioschyzon merolae*. (**a**) An immunoelectron micrograph of an isolated MD/PD machinery complex. A *white* arrowhead indicates the MD machinery, and a *black* arrowhead indicates the PD machinery. Inset shows the localizations of the MD machinery (*red*) and the PD machinery (*green*) in the image. Scale bar, 100 nm. (**b**) Enlarged image of the boxed area in (**a**). Immunogold signals indicate localization of Mda1 protein. Scale bar, 100 nm. (**c**) Immunolocalizations of tubulin-like protein FtsZ1 and GTPase protein Dnm1 in the cells in the interphase (*top*) and M phase (*bottom*). Scale bar, 1 μm. (**b**) Immunolocalizations of WD40 protein Mda1 and coiled-coil protein ZED. Scale bar, 1 μm. Mitochondria (mito) were immunostained with anti-mitochondrial EF-tu antibody (for FtsZ1, Dnm1, and ZED) or anti-mitochondrial porin antibody (for Mda1). Some images were reproduced and modified, with permission, from Yoshida et al. (2009)

more than 80 nm wide with increasing thickness during mitochondrial division (Miyagishima et al. 1999), while the inner MD ring appears to maintain constant width and thickness and disappears in the final phase. Recent studies on yeast cells and human cultured cells indicated the presence of an electron-dense structure at a narrow bridge between two mitochondria, which is presumably executing the process of fission (Yoon et al. 2003; Ingerman et al. 2005). Thus, it is assumed that fungi and animals also contain the MD-ring-like apparatus for mitochondrial division. Elucidation of the detailed structure of the MD ring has been progressing in several projects on *C. merolae*, which showed that the outer MD ring is composed of a bundle of nanofilaments that are narrower than the PD ring filament (Yoshida et al. 2009). At present, the components of both outer and inner MD rings are still unknown, which is currently a particularly focus of investigation.

#### 14.5.1.2 The FtsZ Ring

A series of findings on chloroplast FtsZ have highlighted the need to explore FtsZ homologs involved in mitochondrial division in the genome. A wide range of genomic surveys did not identify any homologous genes that resemble FtsZ in the genomes of classical model organisms such as yeast, animals, and higher plants, while some lower eukaryotes including *C. merolae* do possess FtsZ homologous genes in their genome. It has now been revealed that the primitive red alga *C. merolae* (Takahara et al. 2000), the heterokont alga *Mallomonas splendens* (Beech et al. 2000), and the slime mold *Dictyostelium discoideum* (Gilson and Beech 2001) have an alpha-proteobacteria-type FtsZ gene encoded in their nuclear genome. Interestingly, this FtsZ is specifically expressed during the mitochondrial division phase and localized at the mitochondrial division site in *C. merolae* (Nishida et al. 2003; Fujiwara et al. 2009) (Fig. 14.6c), suggesting that the FtsZ ring was originally involved in mitochondrial division in the same manner as in chloroplasts, but has been lost from higher eukaryotes over the course of evolution.

#### 14.5.1.3 The Dynamin Ring

The dynamin superfamily is involved in various biological processes as follows: cleavage of the budding clathrin-coated vesicles from the plasma membrane, vesicle trafficking from the Golgi, plant cell plate formation, resistance to viral infection, mitochondrial fusion, and organelle division (Praefcke and McMahon 2004). A group of dynamin-related proteins, Dnm1 (also known as Drp1/Dnm1L/DLP1 in mammals and DRP3A/ADL2a and DRP3B/ADL2b in *A. thaliana*), play a critical role in mitochondrial division in various eukaryotes, and homologs of these proteins have been widely identified in the yeast (Bleazard et al. 1999), animal (Smirnova et al. 1998), and plant kingdoms (Arimura and Tsutsumi 2002; Nishida et al. 2003). Similar to classical dynamin, Dnm1 also contains the GTPase domain, the middle domain, and the GED domain but lacks the domains of PH and PRD, which enable interaction with several specific proteins. Instead, a variable domain (also known as insert B) is found between the middle and GED domains and is likely to be required for binding to the target membrane (Mears et al. 2011; Elgass et al. 2013; van der Bliek et al. 2013). The necessity and intracellular function of Dnm1 in mitochondrial division are seemingly universal, as all of them act in similar ways, in which the Dnm1 proteins accumulate at the cytosolic side of the mitochondrial division site (Nishida et al. 2003; Kuroiwa et al. 2006; Friedman and Nunnari 2014). Interestingly, *C. merolae* Dnm1 is probably a multifunctional protein. It is expressed continuously throughout the cell cycle (Nishida et al. 2005, 2007), not similar to a restricted expression pattern of Dnm2 during chloroplast division phase (Miyagishima et al. 2003; Yoshida et al. 2010), and these proteins are transferred between the cytosolic region in interphase and the mitochondrial division site in M phase (Fig. 14.6c). Detected as 10–20 patch-like structures in the cytoplasm during interphase, Dnm1 proteins probably mediate vesicle transport (Nishida et al. 2003). In addition, recent studies have shown that Dnm1 is involved in not only mitochondrial division but also peroxisomal division, which has recently been investigated further utilizing *C. merolae* (see also Chap. 15) (Imoto et al. 2013). Taking these findings together, it is considered highly likely that *C. merolae* Dnm1 plays roles in the following three intracellular functions: mitochondrial division, vesicle transport, and peroxisomal division.

#### 14.5.1.4 Other Components for Mitochondrial Division

In addition to the three fundamental ring structures of the MD, FtsZ, and dynamin rings, several specific proteins for mitochondrial division were identified in studies utilizing *C. merolae*. A WD40-repeat protein, Mda1, was identified as a component of the MD machinery (Nishida et al. 2007) (Fig. 14.6d). Immunocytological studies showed that Mda1 proteins localized on the MD ring in addition to Dnm1 and were subjected to cell-cycle-dependent phosphorylation. Given that Mda1 proteins localize at the mitochondrial division site before localization of Dnm1 at the site, which was further confirmed by a series of cell-cycle inhibition experiments, Mda1 proteins are likely to be mediators or receptor proteins for forming stable linking structures of Dnm1 proteins and the MD ring filaments (Nishida et al. 2007). Other similar WD40-repeat proteins have also been identified, such as Mdv1 (Fekkes et al. 2000; Tieu and Nunnari 2000) and Caf4 (Griffin et al. 2005) in yeast, which are also involved in mitochondrial division and interact with Dnm1 proteins. The mitochondrial outer membrane protein Fis1 is widely conserved in eukaryotes and is also known as a factor involved in mitochondrial division (Mozdy et al. 2000). Recently, a proteomic study of the isolated MD machinery also identified another novel mitochondrial division protein, ZED (Yoshida et al. 2009) (Fig. 14.6d), which partly resembles the bacterial division protein ZapA (Gueiros-Filho and Losick 2002; Adams and Errington 2009) structurally and in terms of its ability to bind to the FtsZ ring. Not found in the mitochondrial division system except the FtsZ ring, the bacterial FtsZ ring appears to be incorporated into the MD machinery, and the other bacterial components might be altered like the ZED protein or eliminated from the mitochondrial division system. The lack of mitochondrial FtsZ in higher organisms supports the assertion that the inner part of the MD machinery was simplified over the course of evolution.

### *14.5.2 Assembly and Dynamics of the MD Machinery*

According to the above findings, the MD machinery is assembled from the inner rings and then the outer rings in *C. merolae*. The mitochondrial FtsZs are expressed in the $G_2$ phase and migrate into the matrix region of the mitochondrion via the guidance of the N-terminal mitochondrial transit peptide sequence. After this translocation, mitochondrial FtsZ proteins polymerize and form a single ring, namely, the mitochondrial FtsZ ring, beneath the inner mitochondrial membrane (Nishida et al. 2003). In this phase, the FtsZ-associated protein ZED is proteolytically processed to a short form by cleavage of part of its protein sequence; thereafter, the short form of ZED interacts with the FtsZ ring (Yoshida et al. 2009). Downregulation of *ZED* showed that mitochondrial FtsZs were expressed but could not form a ring structure in these cells; as a result, only one of the two daughter cells inherited the nondividing mitochondrion. These results indicate that completion of the inner rings of the MD machinery is a critical process for the progression of mitochondrial division. Although the molecular components of the inner MD ring and kinetic properties of the mitochondrial FtsZ ring have not been revealed, current projects such as ex vivo experiments of mitochondrial FtsZs should provide clues to resolving these issues.

Assembly of the outer rings of the MD machinery commences after the accomplishment of inner ring formation. Along the position of the inner rings, the outer MD ring is formed on the outer mitochondrial membrane, and then, Mda1 proteins accumulate on the outer MD ring. Immunofluorescence signals of Mda1 proteins are detected as a ring structure without Dnm1 proteins during the first half of mitochondrial division, and then they are gradually focused into a spot that coincides with the signals of Dnm1 proteins in the second half of the division process. Interestingly, Mda1 proteins were found to be phosphorylated specifically in this phase, suggesting that the activation and/or progression of mitochondrial division is strictly regulated by posttranslational modifications (Nishida et al. 2007). Although various components involved in mitochondrial division have been identified, the molecular components of the outer and inner MD rings have not been revealed. Whereas a protocol for isolating the PD machinery has been developed for these cells, the isolation of the MD machinery was not easy due to the fact that mitochondria are smaller and more frangible than chloroplasts. To overcome this problem, isolation of the MD machinery was carried out by using the binding state of the MD and PD machinery; MD and PD machineries form a complex structure before the contraction phase, which is called the MD/PD machinery complex (Fig. 14.6e) (Yoshida et al. 2009). The isolated MD machinery also displayed circular, super-twisted and spiral forms, like those of the isolated PD machinery, suggesting that the MD machinery can generate the motive force for mitochondrial division in a similar manner to the PD machinery. In addition, the proteomic profile of the fraction revealed approximately 50 proteins, including 20 unknown ones, suggesting that further research on these candidate proteins should disclose the molecular identity of the components of the outer and inner MD rings.

## 14.6 Conclusions and Perspectives

The structure and dynamics of double-membrane-bounded organelles in *C. merolae* are very simple. This simplicity has been available for revealing the mechanism of double-membrane-bounded organelles. Eukaryotic fundamental nuclear structures and components are also conserved in *C. merolae*. The core and linker histones, H2A.Z, and a centromeric histone variant CenH3 are preserved, but H2A variants including H2A.X, H2A.M, and H2A.W are absent. In addition, NPCs and nucleolar proteins are partly preserved, whereas almost all homologs of nuclear lamina and nuclear envelope proteins are lacking. A comparison of the structural characteristics of the nucleus between *A. thaliana* and *C. merolae* is shown in Fig. 14.2. This demonstrates that the nucleus of *C. merolae* has quite a simplified range of components, representing the minimal nuclear proteins in eukaryotes.

Studies utilizing *C. merolae* cells have revealed the system of chloroplast division by the supramolecular complex called the PD machinery (Yoshida et al. 2012) (Fig. 14.4). Multi-omics approaches have also begun to decode the components and molecular mechanisms of the PD machinery at single-molecule resolution. Continuation of such research should reveal more of the molecular details of the mechanism of chloroplast division. The appearance of several candidate proteins from the omics results indicates both significant similarities and differences between the systems of chloroplast and mitochondrial division. The results have also revealed additional uncharacterized proteins that make up the PD machinery. Ultrastructural observations of collapsed PD ring filament bundles revealed that PD ring filaments appeared to be covered by a thin layer structure (Yoshida et al. 2006); thus, not only PDR1 but also additional factors might be involved in the assembly and maintenance of the bundle structures of the PD ring filaments. Furthermore, these candidate proteins should provide clues to elucidating the composition of the inner PD ring, as its components and functional importance have remained completely unknown. The primary structures of the candidate proteins indicate that some of these contain putative transmembrane domains, suggesting that these proteins mediate the formation of a structure that links the inner and outer rings through the chloroplast membranes. Functional characterization of these proteins should reveal the evolutionary relationships of chloroplast division systems between primitive photosynthetic eukaryotes and higher plants. As the development of a technique to isolate the intact PD machinery enables us to perform various types of in vitro experiments, further research on the isolated PD machinery using microfluidic devices should help us to clarify molecular mechanisms in chloroplast division. On the other hand, PDR1 appears to be involved in the formation of the PD ring (ring-shaped polyglucan filaments); however, the molecular mechanism of filament synthesis remains unclear. Therefore, characterization of the enzymatic mechanisms of the PDR1 protein should provide important information about the intracellular dynamics of the polysaccharide filament.

Insights into the mechanisms of mitochondrial and chloroplast division help us to understand the long-standing questions of evolutional origin of them surprisingly; even though these organelles have different endosymbiotic origins, their division systems are strikingly similar. This indicates that very similar evolutionary processes occurred in both mitochondria and chloroplasts during their establishment. However, several considerable differences in their division systems appear to reflect their development into specialized organelles. One significant difference is the lack of mitochondrial FtsZ in higher organisms (Miyagishima et al. 2004; TerBush et al. 2013); therefore, the molecular mechanism behind the pinching of the inner mitochondrial membrane in these organisms is completely unknown. Alongside the loss of FtsZ, a number of alternative Dnm1 receptor proteins have been shown to be involved in mitochondrial division in mammals, such as Mff (Gandre-Babbe and van der Bliek 2008; Liu and Chan 2015), Mid49, Mid51 (Palmer et al. 2011; Losón et al. 2013), and GDAP1 (Niemann et al. 2005) which are only encoded in metazoans. Since mitochondria in higher organisms frequently undergo fusion and fission, simplification of the mechanism of mitochondrial division regarding the interior of this organelle might have been necessary to maintain the homeostasis of mitochondrial functions. This is supported by the discovery of several proteins associated with mitochondrial fusion. Interestingly, fungi and animals contain not only Dnm1 but also some additional dynamin-related proteins for mitochondrial dynamics. Mgm1/Opa1 is a dynamin-related protein that is considered to be an inner membrane fusion component in yeast and animals (Roy et al. 2015). In addition to the finding of Mgm1/Opa1, other dynamin-related proteins were found to be involved in mitochondrial outer membrane fusion, the yeast version of which was named Fzo1 (known as Mfn1 and Mfn2 in humans) (van der Bliek et al. 2013; Roy et al. 2015). Thus, given frequent mitochondrial fusion and fission in these organisms, the divergence of mitochondrial dynamin-related proteins in metazoans provides the impression that the mechanism of mitochondrial dynamics has been significantly developed when these organisms obtained the ability for mitochondrial fusion. On the other hand, some lower eukaryotes including *C. merolae* seem to have retained primitive or fundamental mechanisms for mitochondrial division, such that it exhibits remarkable similarity to the mechanisms of division in chloroplasts. Further studies of mitochondrial proliferation utilizing *C. merolae* should enable us to elucidate how eukaryotic ancestral hosts engulfed endosymbionts and converted them into mitochondria and chloroplasts.

**Acknowledgments** This work was supported by a Human Frontier Science Program Long Term Fellowship (No. LT000356/2011-L) to Y.Y. and a grant for XFEL key technology and the X-ray Free Electron Laser Priority Strategy Program from MEXT to Sa.M. (Tokyo University of Science), MEXT/JSPS KAKENHI No. JP26291067 and JP15H05962 to Sa.M. (Tokyo University of Science), and JP15K18562 to Sh.M. (Tohoku University).

## References

Adams DW, Errington J (2009) Bacterial cell division: assembly, maintenance and disassembly of the Z ring. Nat Rev Microbiol 7(9):642–653. https://doi.org/10.1038/nrmicro2198

Al-Haboubi T, Shumaker DK, Köser J et al (2011) Distinct association of the nuclear pore protein Nup153 with A- and B-type lamins. Nucleus 2(5):500–509. https://doi.org/10.4161/nucl.2.5.17913

Archibald JM (2015) Endosymbiosis and eukaryotic cell evolution. Curr Biol 25:R911–R921. https://doi.org/10.1016/j.cub.2015.07.055

Arimura S, Tsutsumi N (2002) A dynamin-like protein (ADL2b), rather than FtsZ, is involved in *Arabidopsis* mitochondrial division. Proc Natl Acad Sci U S A 99(8):5727–5731. https://doi.org/10.1073/pnas.082663299

Baum DA, Baum B (2014) An inside-out origin for the eukaryotic cell. BMC Biol 12:1. https://doi.org/10.1186/s12915-014-0076-2

Beech PL et al (2000) Mitochondrial FtsZ in a chromophyte alga. Science 287(5456):1276–1279

Berk JM, Tifft KE, Wilson KL (2013) The nuclear envelope LEM-domain protein emerin. Nucleus 4(4):298–314. https://doi.org/10.4161/nucl.25751

Bi EF, Lutkenhaus J (1991) FtsZ ring structure associated with division in *Escherichia coli*. Nature 354(6349):161–164. https://doi.org/10.1038/354161a0

Bleazard W et al (1999) The dynamin-related GTPase Dnm1 regulates mitochondrial fission in yeast. Nat Cell Biol 1(5):298–304. https://doi.org/10.1038/13014

Burma S, Chen BP, Murphy M et al (2001) ATM phosphorylates histone H2AX in response to DNA double-strand breaks. J Biol Chem 276:42462–42467. https://doi.org/10.1074/jbc.C100466200

Cavalier-Smith T (1987) The simultaneous symbiotic origin of mitochondria, chloroplasts, and microbodies. Ann N Y Acad Sci 503:55–71

Chen MS et al (1991) Multiple forms of dynamin are encoded by shibire, a *Drosophila* gene involved in endocytosis. Nature 351(6327):583–586. https://doi.org/10.1038/351583a0

Copenhaver GP, Pikaard CS (1996) RFLP and physical mapping with an rDNA-specific endonuclease reveals that nucleolus organizer regions of *Arabidopsis thaliana* adjoin the telomeres on chromosomes 2 and 4. Plant J 9:259–272. https://doi.org/10.1046/j.1365-313X.1996.09020259.x

De Souza CPC, Osmani SA (2007) Mitosis, not just open or closed. Eukaryot Cell 6:1521–1527. https://doi.org/10.1128/EC.00178-07

DuBois KN et al (2012) NUP-1 is a large coiled-coil nucleoskeletal protein in trypanosomes with lamin-like functions. PLoS Biol 10(3):e1001287. https://doi.org/10.1371/journal.pbio.1001287

Elgass K, Pakay J, Ryan MT, Palmer CS (2013) Recent advances into the understanding of mitochondrial fission. Biochim Biophys Acta, Mol Cell Res 1833(1):150–161. https://doi.org/10.1016/j.bbamcr.2012.05.002

Faelber K et al (2011) Crystal structure of nucleotide-free dynamin. Nature 477(7366):556–560. https://doi.org/10.1038/nature10369

Fekkes P, Shepard KA, Yaffe MP (2000) Gag3p, an outer membrane protein required for fission of mitochondrial tubules. J Cell Biol 151(2):333–340

Ford MGJ, Jenni S, Nunnari J (2011) The crystal structure of dynamin. Nature 477(7366):561–566. https://doi.org/10.1038/nature10441

Friedman JR, Nunnari J (2014) Mitochondrial form and function. Nature 505(7483):335–343. https://doi.org/10.1038/nature12985

Fujiwara T et al (2009) Periodic gene expression patterns during the highly synchronized cell nucleus and organelle division cycles in the unicellular red alga *Cyanidioschyzon merolae*. DNA Res 16(1):59–72. https://doi.org/10.1093/dnares/dsn032

Fujiwara T, Tanaka K, Kuroiwa T, Hirano T (2013) Spatiotemporal dynamics of condensins I and II: evolutionary insights from the primitive redalga *Cyanidioschyzon merolae*. Mol Biol Cell 24(16):2515–2527. https://doi.org/10.1091/mbc.E13-04-0208

Galy V et al (2008) A role for gp210 in mitotic nuclear-envelope breakdown. J Cell Sci 121:317–328. https://doi.org/10.1242/jcs.022525

Gandre-Babbe S, van der Bliek AM (2008) The novel tail-anchored membrane protein Mff controls mitochondrial and peroxisomal fission in mammalian cells. Mol Biol Cell 19(6):2402–2412. https://doi.org/10.1091/mbc.E07-12-1287

Gao H, Kadirjan-Kalbach D, Froehlich JE, Osteryoung KW (2003) ARC5, a cytosolic dynamin-like protein from plants, is part of the chloroplastdivision machinery. Proc Natl Acad Sci U S A 100(7):4328–4333. https://doi.org/10.1073/pnas.0530206100

Gillham NW, Boynton JE, Hauser CR (1994) Translational regulation of gene expression in chloroplasts and mitochondria. Annu Rev Genet 28:71–93. https://doi.org/10.1146/annurev.ge.28.120194.000443

Gilson PR, Beech PL (2001) Cell division protein FtsZ: running rings around bacteria, chloroplasts and mitochondria. Res Microbiol 152(1):3–10

Glynn JM, Froehlich JE, Osteryoung KW (2008) *Arabidopsis* ARC6 coordinates the division machineries of the inner and outer chloroplast membranes through interaction with PDV2 in the intermembrane space. Plant Cell 20(9):2460–2470. https://doi.org/10.1105/tpc.108.061440

Glynn JM et al (2009) PARC6, a novel chloroplast division factor, influences FtsZ assembly and is required for recruitment of PDV1 during chloroplast division in *Arabidopsis*. Plant J 59(5):700–711. https://doi.org/10.1111/j.1365-313X.2009.03905.x

Griffin EE, Graumann J, Chan DC (2005) The WD40 protein Caf4p is a component of the mitochondrial fission machinery and recruits Dnm1p to mitochondria. J Cell Biol 170(2):237–248. https://doi.org/10.1083/jcb.200503148

Gueiros-Filho FJ, Losick R (2002) A widely conserved bacterial cell division protein that promotes assembly of the tubulin-like protein FtsZ. Genes Dev 16(19):2544–2556. https://doi.org/10.1101/gad.1014102

Hashimoto H (1986) Double ring structure around the constricting neck of dividing plastids of *Avena sativa*. Protoplasma 135(2–3):166–172

Hayashi T, Rizzuto R, Hajnoczky G, T-P S (2009) MAM: more than just a housekeeper. Trends Cell Biol 19:81–88. https://doi.org/10.1016/j.tcb.2008.12.002

Hoppins S (2014) The regulation of mitochondrial dynamics. Curr Opin Cell Biol 29:46–52. https://doi.org/10.1016/j.ceb.2014.03.005

Imoto Y et al (2013) Single-membrane-bounded peroxisome division revealed by isolation of dynamin-based machinery. Proc Natl Acad Sci U S A 110(23):9583–9588. https://doi.org/10.1073/pnas.1303483110

Ingerman E et al (2005) Dnm1 forms spirals that are structurally tailored to fit mitochondria. J Cell Biol 170(7):1021–1027. https://doi.org/10.1083/jcb.200506078

Kalashnikova AA, Rogge RA, Hansen JC (2016) Linker histone H1 and protein-protein interactions. Biochim Biophys Acta 1859(3):455–461. https://doi.org/10.1016/j.bbagrm.2015.10.004

Kato S, Imoto Y, Ohnuma M et al (2011) Aurora kinase of a red alga *Cyanidioschyzon merolae* is related to both mitochondrial division and mitotic spindle formation. Cytologia 76:465–472. https://doi.org/10.1508/cytologia.76.455

Kawashima T et al (2015) Diversification of histone H2A variants during plant evolution. Trends Plant Sci 20(7):419–425. https://doi.org/10.1016/j.tplants.2015.04.005

Koenig JH, Ikeda K (1989) Disappearance and reformation of synaptic vesicle membrane upon transmitter release observed under reversible blockage of membrane retrieval. J Neurosci 9(11):3844–3860

Krüger A (2012) Characterization of NE81, the first lamin-like nucleoskeleton protein in a unicellular organism. Mol Biol Cell 23(2):360–370. https://doi.org/10.1091/mbc.E11-07-0595

Kuroiwa T (1998) The primitive red algae *Cyanidium caldarium* and *Cyanidioschyzon merolae* as model system for investigating the dividing apparatus of mitochondria and plastids. BioEssays 20(4):344–354. https://doi.org/10.1002/(SICI)1521-1878(199804)20:4<344::AIDBIES11>3.0.CO;2-2

Kuroiwa T, Ohta T, Kuroiwa H, Shigeyuki K (1994) Molecular and cellular mechanisms of mitochondrial nuclear division and mitochondriokinesis. Microsc Res Tech 27(3):220–232. https://doi.org/10.1002/jemt.1070270304

Kuroiwa T et al (1995) Mitochondria-dividing ring: ultrastructural basis for the mechanism of mitochondrial division in *Cyanidioschyzon merolae*. Protoplasma 186:12–23

Kuroiwa T et al (1998) The division apparatus of plastids and mitochondria. Int Rev Cytol 181:1–41

Kuroiwa H, Mori T, Takahara M et al (2002) Chloroplast division machinery as revealed by immunofluorescence and electron microscopy. Planta 215(2):185–190. https://doi.org/10.1007/s00425-002-0734-4

Kuroiwa T et al (2006) Structure, function and evolution of the mitochondrial division apparatus. Biochim Biophys Acta 1763(5–6):510–521. https://doi.org/10.1016/j.bbamcr.2006.03.007

Lane N, Martin W (2010) The energetics of genome complexity. Nature 467(7318):929–934. https://doi.org/10.1038/nature09486

Lee J, Kim DH, Hwang I (2014) Specific targeting of proteins to outer envelope membranes of endosymbiotic organelles, chloroplasts, and mitochondria. Front Plant Sci. https://doi.org/10.3389/fpls.2014.00173

Liu R, Chan DC (2015) The mitochondrial fission receptor Mff selectively recruits oligomerized Drp1. Mol Biol Cell:1–27. https://doi.org/10.1091/mbc.E15-08-0591

Losón OC, Song Z, Chen H, Chan DC (2013) Fis1, Mff, MiD49, and MiD51 mediate Drp1 recruitment in mitochondrial fission. Mol Biol Cell 24(5):659–667. https://doi.org/10.1091/mbc.E12-10-0721

Löwe J, Amos LA (1998) Crystal structure of the bacterial cell-division protein FtsZ. Nature 391(6663):203–206. https://doi.org/10.1038/34472

Maple J et al (2004) GIANT CHLOROPLAST 1 is essential for correct plastid division in *Arabidopsis*. Curr Biol 14(9):776–781. https://doi.org/10.1016/j.cub.2004.04.031

Martin WF, Garg S, Zimorski V (2015) Endosymbiotic theories for eukaryote origin. Phil Trans R Soc B 370:20140330. https://doi.org/10.1098/rstb.2014.0330

Maruyama S et al (2004) The minimal eukaryotic ribosomal DNA units in the primitive red alga *Cyanidioschyzon merolae*. DNA Res 11(2):83–91

Maruyama S, Kuroiwa H, Miyagishima S-Y et al (2007) Centromere dynamics in the primitive red alga *Cyanidioschyzon merolae*. Plant J 49:1122–1129. https://doi.org/10.1111/j.1365-313X.2006.03024.x

Masuda K et al (1997) Peripheral framework of carrot cell nucleus contains a novel protein predicted to exhibit a long alpha-helical domain. Exp Cell Res 232:173–181. https://doi.org/10.1006/excr.1997.3531

Matsuanga S, Katagiri Y, Nagashima Y et al (2013) New insights into the dynamics of plant cell nuclei and chromosomes. Int Rev Cell Mol Biol 305:253–301. https://doi.org/10.1016/B978-0-12-407695-2.00006-8

Matsuzaki M et al (2004) Genome sequence of the ultrasmall unicellular red alga *Cyanidioschyzon merolae* 10D. Nature 428(6983):653–657. https://doi.org/10.1038/nature02398

McKeon FD, Kirschner MW, Caput D (1986) Homologies in both primary and secondary structure between nuclear envelope and intermediate filament proteins. Nature 319(6053):463–468. https://doi.org/10.1038/319463a0
Mears JA et al (2011) Conformational changes in Dnm1 support a contractile mechanism for mitochondrial fission. Nat Struct Mol Biol 18(1):20–26. https://doi.org/10.1038/nsmb.1949
Meinke P, Schirmer EC (2015) LINC'ing form and function at the nuclear envelope. FEBS Lett 589:2514–2521. https://doi.org/10.1016/j.febslet.2015.06.011
Misumi O et al (2005) *Cyanidioschyzon merolae* genome. A tool for facilitating comparable studies on organelle biogenesis in photosynthetic eukaryotes. Plant Physiol 137(2):567–585. https://doi.org/10.1104/pp.104.053991
Mita T, Kanbe T, Tanaka K, Kuroiwa T (1986) A ring structure around the dividing plane of the *Cyanidium caldarium* chloroplast. Protoplasma 130(2–3):211–213
Miyagishima S-Y, Kabeya Y (2010) Chloroplast division: squeezing the photosynthetic captive. Curr Opin Microbiol 13(6):738–746. https://doi.org/10.1016/j.mib.2010.10.004
Miyagishima S-Y et al (1998) Identification of a triple ring structure involved in plastid division in the primitive red alga *Cyanidioschyzon merolae*. J Electron Microsc 47(3):269–272. https://doi.org/10.1093/oxfordjournals.jmicro.a023589
Miyagishima S, Itoh R, Toda K et al (1999) Real-time analyses of chloroplast and mitochondrial division and differences in the behavior of their dividing rings during contraction. Planta 207(3):343–353. https://doi.org/10.1007/s004250050491
Miyagishima S, Takahara M, Kuroiwa T (2001a) Novel filaments 5 nm in diameter constitute the cytosolic ring of the plastid division apparatus. Plant Cell 13(3):707–721
Miyagishima SY et al (2001b) Plastid division is driven by a complex mechanism that involves differential transition of the bacterial and eukaryotic division rings. Plant Cell 13(10):2257–2268. https://doi.org/10.1105/tpc.010185
Miyagishima S et al (2003) A plant-specific dynamin-related protein forms a ring at the chloroplast division site. Plant Cell Online 15(3):655–665
Miyagishima S et al (2004) Two types of FtsZ proteins in mitochondria and red-lineage chloroplasts: the duplication of FtsZ is implicated in endosymbiosis. J Mol Evol 58(3):291–303. https://doi.org/10.1007/s00239-003-2551-1
Miyagishima S, Kuwayama H, Urushihara H, Nakanishi H (2008) Evolutionary linkage between eukaryotic cytokinesis and chloroplast division by dynamin proteins. Proc Natl Acad Sci U S A 105(39):15202–15207. https://doi.org/10.1073/pnas.0802412105
Miyagishima S-Y, Kabeya Y, Sugita C et al (2014) DipM is required for peptidoglycan hydrolysis during chloroplast division. BMC Plant Biol 14:57. https://doi.org/10.1186/1471-2229-14-57.
Mori T, Kuroiwa H, Takahara M et al (2001) Visualization of an FtsZ ring in chloroplasts of *Lilium longiflorum* leaves. Plant Cell Physiol 42:555–559
Mozdy AD, McCaffery JM, Shaw JM (2000) Dnm1p GTPase-mediated mitochondrial fission is a multi-step process requiring the novel integral membrane component Fis1p. J Cell Biol 151(2):367–379. https://doi.org/10.1083/jcb.151.2.367
Muñoz-Gómez SA, Slamovits CH, Dacks JB et al (2015) Ancient homology of the mitochondrial contact site and cristae organizing system points to an endosymbiotic origin of mitochondrial cristae. Curr Biol 25:1489–1495. https://doi.org/10.1016/j.cub.2015.04.006
Niemann A, Ruegg M, La Padula V et al (2005) Ganglioside-induced differentiation associated protein 1 is a regulator of the mitochondrial network: new implications for Charcot-Marie-tooth disease. J Cell Biol 170(7):1067–1078. https://doi.org/10.1083/jcb.200507087
Nishida K et al (2003) Dynamic recruitment of dynamin for final mitochondrial severance in a primitive red alga. Proc Natl Acad Sci U S A 100(4):2146–2151. https://doi.org/10.1073/pnas.0436886100
Nishida K, Yagisawa F, Kuroiwa H et al (2005) Cell cycle-regulated, microtubule-independent organelle division in *Cyanidioschyzon merolae*. Mol Biol Cell 16(May):2493–2502. https://doi.org/10.1091/mbc.E05-01-0068
Nishida K, Yagisawa F, Kuroiwa H et al (2007) WD40 protein Mda1 is purified with Dnm1 and forms a dividing ring for mitochondria before Dnm1 in *Cyanidioschyzon merolae*. Proc Natl Acad Sci U S A 104(11):4736–4741. https://doi.org/10.1073/pnas.0609364104

Nogales E, Wolf SG, Downing KH et al (1998) Structure of the alpha beta tubulin dimer by electron crystallography. Nature 393:199–203. https://doi.org/10.1038/34465

Okamura E, Sakamoto T, Sasaki T, Matsunaga S (2017) A plant ancestral polo-like kinase sheds light on the mystery of the evolutionary disappearance of polo-like kinases in the plant kingdom. Cytologia 82(3):261–266. https://doi.org/10.1508/cytologia.82.261

Olins AL, Rhodes G, Welch DB et al (2010) Lamin B receptor: multi-tasking at the nuclear envelope. Nucleus 1(1):53–70. https://doi.org/10.4161/nucl.1.1.10515

Oliva MA, Cordell SC, Löwe J (2004) Structural insights into FtsZ protofilament formation. Nat Struct Mol Biol 11(12):1243–1250. https://doi.org/10.1038/nsmb855

Osawa M, Anderson DE, Erickson HP (2008) Reconstitution of contractile FtsZ rings in liposomes. Science 320(5877):792–794. https://doi.org/10.1126/science.1154520

Osteryoung KW, Vierling E (1995) Conserved cell and organelle division. Nature 376(6540): 473–474. https://doi.org/10.1038/376473b0

Palmer CS et al (2011) MiD49 and MiD51, new components of the mitochondrial fission machinery. EMBO Rep 12(6):565–573. https://doi.org/10.1038/embor.2011.54

Paull TT et al (2000) A critical role for histone H2AX in recruitment of repair factors to nuclear foci after DNA damage. Curr Biol 10(15):886–895

Praefcke GJK, McMahon HT (2004) The dynamin superfamily: universal membrane tubulation and fission molecules? Nat Rev Mol Cell Biol 5(2):133–147. https://doi.org/10.1038/nrm1313

Roy M, Reddy PH, Iijima M, Sesaki H (2015) Mitochondrial division and fusion in metabolism. Curr Opin Cell Biol 33:111–118. https://doi.org/10.1016/j.ceb.2015.02.001

Sakamoto Y, Takagi S (2013) LITTLE NUCLEI 1 and 4 regulate nuclear morphology in *Arabidopsis thaliana*. Plant Cell Physiol 54(4):622–633. https://doi.org/10.1093/pcp/pct031

Schmitz AJ, Glynn JM, Olson BJSC et al (2009) *Arabidopsis* FtsZ2-1 and FtsZ2-2 are functionally redundant, but FtsZ-based plastid division is not essential for chloroplast partitioning or plant growth and development. Mol Plant 2(6):1211–1222. https://doi.org/10.1093/mp/ssp077

Shibata S, Matsuoka Y, Yoneda Y (2002) Nucleocytoplasmic transport of proteins and poly (A) RNA in reconstituted Tpr-less nuclei in living mammalian cells. Genes Cells 7:421–434

Shimada H et al (2004) ARC3, a chloroplast division factor, is a chimera of prokaryotic FtsZ and part of eukaryotic phosphatidylinositol-4-phosphate 5-kinase. Plant Cell Physiol 45(8):960–967. https://doi.org/10.1093/pcp/pch130

Shpetner HS, Vallee RB (1989) Identification of dynamin, a novel mechanochemical enzyme that mediates interactions between microtubules. Cell 59(3):421–432

Smirnova E, Shurland D-L, Ryazantsev SN et al (1998) A human dynamin-related protein controls the distribution of mitochondria. J Cell Biol 143(2):351–358

Smythe C, Jenkins HE, Hutchison CJ (2000) Incorporation of the nuclear pore basket protein nup153 into nuclear pore structures is dependent upon lamina assembly: evidence from cell-free extracts of *Xenopus* eggs. EMBO J 19:3918–3931. https://doi.org/10.1093/emboj/19.15.3918

Spang A, Saw JH, Jorgensen SL et al (2015) Complex archaea that bridge the gap between prokaryotes and eukaryotes. Nature 521:173–179. https://doi.org/10.1038/nature14447

Stiff T, O'Driscoll M, Rief N et al (2004) ATM and DNA-PK function redundantly to phosphorylate H2AX after exposure to ionizing radiation. Cancer Res 64:2390–2396

Sumiya N, Hirata A, Kawano S (2008) Multiple FtsZ ring formation and reduplicated chloroplast DNA in *Nannochloris bacillaris* (Chlorophyta, Trebouxiophyceae) under phosphate-enriched culture. J Phycol 44(6):1476–1489. https://doi.org/10.1111/j.1529-8817.2008.00589.x

Takahara M et al (2000) A putative mitochondrial ftsZ gene is present in the unicellular primitive red alga *Cyanidioschyzon merolae*. Mol Gen Genet 264(4):452–460

Tamura K, Hara-Nishimura I (2013) The molecular architecture of the plant nuclear pore complex. J Exp Bot 64(4):823–832. https://doi.org/10.1093/jxb/ers258

Tamura K, Fukao Y, Iwamoto M et al (2010) Identification and characterization of nuclear pore complex components in *Arabidopsis thaliana*. Plant Cell 22:4084–4097. https://doi.org/10.1105/tpc.110.079947

TerBush AD, Osteryoung KW (2012) Distinct functions of chloroplast FtsZ1 and FtsZ2 in Z-ring structure and remodeling. J Cell Biol 199(4):623–637. https://doi.org/10.1083/jcb.201205114

TerBush AD, Yoshida Y, Osteryoung KW (2013) FtsZ in chloroplast division: structure, function and evolution. Curr Opin Cell Biol 25(4):461–470. https://doi.org/10.1016/j.ceb.2013.04.006

Tieu Q, Nunnari J (2000) Mdv1p is a WD repeat protein that interacts with the dynamin-related GTPase, Dnm1p, to trigger mitochondrial division. J Cell Biol 151(2):353–365

van der Bliek AM, Meyerowitz EM (1991) Dynamin-like protein encoded by the *Drosophila* shibire gene associated with vesicular traffic. Nature 351(6325):411–414. https://doi.org/10.1038/351411a0

van der Bliek AM, Shen Q, Kawajiri S (2013) Mechanisms of mitochondrial fission and fusion. Cold Spring Harb Perspect Biol 5(6):1–16. https://doi.org/10.1101/cshperspect.a011072

Vitha S, McAndrew RS, Osteryoung KW (2001) FtsZ ring formation at the chloroplast division site in plants. J Cell Biol 153:111–120

Wang H, Wang M, Wang H et al (2005) Complex H2AX phosphorylation patterns by multiple kinases including ATM and DNA-PK in human cellsexposed to ionizing radiation and treated with kinase inhibitors. J Cell Physiol 202:492–502. https://doi.org/10.1002/jcp.20141

XM X et al (2007) NUCLEAR PORE ANCHOR, the *Arabidopsis* homolog of Tpr/Mlp1/Mlp2/megator, is involved in mRNA export and SUMO homeostasis and affects diverse aspects of plant development. Plant Cell 19:1537–1548. https://doi.org/10.1105/tpc.106.049239

Yaffe MP (1999) The machinery of mitochondrial inheritance and behavior. Science 283 (5407):1493–1497

Yagisawa F et al (2012) Mitotic inheritance of endoplasmic reticulum in the primitive red alga *Cyanidioschyzon merolae*. Protoplasma 249(4):1129–1135. https://doi.org/10.1007/s00709-011-0359-1

Yoon Y, Krueger EW, Oswald BJ, McNiven MA (2003) The mitochondrial protein hFis1 regulates mitochondrial fission in mammalian cells through an interaction with the dynamin-like protein DLP1. Mol Cell Biol 23(15):5409–5420

Yoshida Y, Mogi Y (2016) Reconstitution of the chloroplast FtsZ ring using the heterologous yeast system *Pichia pastoris*. Cytologia (Tokyo) 81(3):249–250. https://doi.org/10.1508/cytologia.81.249

Yoshida Y et al (2006) Isolated chloroplast division machinery can actively constrict after stretching. Science 313(5792):1435–1438. https://doi.org/10.1126/science.1129689

Yoshida Y, Nishida K, Kawano S, Kuroiwa T (2008) Novel dynamics of FtsZ ring before plastid abscission. Cytologia 73:197–201

Yoshida Y et al (2009) The bacterial ZapA-like protein ZED is required for mitochondrial division. Curr Biol 19(17):1491–1497. https://doi.org/10.1016/j.cub.2009.07.035

Yoshida Y et al (2010) Chloroplasts divide by contraction of a bundle of nanofilaments consisting of polyglucan. Science 329(5994):949–953. https://doi.org/10.1126/science.1190791

Yoshida Y, Miyagishima S-Y, Kuroiwa H, Kuroiwa T (2012) The plastid-dividing machinery: formation, constriction and fission. Curr Opin Plant Biol 15(6):714–721. https://doi.org/10.1016/j.pbi.2012.07.002

Yoshida Y, Fujiwara T, Imoto Y et al (2013) The kinesin-like protein TOP promotes aurora localisation and induces mitochondrial, chloroplast and nuclear division. J Cell Sci 126:2392–2400. https://doi.org/10.1242/jcs.116798

Yoshida Y, Mogi Y, TerBush AD, Osteryoung KW (2016) Chloroplast FtsZ assembles into a contractible ring via tubulin-like heteropolymerization. Nat Plants 2(7):16095. https://doi.org/10.1038/nplants.2016.95

Zaremba-Niedzwiedzka K, Caceres EF, Saw JH et al (2017) Asgard archaea illuminate the origin of eukaryotic cellular complexity. Nature 541:353–358. https://doi.org/10.1038/nature21031

Zhang M, Chen C, Froehlich JE et al (2015) Roles of *Arabidopsis* PARC6 in coordination of the chloroplast division complex and negative regulation of FtsZ assembly. Plant Physiol 170(517):01460.2015. https://doi.org/10.1104/pp.15.01460

Zhou X, Groves NR, Meier I (2015) Plant nuclear shape is independently determined by the SUN-WIP-WIT2-myosin XI-i complex and CRWN1. Nucleus 6(2):144–153. https://doi.org/10.1080/19491034.2014.1003512

# Chapter 15
# Single-Membrane-Bound Organelles: Division and Inheritance

**Fumi Yagisawa, Yuuta Imoto, Takayuki Fujiwara, and Shinya Miyagishima**

**Abstract** *C. merolae* cells contain four types of single-membrane-bound organelles—the ER, Golgi bodies, vacuoles/lysosomes, and peroxisomes—the dynamics of which have been studied during the cell cycle by fluorescence microscopy and electron microscopy. The ER retains some integrity throughout the cell cycle and is apportioned to daughter cells by spindle elongation. Golgi bodies proliferate prior to M phase and lie close to the spindle poles, with which they are inherited. Vacuoles proliferate and migrate toward mitochondria before M phase, and by binding to the mitochondria, they become inherited by daughter cells. This binding requires a coiled-coil protein, vacuole inheritance gene 1 (VIG1). Peroxisomes associate with the division planes of mitochondria before mitochondrial division, after which they divide—a process that is dependent on dynamin and peroxisome-dividing (POD) machinery—and are inherited with the divided mitochondria. The inheritance of vacuoles and peroxisomes relies on the spindle, because it is required for mitochondrial inheritance. Because the spindle segregates chromosomes, the inheritance of all single-membrane-bound organelles is coupled to chromosomal

F. Yagisawa (✉)
Center for Research Advancement and Collaboration, University of the Ryukyus, 1 Senbaru, Nishihara-cho, Nakagami-gun, Okinawa 903-0213, Japan
e-mail: yagisawa@lab.u-ryukyu.ac.jp

Y. Imoto
Division of Organelle Homeostasis, Medical Institute of Bioregulation, Kyushu University, 3-1-1 Maidashi, Higashiku Fukuoka 812-8582, Japan

T. Fujiwara
Department of Cell Genetics, National Institute of Genetics, 1111 Yata, Mishima, Shizuoka 411-8540, Japan

Department of Genetics, Graduate University for Advanced Studies (SOKENDAI), 1111 Yata, Mishima, Shizuoka 411-8540, Japan

Core Research for Evolutional Science and Technology Program, Japan Science and Technology Agency, Saitama 332-0012, Japan

S. Miyagishima
Department of Cell Genetics, National Institute of Genetics, 1111 Yata, Mishima, Shizuoka 411-8540, Japan

 
T. Kuroiwa et al. (eds.), *Cyanidioschyzon merolae*,
https://doi.org/10.1007/978-981-10-6101-1_15

inheritance via the spindle. This coupling allows cells to retain organelles and remain functional after cell division.

**Keywords** Endoplasmic reticulum (ER) • Golgi body • Vacuole • Peroxisome • Spindle poles • Proliferation • Inheritance

## 15.1 Introduction

The number (or amount) of organelles must be controlled to maintain proper cell function. Eukaryotic cells adjust the number of organelles in various situations, including cell division, differentiation, and changes in environment. In *C. merolae*, the number of organelles is tightly regulated by the cell cycle. Most organelles divide or proliferate prior to cytokinesis and are evenly distributed to daughter cells. Recent studies have demonstrated the presence of a cell cycle checkpoint, based on the status of organelle proliferation (Chap. 13). This type of regulation would be indispensable for the normal functioning of cells that contain small numbers of organelles after cytokinesis.

During eukaryotic evolution, various mechanisms of organelles inheritance developed, presumably to adopt several routes of cell division and changes in the morphologies of organelles and in the numbers of organelles that are contained in a cell. In higher eukaryotes, certain organelles break down at the onset of mitosis and are reconstructed in daughter cells. These morphological changes are atypical in simple eukaryotes, including *C. merolae*. Also, in several organisms, such as the budding yeast *S. cerevisiae*, most organelles are inherited in actin- and myosin-dependent mechanisms. In contrast, *C. merolae* cells lack a conventional actomyosin system (Chap. 16). Organelle inheritance in *C. merolae* is linked to chromosomal inheritance through spindle poles. Coupling of inheritance of various types of organelles is seen in other eukaryotes, such as *Trypanosoma brucei* and *Toxoplasma gondii*. Thus, we consider the mechanisms for organelle inheritance in *C. merolae* to be partially conserved in other eukaryotes.

*C. merolae* cells contain four types of single-membrane-bound organelles: Golgi bodies, the ER, vacuole/lysosomes, and peroxisomes. The ER retains some integrity throughout the cell cycle and is segregated into two parts as cells divide. Golgi bodies lie close to the spindle poles and proliferate before mitosis. Vacuoles bind to mitochondria for transmission to daughter cells. Peroxisomes proliferate by fission just prior to cytokinesis. In this chapter, we will discuss our current understanding of the dynamics of single-membrane-bound organelles during the cell cycle and their underlying molecular mechanisms in *C. merolae*.

## 15.2 Endoplasmic Reticulum

In *C. merolae*, the behavior of the ER has been studied by immunofluorescence microscopy of calnexin, a well-conserved ER chaperone (Yagisawa et al. 2012). The ER consists of nuclear ER (nuclear membrane) that surrounds the nucleus and a single layer of peripheral ER that extends from the nuclear ER. Unlike mammalian and plant ER, nuclear ER in *C. merolae* does not break down during mitosis. However, in metaphase, it partially dissolves, and the nuclear pore complex (NPC) that is associated with it disassembles (semi-open mitosis; Fujiwara et al. 2013). Incomplete disassembly of the NPC also occurs in certain yeasts and fungi, such as *Schizosaccharomyces japonicus* and *Aspergillus nidulans* (Aoki et al. 2011; De Souza et al. 2004). Subsequently, the mitotic spindle forms in the nucleus, and the nuclear ER is distributed to daughter cells with the chromosomes as the spindle stretches. The nuclear ER then divides into two during telophase.

Whereas the peripheral ER has a complicated, mesh-like structure in typical model eukaryotes, including *S. cerevisiae*, its structure is simpler in *C. merolae*. In most *C. merolae* cells, the peripheral ER fits into a single focal plane under a conventional fluorescence microscope. The peripheral ER forms ring-like structures during G1/S phase and extends toward the mitochondria and cell division planes during M phase. This extension depends on tubulin. The peripheral ER is inherited with nuclear ER. Interactions between the ER and other organelles have recently been identified in yeast and animals. The sites of ER membrane contact with other organelles are known as membrane contact sites (MCSs) (Phillips and Voeltz 2015). MCSs may perform specialized functions such as transferring chemical signals and lipids, sensing intracellular sterol levels, and participating in organellar biogenesis and dynamics. Interorganellar interactions between the ER and mitochondria are believed to be important for mitochondrial division in yeast and animals. During the process described as ER-associated mitochondrial division (ERMD), ER tubules cover part of the mitochondrion; subsequently, this site on the mitochondrial outer membrane becomes enriched with the membrane fission GTPase DRP1 and becomes the division site (Friedman et al. 2011). Indeed, the ER is in close proximity to the mitochondria in *C. merolae*, indicating that ER-mitochondrial interactions might play important roles for organelle biogenesis in primitive eukaryotes.

In several eukaryotes, the mechanisms for inheriting peripheral ER differ from those for nuclear ER. In *S. cerevisiae*, tubules of cortical ER (equivalent to peripheral ER) extend into daughter cells, based on actin and type V myosin Myo4 activity (Estrada et al. 2003), whereas nuclear ER inheritance requires microtubules (Du et al. 2004). In higher eukaryotes, including plant and mammals, nuclear ER breaks down and is reconstructed in daughter cells, and peripheral ER remains intact. Whether *C. merolae* regulates the inheritance of the two types of ER separately is unknown. Given the simple structure of peripheral ER in *C. merolae*, its association with nuclear ER might be sufficient for inheritance.

## 15.3 Golgi Bodies

Okuwaki et al. (1996) reported that *C. merolae* cells contain one Golgi body that consists of two layers and abuts the cell nucleus by electron microscopy. A more recent immunofluorescence microscopy study, labeling a homolog of the yeast SNARE protein Sed5, determined when Golgi bodies proliferate and how they are inherited (Yagisawa et al. 2013). During interphase, the Golgi body lies close to the cell nucleus. Although most interphase cells contain one Golgi body, some contain two or more. No M phase cells with fewer than two Golgi bodies have been observed, suggesting that Golgi bodies amplify before mitosis—at least when cells contain only one. During M phase, Golgi bodies near both spindle poles are inherited together with them. When cells are treated with oryzalin, an inhibitor of tubulin polymerization, the bipolar localization of Golgi bodies is disrupted; the Golgi bodies are elongated, or two Golgi bodies are localized next to each other. Thus, the microtubule is likely to function in Golgi segregation and inheritance.

The mechanisms of Golgi proliferation are poorly understood in *C. merolae*. In eukaryotes, Golgi proliferation occurs through several pathways (Munro 2002). In yeasts and the unicellular parasite *Toxoplasma brucei*, Golgi bodies are generated de novo. They are generated distally from old bodies in the budding yeast *Pichia pastoris* (Bevis et al. 2002) but next to them in *T. brucei* (He et al. 2005). In seed plants, green algae, and *T. gondii*, Golgi bodies proliferate by fission (Staehelin and Kang 2008; Noguchi 1978; Noguchi and Kakami 1999; Pelletier et al. 2002). In *C. merolae*, because Golgi bodies elongate in the presence of oryzalin, they might also proliferate by fission.

The timing of Golgi proliferation and how they are inherited in *C. merolae* are somewhat similar to those in other eukaryotes. In mammalian cells, Golgi bodies exist as a single perinuclear Golgi ribbon and associate with centrosomes (Rios and Bornens 2003). During mitosis, fragmented Golgi bodies are inherited by daughter cells in a process that is dependent on the spindle (Warren and Wickner 1996; Persico et al. 2009; Lippincott-Schwartz and Zaal 2000; Wei and Seemann 2009). In *T. brucei* and *T. gondii*, Golgi bodies duplicate prior to mitosis and are inherited with centrioles and basal bodies—two structures that are related to the microtubule organizing center (MTOC), like spindle poles. Also, Golgi bodies proliferate before mitosis in seed plants and green algae (Noguchi 1978; Ueda 1997; Segui-Simarro and Staehelin 2006). Thus, some of the molecular mechanisms of Golgi inheritance in *C. merolae* might be shared by other eukaryotes.

## 15.4 Vacuoles

*C. merolae* cells contain several vacuoles that are ~500 nm in diameter. These vacuoles can be visualized with several fluorescent dyes, such as quinacrine, which accumulates in acidic compartments, and a high concentration of DAPI, which stains polyphosphate that is stored in vacuoles and fluoresces yellow. They can also be examined by immunofluorescence microscopy using antibodies against vacuolar markers, such as vacuolar-type proton ATPase (Yagisawa et al. 2009; Fujiwara et al. 2010). By electron microscopy, vacuoles are observed as single-membrane-bound compartments with electron-dense substances that vary in amount. This substance has been suggested to contain polyphosphate and cations, based on observations of vacuoles in other organisms, and is consistent with the finding that it disappears when cells are depleted of phosphate (Yagisawa et al. 2016). Thick layers are observed around the membranes of some vacuoles.

Vacuolar proliferation is governed by environmental factors and the cell cycle. Vacuoles increase in number when cells are transplanted from a stock culture to fresh medium and aerated vigorously by bubbling to effect synchronization (Yagisawa et al. unpublished data). The resulting interphase cells contain approximately five vacuoles per cell on average, which increases to roughly six when cells enter mitosis; this value is halved after cytokinesis (Yagisawa et al. 2007). In the same culture, *C. merolae* can undergo second and third rounds of synchronized cell division, in which the number of vacuoles nearly doubles just before mitosis and is halved after cytokinesis.

Vacuoles are almost evenly inherited between daughter cells (Yagisawa et al. 2007). At the beginning of mitosis, they migrate from the cytoplasm, close to the nucleus, to near mitochondria (Fig. 15.1a-c). Subsequently, vacuoles bind to mitochondria through filamentous structures (Fig. 15.1d, f–g). This movement toward the mitochondria is unlikely to depend on microtubules, because it is unaffected by oryzalin. Because conventional actomyosin system is presumably absent from *C. merolae* (Chap. 16), this translocation is likely to be independent of the conventional cytoskeleton. During cytokinesis, vacuoles approach the cell division plane and start to dissociate from the mitochondria (Fig. 15.1e).

To identify the proteins that are involved in vacuolar migration and the association between vacuoles and mitochondria, Fujiwara et al. (2009, 2010) screened genes that were expressed specifically when vacuoles bound to mitochondria during the cell cycle in *C. merolae* by transcriptome analysis. One such gene, *VIG1*, encoded a 30-kDa coiled-coil protein. VIG1 began to localize to the surface of vacuoles around the nucleus in late S phase (Fig. 15.2a). Then, the vacuoles that were associated with VIG1 migrated to the surface of the mitochondrion at the outset of mitosis (Fig. 15.2a–c). In telophase, VIG1 became degraded, and the vacuoles dissociated from the mitochondrion (Fig. 15.2a). When *VIG1* was downregulated by antisense RNA, the vacuoles did not migrate or bind to mitochondria (Fujiwara et al. 2010).

Although filamentous structures are observed between vacuoles and mitochondria by transmission electron microscopy, the relationship between them and VIG1 remains poorly understood. VIG1 is homologous to CHMP5/VPS60, which is a component of the endosomal sorting complex required for transport (ESCRT)

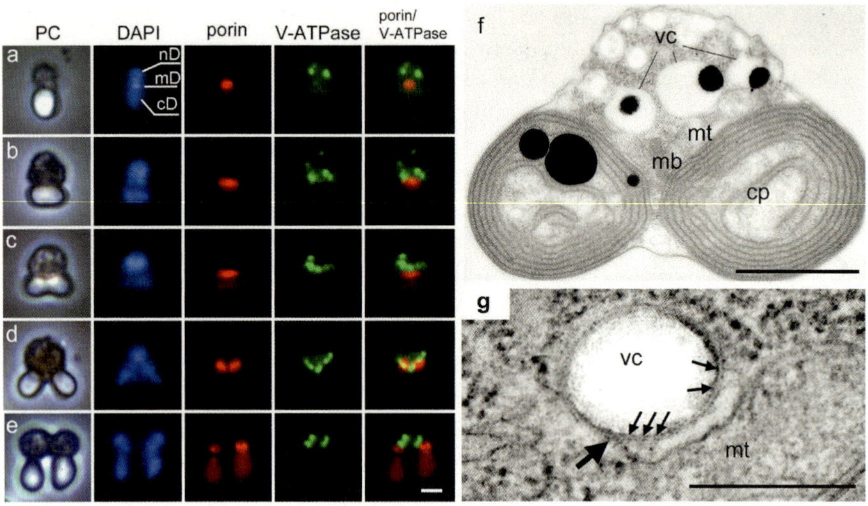


**Fig. 15.1** *Contact between vacuoles and mitochondria.* **(a–e)** Immunofluorescence images of cells in interphase (**a**), G2 phases (**b, c**), M phase (**d**), and cytokinesis (**e**) labeled with DAPI, anti-porin (mitochondria), and anti-V-ATPase (vacuole). The vacuoles localize close to the mitochondria during M phase and start to dissociate from the mitochondria after being distributed to daughter cells. (**f**) EM image of an M phase cell. (**g**) Filamentous structures between the vacuole and mitochondria during M phase. The mitochondrial membrane is stretched toward the vacuole (*large arrow*) and binds it through filamentous structures (*small arrows*) (*nD* nuclear DNA, *mD* mitochondrial DNA, *cD* chloroplast DNA, *vc* vacuole, *mb* microbody, *mt* mitochondria). Bars = 2 μm (*e*), 1 μm (*f*), 250 nm (*g*). Panels *a–f* and *g* are reproduced from Yagisawa et al. (2012) Plant Morphology 24: 105–109 and Yagisawa et al. (2007), respectively

complex. Although the ESCRT complex was originally shown to mediate protein sorting in endosomes, it is now known to be involved in other physiological processes, such as cytokinetic membrane scission, reconstruction of the nuclear membrane after mitosis, and viral budding (Campsteijn et al. 2016).

It is unknown whether vacuole-mitochondria contact through VIG1 occurs in other eukaryotes. In *S. cerevisiae*, vacuoles associate with mitochondria, a process that is dependent on the homotypic fusion and vacuolar sorting (HOPS) complex protein Vps39/Vam6 and the Rab GTPase Ypt7 (Elbaz-Alon et al. 2014 Dev Cell; Honscher et al. 2014 Dev Cell), which also participates in membrane fusion of vacuoles. Homologs of these proteins have not been found in *C. merolae* by BLAST search. Studies on the molecular mechanisms of the vacuole-mitochondrion association in *C. merolae* would increase our understanding of organelle-organelle interactions in eukaryotic cells.

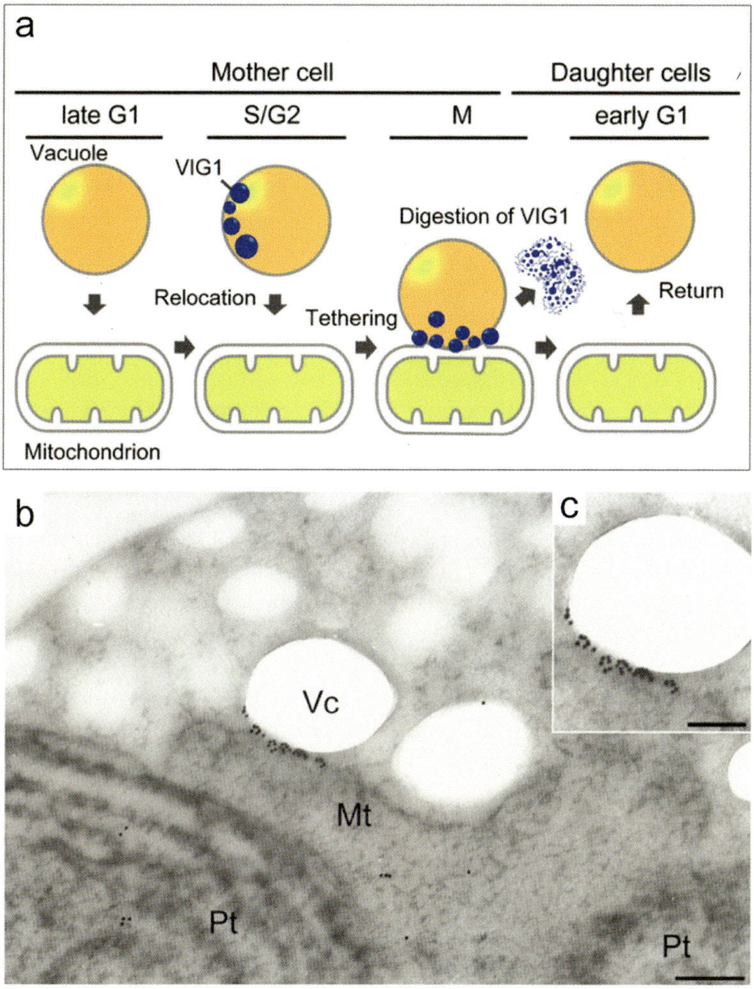


**Fig. 15.2** *Involvement of VIG1 in the binding between vacuoles and mitochondria.* (**a**) Model for vacuole inheritance in *C. merolae*. VIG1 accumulates at patch structures on the vacuolar surface in the cytosol. The patch structures have an asymmetric distribution during S phase. During M phase, the VIG1 patches localizes to between the vacuolar membrane and mitochondrial surface and tether the vacuole to the mitochondria. After cytokinesis, digestion of VIG1 allows the vacuole to return to the cytosol. (**b**) Section showing VIG1 localization when the vacuole is bound to the mitochondria. VIG1 localizes to between the vacuoles and mitochondria. The section was immunolabeled with anti-VIG1 and 15-nm gold particle-conjugated secondary antibodies. (**c**) A magnified image of the vacuole binding to the mitochondria in *b*. *Mt* mitochondrion, *Pt* plastid, *Vc* vacuole. Bars = 200 nm (*b*), 100 nm (*c*). Panels are reproduced (Fujiwara et al. (2010) with permission from the American Society of Plant Biologists)

## 15.5 Peroxisomes

*C. merolae* is an excellent model organism for examining peroxisomal division and inheritance, because it contains only one peroxisome that proliferates once per cell cycle. A three-dimensional quantitative analysis by thin-section EM has shown that the peroxisome doubles its volume in M phase and increases in number on constriction of the equator plane just after mitochondrial division (Miyagishima et al. 1999). By thin-section EM, both tips of the dividing peroxisome are tethered to daughter mitochondria via an electron-dense connector that is 30–50 nm in diameter. This structure remains after peroxisomal division, and the daughter peroxisomes are inherited by progeny cells with the mitochondria. The linked peroxisomes and mitochondria are anchored by centrosomes and microtubules at the tip of the daughter mitochondrion (Nishida et al. 2005); thus, the peroxisome proliferates by binary division, and inheritance of the peroxisome is regulated by spindle pole bodies and microtubules (Imoto et al. 2010).

Consistent with these observations, peroxisomal division can be synchronized by light/dark stimulation and treatment with oryzalin. Oryzalin delays the division of peroxisomes, and cells that contain dividing peroxisomes can be collected in bulk. This method allows one to isolate dividing peroxisomes and perform an ultrastructural analysis of its division plane. By whole-mount EM analysis, a ring-shaped division apparatus of the peroxisome has been identified—termed peroxisome-dividing (POD) machinery—which encircles the division plane of the peroxisome (Imoto et al. 2012, 2013). The diameter of POD machinery is approximately 600 nm at its maximum and contracts to 50 nm during peroxisomal division.

POD machinery comprises two structures: a dynamin-based (DB) ring and a filamentous ring (Fig. 15.3a, b). The DB ring consists of amorphous strings and contains the dynamin-related protein Dnm1, which is well conserved in eukaryotes and belongs to a family of large GTPases that regulate membrane remodeling. The filamentous ring is the skeletal structure of the machinery and is composed of numerous fine filaments that are 4–5 nm in width, formed between the DB ring and the cytoplasmic side of the peroxisomal membrane. Motive force for the contraction of POD machinery is likely to be generated by the DB ring. Initial constriction of the peroxisome occurs concomitantly with disassembly of the filamentous ring and contraction of the DB ring. In the final step of peroxisomal division (the membrane fission phase), the filamentous ring disassembles completely, and the DB ring interacts directly with the membrane and is believed to pinch the membrane off (Fig. 15.3c).

Dnm1 is constitutively expressed during the cell cycle and localizes to punctate foci in the cytosol (cytosolic Dnm1 patches) and the division plane of mitochondria and peroxisomes (Nishida et al. 2003; Imoto et al. 2013). On isolation of cytosolic Dnm1 patches by Optiprep density floatation and the development of their nucleotide statement analysis by liquid chromatography-electrospray ionization tandem mass spectrometry (LC ESI-MS/MS), nearly all Dnm1 in cytosolic patches has

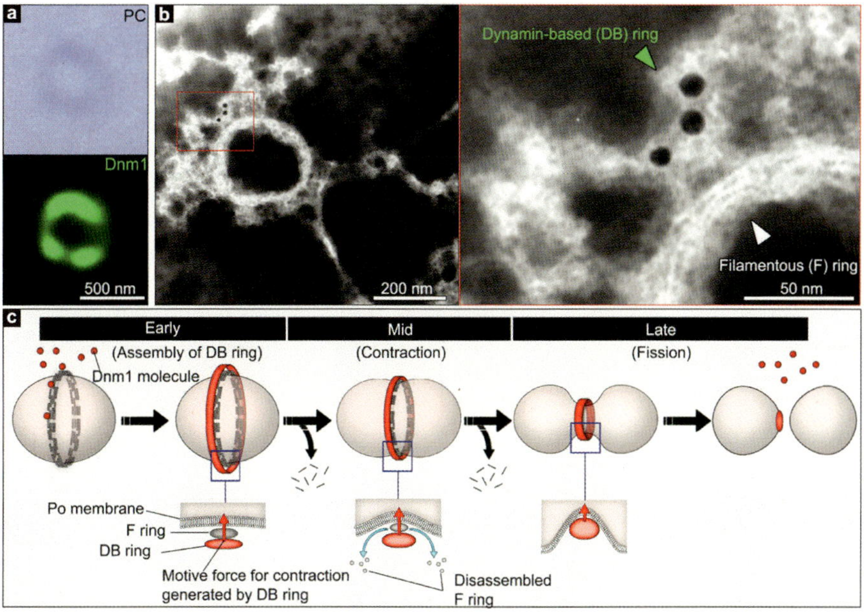


**Fig. 15.3** *Structure and function of peroxisome-dividing (POD) machinery.* (**a**) Isolated POD machinery immunostained with anti-Dnm1. (**b**) Whole-mount negative staining EM images of isolated POD machinery. The DB ring is partially separated by detergent. *Right panel* shows enlarged image of the *red* boxed area in the *left panel*. (**c**) Working model of POD machinery. The POD machinery consists of an outer DB ring and an inner filamentous ring that are assembled on the cytosolic side of the peroxisome (Po) membrane. The DB ring (Dnm1 molecules and amorphous string) forms around the filamentous ring (early). The DB ring then generates a contraction force for the POD machinery, with partial digestion of the filamentous ring (*middle*). The filamentous ring disassembles, and then, the single-membrane bridge between daughter peroxisomes is pinched off directly by the DB ring (*late*) (Figures are referred and modified from Imoto et al. 2013)

been demonstrated to exist in GTP-unbound form, whereas <90% of Dnm1 in the POD machinery is GTP-bound (Imoto et al. 2017). During peroxisomal division, Dnm1 in cytosolic patches transits to its GTP-bound form and relocates immediately to the division plane (Imoto et al. 2017). Around the division plane, the DB ring organizing center (DOC) mediates the assembly of the GTP-bound form of Dnm1 and initiates DB ring formation (Imoto et al. 2017). During initial constriction of the peroxisome, a single DOC is formed on the ring. The DB ring elongates from the DOC by unidirectional assembly of Dnm1 in a mechanism that relies on the GTP concentration but is independent of GTP hydrolysis (Fig. 15.4) (Imoto et al. 2017). Notably, the GTP-bound form of Dnm1 is integrated into the DB ring, while the same amount of Dnm1 dissociates from the DB ring.

According to an in vitro study in which Dnm1 dissociated rapidly from lipid tubules on GTP hydrolysis (Mears et al. 2011) and a transcriptomic analysis of *C. merolae* that showed constitutive transcription of Dnm1 throughout the cell

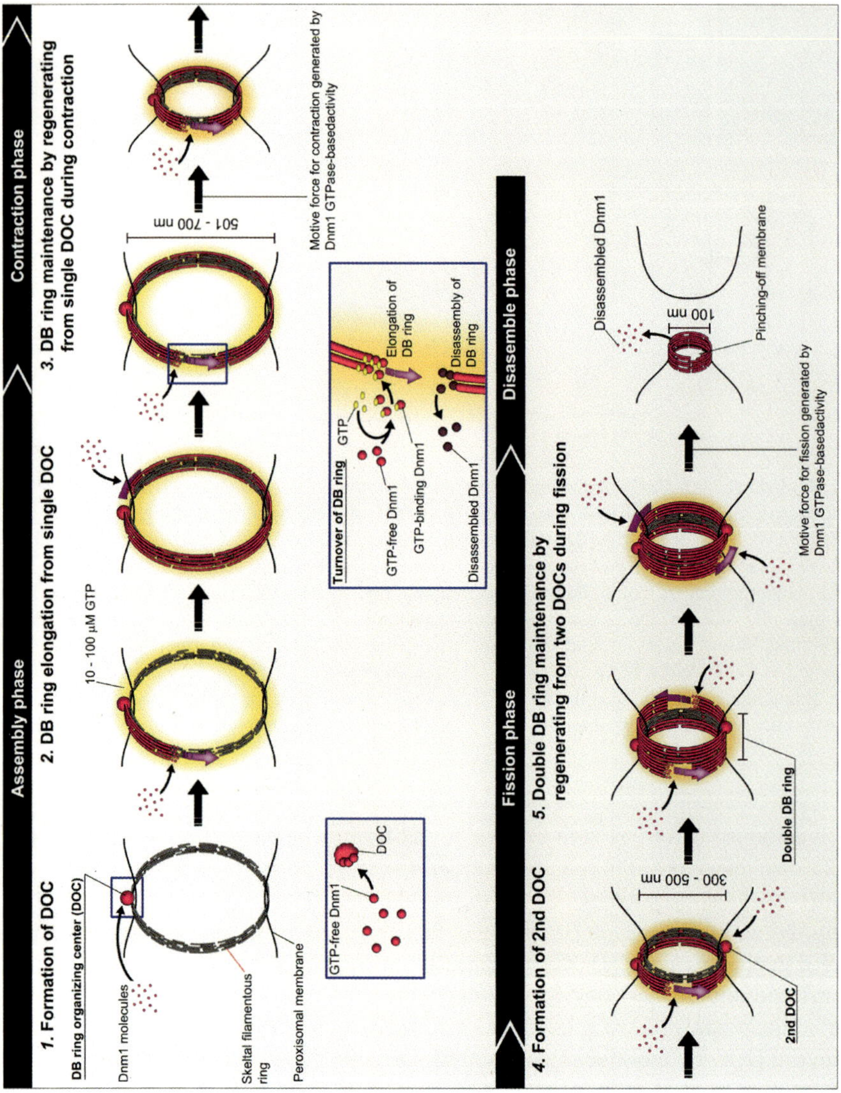
Assembly phase
Contraction phase
1. Formation of DOC
DB ring organizing center (DOC)
Dnm1 molecules
Skeltal filamentous ring
Peroxisomal membrane
GTP-free Dnm1
DOC
2. DB ring elongation from single DOC
10 - 100 μM GTP
3. DB ring maintenance by regenerating from single DOC during contraction
501 - 700 nm
Motive force for contraction generated by Dnm1 GTPase-basedactivity
Turnover of DB ring
GTP
GTP-free Dnm1
GTP-binding Dnm1
Elongation of DB ring
Disassembly of DB ring
Disassembled Dnm1
Fission phase
Disasemble phase
4. Formation of 2nd DOC
5. Double DB ring maintenance by regenerating from two DOCs during fission
300 - 500 nm
2nd DOC
Double DB ring
Motive force for fission generated by Dnm1 GTPase-basedactivity
100 nm
Disassembled Dnm1
Pinching-off membrane

**Fig. 15.4** *Working model of DB ring formation via DB ring organizing center (DOC)*. DB ring formation during peroxisomal division takes place as follows: (*1*) During the assembly phases, GTP-free Dnm1 molecules form a single DOC on the peroxisome division plane. (*2*) The DB ring is formed by unidirectional elongation from the DOC by the assembly of Dnm1 from cytosolic Dnm1 patches in a GTP concentration-dependent manner. (*3*) DB ring turnover occurs continuously during the contraction phase. (*4*) During the fission phase, a second DOC is formed on the opposite side of the first DOC. (*5*) A double DB ring forms by unidirectional elongation from both DOCs. The POD machinery is believed to sever the membrane through the GTPase activity of Dnm1 in the DB ring (Figure is referred and modified from Imoto et al. 2017)

cycle (Fujiwara et al. 2009), the DB ring is likely turning over to maintain its function in constriction. During the later phases of peroxisomal division (membrane fission), a second DOC forms on the opposite side of the first species (Fig. 15.4). A DB ring is formed via the second DOC independently from that of the first DOC. Hence, a double DB ring exists during the membrane fission phase. An in vitro kinetic assay of dynamin family members showed that increased attachment of dynamin or Dnm1 string to the lipid surface constricted lipid tubules more efficiently (Mears et al. 2011; Sundborger et al. 2014). Thus, double DB ring formation possibly generates strong constriction forces during membrane fission.

As discussed, the DB ring is an important component of the POD machinery. Notably, Dnm1 in the DB ring was originally identified as a constituent of mitochondrial division (MD) machinery (Nishida et al. 2003: see Chap. 14). Molecular genetic studies in yeast, mammals, and land plant cells have shown that defects in Dnm1 impair peroxisomal and mitochondrial division (Bleazard et al. 1999; Koch et al. 2003; Mano et al. 2004; Motley and Hettema 2007). Moreover, Dnm1-associated proteins, such as Fis1 (Mozdy et al. 2000), Mdv1 (Tieu and Nunnari 2000), and Mff (Gandre-Babbe and van der Bliek 2008), participate in peroxisomal and mitochondrial division (Motley and Hettema 2007; Motley et al. 2008). Thus, Dnm1-based organelle division machinery is likely to be shared between peroxisomal and mitochondrial division in eukaryotes. Peroxisomes and mitochondria are believed to have different origins (de Duve 2007), raising the question of why they share membrane fission machinery for division.

To this end, it has been hypothesized that the peroxisome was acquired originally via budding from the mitochondrial outer membrane using Dnm1-based MD machinery (Imoto et al. 2012). Notably, there is emerging evidence in mammalian cells that suggests that mitochondrial outer membrane-derived vesicles transform into peroxisomes under certain conditions. This biogenesis of peroxisomes has been suggested to involve peroxisomal biogenesis factor 3 (Pex3), which is also conserved in *C. merolae* (Sugiura et al. 2014). Studies on the molecular composition and mechanics of the POD machinery continue and should shed light on the biogenesis and origin of single-membrane-bound organelles, such as peroxisomes.

## 15.6 Link Between Inheritance of Single-Membrane-Bound Organelles and Chromosome Segregation

We have proposed that organelle inheritance in *C. merolae* is linked to chromosome inheritance through spindle poles. Golgi bodies lie near the spindle poles and are inherited in spindle-dependent mechanisms. The ER does not break down completely and is inherited with chromosomes by spindle elongation. Vacuoles and peroxisomes bind to mitochondria, which are connected to the edges of spindle poles and are segregated in a spindle-dependent process. This coupling ensures proper organelle inheritance, especially in simple eukaryotes with few organelles.

**Acknowledgments** We thank Dr. Yamato Yoshida for a discussion on the manuscript. Our study was partly supported by MEXT KAKENHI (16 K14770 to F.Y.) and Grants-in-Aid for JSPS fellows (1044 to F.Y.).

## References

Aoki K, Hayashi H, Furuya K, Sato M, Takagi T, Osumi M, Kimura A, Niki H (2011) Breakage of the nuclear envelope by an extending mitotic nucleus occurs during anaphase in *Schizosaccharomyces japonicus*. Genes Cells 16:911–926

Bevis BJ, Hammond AT, Reinke CA, Glick BS (2002) De novo formation of transitional ER sites and Golgi structures in *Pichia Pastoris*. Nat Cell Biol 4:750–756

Bleazard W, McCaffery JM, King EJ, Bale S, Mozdy A, Tieu Q, Nunnari J, Shaw JM (1999) The dynamin-related GTPase Dnm1 regulates mitochondrial fission in yeast. Nat Cell Biol 1:298–304

Campsteijn C, Vietri M, Stenmark H (2016) Novel ESCRT functions in cell biology: spiraling out of control? Curr Opin Cell Biol 41:1–8

de Duve C (2007) The origin of eukaryotes: a reappraisal. Nat Rev Genet 8(5):395–403

De Souza CP, Osmani AH, Hashmi SB, Osmani SA (2004) Partial nuclear pore complex disassembly during closed mitosis in *Aspergillus nidulans*. Curr Biol 14:1973–1984

Du Y, Ferro-Novick S, Novick P (2004) Dynamics and inheritance of the endoplasmic reticulum. J Cell Sci 117:2871–2878

Elbaz-Alon Y, Rosenfeld-Gur E, Shinder V, Futerman AH, Geiger T, Schuldiner M (2014) A dynamic interface between vacuoles and mitochondria in yeast. Dev Cell 30:95–102

Estrada P, Kim J, Coleman J, Walker L, Dunn B, Takizawa P, Novick P, Ferro-Novick S (2003) Myo4p and She3p are required for cortical ER inheritance in *Saccharomyces Cerevisiae*. J Cell Biol 163:1255–1266

Friedman JR, Lackner LL, West M, DiBenedetto JR, Nunnari J, Voeltz GK (2011) ER tubules mark sites of mitochondrial division. Science 334:358–362

Fujiwara T, Misumi O, Tashiro K, Yoshida Y, Nishida K, Yagisawa F, Imamura S, Yoshida M, Mori T, Tanaka K, Kuroiwa H, Kuroiwa T (2009) Periodic gene expression patterns during the highly synchronized cell nucleus and organelle division cycles in the unicellular red alga *Cyanidioschyzon merolae*. DNA Res 16:59–72

Fujiwara T, Kuroiwa H, Yagisawa F, Ohnuma M, Yoshida Y, Yoshida M, Nishida K, Misumi O, Watanabe S, Tanaka K, Kuroiwa T (2010) The coiled-coil protein VIG1 is essential for tethering vacuoles to mitochondria during vacuole inheritance of *Cyanidioschyzon merolae*. Plant Cell 22:772–781

Fujiwara T, Tanaka K, Kuroiwa T, Hirano T (2013) Spatiotemporal dynamics of condensins I and II: evolutionary insights from the primitive red alga *Cyanidioschyzon merolae*. Mol Biol Cell 24(16):2515–2527

Gandre-Babbe S, van der Bliek AM (2008) The novel tail-anchored membrane protein Mff controls mitochondrial and peroxisomal fission in mammalian cells. Mol Biol Cell 19:2402–2412

He CY, Pypaert M, Warren G (2005) Golgi duplication in *Trypanosoma brucei* requires Centrin2. Science 310:1196–1198

Honscher C, Mari M, Auffarth K, Bohnert M, Griffith J, Geerts W, van der Laan M, Cabrera M, Reggiori F, Ungermann C (2014) Cellular metabolism regulates contact sites between vacuoles and mitochondria. Dev Cell 30:86–94

Imoto Y, Fujiwara T, Yoshida Y, Kuroiwa H, Maruyama S, Kuroiwa T (2010) Division of cell nuclei, mitochondria, plastids, and microbodies mediated by mitotic spindle poles in the primitive red alga *Cyanidioschyzon merolae*. Protoplasma 241:63–74

Imoto Y, Kuroiwa H, Ohnuma M, Kawano S, Kuroiwa T (2012) Identification of peroxisome-dividing ring in *Cyanidioschyzon merolae* based on organelle partner hypothesis. Cytologia 77:515–522

Imoto Y, Kuroiwa H, Yoshida Y, Ohnuma M, Fujiwara T, Yoshida M, Nishida K, Yagisawa F, Hirooka S, Miyagishima SY, Misumi O, Kawano S, Kuroiwa T (2013) Single-membrane-bounded peroxisome division revealed by isolation of dynamin-based machinery. Proc Natl Acad Sci U S A 110:9583–9588

Imoto Y, Abe Y, Okumoto K, Honsho M, Kuroiwa H, Kuroiwa T, Fujiki Y (2017) Defining dynamin-based ring organizing center on the peroxisome-dividing machinery isolated from *Cyanidioschyzon merolae*. J Cell Sci 130:853–867

Koch A, Thiemann M, Grabenbauer M, Yoon Y, McNiven MA, Schrader M (2003) Dynamin-like protein 1 is involved in peroxisomal fission. J Biol Chem 278:8597–8605

Lippincott-Schwartz J, Zaal KJ (2000) Cell cycle maintenance and biogenesis of the Golgi complex. Histochem Cell Biol 114:93–103

Mano S, Nakamori C, Kondo M, Hayashi M, Nishimura M (2004) An Arabidopsis dynamin-related protein, DRP3A, controls both peroxisomal and mitochondrial division. Plant J 38:487–498

Mears JA, Lackner LL, Fang S, Ingerman E, Nunnari J, Hinshaw JE (2011) Conformational changes in Dnm1 support a contractile mechanism for mitochondrial fission. Nat Struct Mol Biol 18:20–26

Miyagishima S-y, Itoh R, Kuroiwa H, Nishimura M, Kuroiwa T, Toda K (1999) Microbody proliferation and segregation cycle in the single-microbody alga *Cyanidioschyzon merolae*. Planta 208(3):326–336

Motley AM, Hettema EH (2007) Yeast peroxisomes multiply by growth and division. J Cell Biol 178:399–410

Motley AM, Ward GP, Hettema EH (2008) Dnm1p-dependent peroxisome fission requires Caf4p, Mdv1p and Fis1p. J Cell Sci 121:1633–1640

Mozdy AD, McCaffery JM, Shaw JM (2000) Dnm1p GTPase-mediated mitochondrial fission is a multi-step process requiring the novel integral membrane component Fis1p. J Cell Biol 151:367–380

Munro S (2002) More than one way to replicate the Golgi apparatus. Nat Cell Biol 4:E223–E224

Nishida K, Takahara M, Miyagishima SY, Kuroiwa H, Matsuzaki M, Kuroiwa T (2003) Dynamic recruitment of dynamin for final mitochondrial severance in a primitive red alga. Proc Natl Acad Sci U S A 100:2146–2151

Nishida K, Yagisawa F, Kuroiwa H, Nagata T, Kuroiwa T (2005) Cell cycle-regulated, microtubule-independent organelle division in *Cyanidioschyzon merolae*. Mol Biol Cell 16:2493–2502

Noguchi T (1978) Transformation of the Golgi apparatus in the cell cycle especially at the resting and earliest developmental stages of a green alga, *Micrasterias americana*. Protoplasma 95:73–88

Noguchi T, Kakami F (1999) Transformation of trans-Golgi network during the cell cycle in a green alga, *Botryococcus braunii*. J Plant Res 112:175–186

Okuwaki T, Takahashi H, Itoh R, Toda K, Kawazu T, Kuroiwa H, Kuroiwa T (1996) Ultrastructures of the Golgi body and cell surface in *Cyanidioschyzon merolae*. Cytologia 61:69–74

Pelletier L, Stern CA, Pypaert M, Sheff D, Ngo HM, Roper N, He CY, Hu K, Toomre D, Coppens I, Roos DS, Joiner KA, Warren G (2002) Golgi biogenesis in *Toxoplasma gondii*. Nature 418:548–552

Persico A, Cervigni RI, Barretta ML, Colanzi A (2009) Mitotic inheritance of the Golgi complex. FEBS Lett 583:3857–3862

Phillips MJ, Voeltz GK (2015) Structure and function of ER membrane contact sites with other organelles. Nat Rev Mol Cell Biol 17:1–14

Rios RM, Bornens M (2003) The Golgi apparatus at the cell centre. Curr Opin Cell Biol 15:60–66

Segui-Simarro JM, Staehelin LA (2006) Cell cycle-dependent changes in Golgi stacks, vacuoles, clathrin-coated vesicles and multivesicular bodies in meristematic cells of *Arabidopsis thaliana*: a quantitative and spatial analysis. Planta 223:223–236

Staehelin LA, Kang BH (2008) Nanoscale architecture of endoplasmic reticulum export sites and of Golgi membranes as determined by electron tomography. Plant Physiol 147:1454–1468

Sugiura A, McLelland GL, Fon EA, McBride HM (2014) A new pathway for mitochondrial quality control: mitochondrial-derived vesicles. EMBO J 33:2142–2156

Sundborger AC, Fang S, Heymann JA, Ray P, Chappie JS, Hinshaw JE (2014) A dynamin mutant defines a superconstricted prefission state. Cell Rep 8:734–742

Tieu Q, Nunnari J (2000) Mdv1p is a WD repeat protein that interacts with the dynamin-related GTPase, Dnm1p, to trigger mitochondrial division. J Cell Biol 151:353–366

Ueda K (1997) The synchronous division of dictyosomes at the premitotic stage. Ann Bot 80(1):29–33

Warren G, Wickner W (1996) Organelle inheritance. Cell 84(3):395–400

Wei JH, Seemann J (2009) Spindle-dependent partitioning of the Golgi ribbon. Commun Integr Biol 2:406–407

Yagisawa F, Nishida K, Kuroiwa H, Nagata T, Kuroiwa T (2007) Identification and mitotic partitioning strategies of vacuoles in the unicellular red alga *Cyanidioschyzon merolae*. Planta 226:1017–1029

Yagisawa F, Nishida K, Yoshida M, Ohnuma M, Shimada T, Fujiwara T, Yoshida Y, Misumi O, Kuroiwa H, Kuroiwa T (2009) Identification of novel proteins in isolated polyphosphate vacuoles in the primitive red alga *Cyanidioschyzon merolae*. Plant J 60:882–893

Yagisawa F, Fujiwara T, Kuroiwa H, Nishida K, Imoto Y, Kuroiwa T (2012) Mitotic inheritance of endoplasmic reticulum in the primitive red alga *Cyanidioschyzon merolae*. Protoplasma 249:1129–1135

Yagisawa F, Fujiwara T, Ohnuma M, Kuroiwa H, Nishida K, Imoto Y, Yoshida Y, Kuroiwa T (2013) Golgi inheritance in the primitive red alga, *Cyanidioschyzon merolae*. Protoplasma 250:943–948

Yagisawa F, Kuroiwa H, Fujiwara T, Kuroiwa T (2016) Intracellular structure of the unicellular red alga *Cyanidioschyzon merolae* in response to phosphate depletion and resupplementation. Cytologia 81:341–347

# Chapter 16
# Regulation of Organelle and Cell Division by Cytoskeletal and Motor Proteins in *Cyanidioschyzon merolae*

**Yamato Yoshida and Yuuta Imoto**

**Abstract** During cell proliferation, many types of organelles must be correctly distributed into daughter cells before cytokinesis. Pioneering microscopic and molecular physiological experiments revealed that organelle proliferation and segregation into daughter cells of *Cyanidioschyzon merolae* are tightly regulated. The first breakthroughs were achieved using synchronized cell cultures treated with cell-cycle inhibitors, which enabled elucidation of the sequential patterns of organelle divisions within the cell division cycle. In addition, the dynamics of organelle movements during cell division suggested that they were mediated by the cytoskeletal network and motor proteins. Recent state-of-the-art investigations into the molecular functions of a kinesin-like protein (TOP) and a mitotic kinase (AUR) have revealed that an integrated regulatory mechanism controls the division of many types of organelles. Transcriptomic and proteomic analyses suggest that *C. merolae* cells lack a classical actomyosin-based cytokinesis system. However, eukaryotic elongation factor 1 alpha (eEF-1α) is specifically and spatiotemporally enriched at the cell division site during cytokinesis, suggesting that *C. merolae* cells utilize an evolutionarily ancient cell division system.

**Keywords** Aurora serine/threonine mitotic kinase (AUR) • *Cyanidioschyzon merolae* • Cytokinesis • Eukaryotic elongation factor 1 alpha (eEF-1α) • Microtubules • Spindle apparatus • Kinesin superfamily protein • Organelle division • Organelle segregation

Y. Yoshida (✉)
Laboratory for Single Cell Gene Dynamics, Quantitative Biology Center, RIKEN, 6-2-3 Furuedai, Suita, Osaka 565-0874, Japan
e-mail: yamato.yoshida@riken.jp

Y. Imoto
Division of Organelle Homeostasis, Medical Institute of Bioregulation, Kyushu University, 3-1-1 Maidashi, Higashiku Fukuoka 812-0053, Japan



T. Kuroiwa et al. (eds.), *Cyanidioschyzon merolae*,
https://doi.org/10.1007/978-981-10-6101-1_16

## 16.1 Introduction

Mitochondria and plastids, which evolved from bacterial origins, possess genomic DNA and do not generate de novo. Therefore, photosynthetic eukaryotes contain three types of organelles with their own DNA genomes: nuclei, mitochondria, and chloroplasts (plastids). These genomes must be correctly distributed into daughter cell nuclei, mitochondria, and plastids, respectively, during cell proliferation (Fig. 16.1). The behavior of these DNA genomes suggests that the eukaryotic cell division cycle must include both a mitotic (nuclear) division cycle and mitochondrial and plastid division cycles. Mitochondrial and plastid DNA replication occurs before nuclear DNA replication in many eukaryotes. Recent studies using *C. merolae* and tobacco BY-2 cells indicate that mitochondrial and plastid DNA replication activates cyclin-dependent kinase A, which leads to the accumulation of a tetrapyrrole intermediate and subsequent initiation of nuclear DNA replication (Kobayashi et al. 2009, 2011) (see Chap. 13). In addition, other recent studies have elucidated the molecular mechanism that synchronizes the cell division and plastid division cycles (Sumiya et al. 2016).

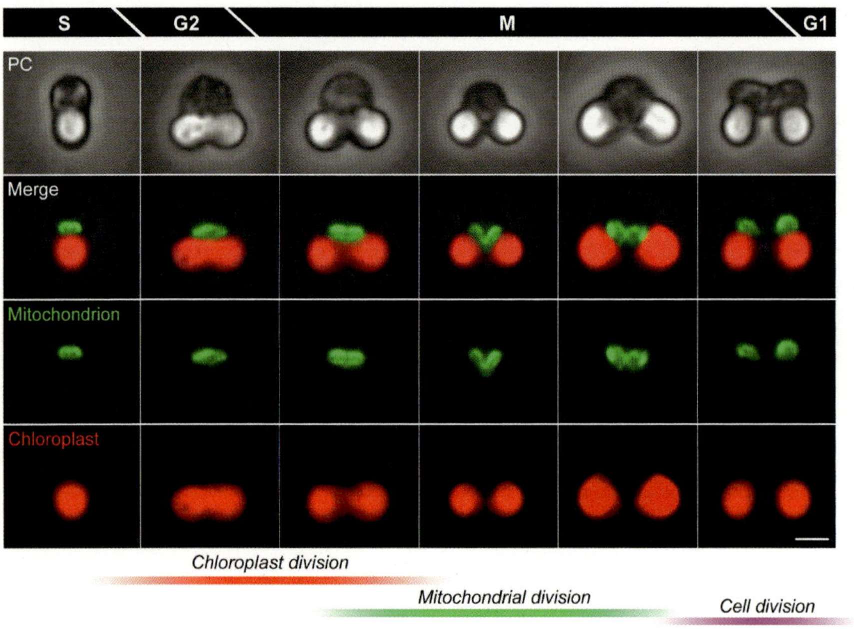


**Fig. 16.1** Sequential process of chloroplast, mitochondrion, and cell division in *Cyanidioschyzon merolae*. Immunofluorescence images of *C. merolae* cells in each phase of the cell cycle. Mitochondria were visualized by immunostaining using antimitochondrial porin antibody, and chloroplasts emit autofluorescence. Scale bar, 1 μm (Images were reproduced and modified, with permission, from Yoshida et al. 2013)

On the other hand, a new model has been proposed in which an integrated system regulates the sequential division of DNA-containing organelles during the cell cycle because several cytoskeletal and motor proteins have been implicated in cell division and organelle division cycles in *C. merolae* cells. In this chapter, we review the current knowledge of cytoskeletal and motor proteins that are involved in cell and organelle proliferation in *C. merolae*. *C. merolae* cells have an ancient genome that encodes only a few cytoskeletal and motor proteins compared with other model eukaryotes. This relative simplicity facilitates an exploration of the molecular mechanisms controlling intracellular motility, organelle transport, organelle division, and cytokinesis.

## 16.2 Tubulin, Microtubules, and Spindle Apparatus

The *C. merolae* genome encodes a minimal set of tubulin family proteins, including alpha-tubulin (gene ID: CMT504C), beta-tubulin (CMN263C), and gamma-tubulin (CMN304C). Transcriptomic analyses using synchronized cell cultures showed that these three tubulin genes are expressed only during the S to M phase of the cell cycle (Fig. 16.2), suggesting that tubulin proteins might enable dynamic intracellular movements during S to M phases. The tubulin distribution in *C. merolae* cells was analyzed by performing immunocytochemical studies using anti-alpha-tubulin antibody (Nishida et al. 2005; Imoto et al. 2010) (Fig. 16.3). This analysis showed that the alpha-tubulin proteins appear in the cytoplasmic region during S phase and then accumulate on mitochondria during G2. At the start of the mitotic phase, tubulin is localized along mitochondria, and then two symmetrical spindle poles appear in the cytosol. A series of electron microscopic analyses and experimental results revealed that the two spindle poles play a key role in the division process of the cell nucleus (Nishida et al. 2005; Maruyama et al. 2007). Interestingly, the two spindle poles are associated with nuclear envelope and physically tethered to the two ends of the mitochondria. Then, microtubules elongate from the two spindle poles and then form astral-like structures. After plastid binary fission is complete, the two spindle poles move closer and assemble into a spindle apparatus in the cell nucleus. During this phase, nuclear DNA appears to be compacted and shifts into the mid-region between the spindle poles. Mitochondrial division occurs at approximately the same time. Subsequently, elongated microtubules from the two spindle poles form a thick spindle apparatus. The spindle apparatus is immediately disassembled into tubulin subunits, and the mitotic phase is terminated upon completion of nuclear division and nuclear DNA segregation into the daughter cells. Tubulin gene expression and tubulin proteins promptly vanish at the transition to G1 phase. This disappearance suggests that the tubulin cytoskeleton does not have a role in cytoskeletal architecture and intracellular dynamics during the G1 phase in *C. merolae* cells.

| Description | Annotation | Gene ID |
|---|---|---|
| Tubulin superfamily | Alpha-tubulin | CMT504C |
| | Beta-tubulin | CMN263C |
| | Gamma-tubulin | CMN304C |
| Kinesin superfamily | Kinesin-related protein BimC subfamily | CMC157C |
| | Kinesin-related protein | CMO070C |
| | Kinesin-related protein CENP-E | CMQ429C |
| | Kinesin-related protein TOP | CMR497C |
| | Kinesin-related protein | CMT097C |
| Chloroplast division | FtsZ2-1 | CMS004C |
| | FtsZ2-2 | CMO089C |
| | PDR1 | CMR358C |
| | Dnm2 | CMN262C |
| Mitochondrial division | FtsZ1-1 | CMS361C |
| | FtsZ1-2 | CMT493C |
| | ZED | CMQ228C |
| | Mda1 | CMR185C |
| | Dnm1 | CME019C |

**Fig. 16.2** Gene expression profiles of cytoskeletal proteins, motor proteins, and organelle division proteins in *C. merolae*. *Z* scores of expression levels through the cell cycle were calculated from the data set of a *C. merolae* DNA microarray analysis (Fujiwara et al. 2009)

Microtubules in *C. merolae* are generally assembled in the cytosol near the nucleus; however, specific microtubule structures project from the two spindle poles to specific organelles such as a dividing mitochondrion (Fig. 16.3). This connection suggests that the spindle apparatus regulates both the nuclear DNA segregation and the segregation of dividing organelles into the daughter cells. Such a hypothesis is supported by the observation that inhibition of microtubule organization blocks the segregation of divided mitochondria into the daughter cells and arrests the cell division cycle (Nishida et al. 2005). Recent studies in *C. merolae* reported that several types of organelles are segregated into daughter cells via microtubule dynamics. Golgi are connected to the spindle poles and thereby segregated into daughter cells (Yagisawa et al. 2013). The endoplasmic reticulum (ER) is not dispersed during M phase and is equipartitioned into daughter cells via the nuclear division machinery (Yagisawa et al. 2012) (see Chap. 15). Vacuoles and peroxisomes also are tethered to the mitochondrial surface during the S to M phase and are subsequently transported into daughter cells (see Chap. 15). Together, these studies indicate that the spindle apparatus has a crucial role in the segregation of genomic DNA and divided organelles into daughter cells in *C. merolae*.

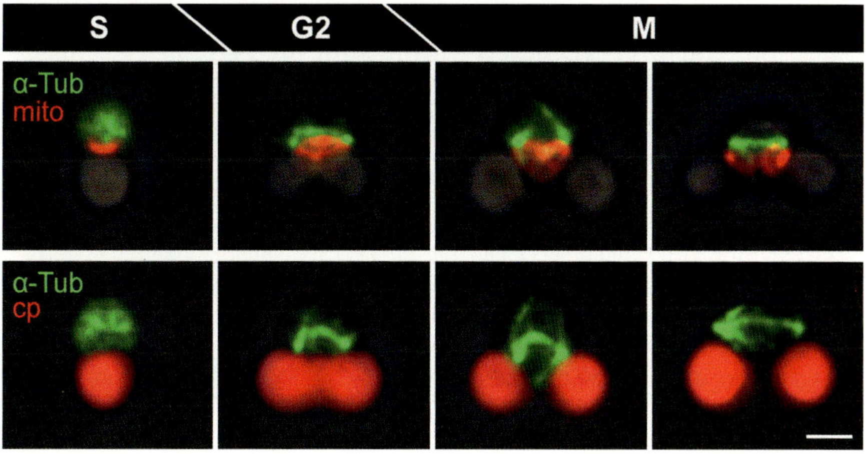


**Fig. 16.3** Protein distribution of tubulin during organelle and cell divisions. Immunofluorescence images of *C. merolae* cells. Alpha-tubulin (α-Tub) was immunostained with anti-alpha-tubulin antibody. Mitochondria (mito) were immunostained with antimitochondrial porin antibody, and chloroplast (cp) emitted red autofluorescence. Scale bar, 1 μm (Images were reproduced and modified, with permission, from Yoshida et al. 2013)

## 16.3 Kinesin Superfamily

There are two current approaches for functional analyses of microtubule dynamics in organelle proliferation and cell division in *C. merolae*. One approach investigates common molecular regulatory mechanisms in cell division and organelle division. The second approach focuses on identifying functional partner molecules of tubulin and kinesin superfamily proteins which drive intracellular transport of cargos such as protein complexes, mRNAs, and organelles (Hirokawa 1998; Hirokawa et al. 2009). The kinesin superfamily members are generally classified into three subgroups depending on the position of the motor domain within the protein sequence. N-kinesins contain an amino-terminal motor domain, which drives microtubule plus-ended microtubule motility. M-kinesins contain a motor domain in the middle of the sequence, where the domain is considered to function in microtubule depolymerization. C-kinesins contain a carboxyl-terminal motor domain that drives minus-ended microtubule motility. Eukaryotic genomes generally contain abundant numbers of kinesin superfamily genes in their genome. By contrast, the *C. merolae* genome contains only five types of kinesin superfamily genes (CMC157C, CMO070C, CMQ429C, CMR497C, and CMT097; Fig. 16.2) (Matsuzaki et al. 2004), suggesting that fundamental functions driven by kinesin superfamily proteins might be performed by these five kinesins in the evolutionarily ancient *C. merolae*.

## 16.4 Regulation of Organelle Division by the Carboxy-Terminal Motor Kinesin TOP

The biological functions of kinesins have not been completely elucidated in *C. merolae* cells, although a recent study reported that a kinesin superfamily protein functions in organelle proliferation (Yoshida et al. 2013). Mitochondrial and plastid division in *C. merolae* proceeds as follows: the division machinery for each mitochondrion (MD machinery) and plastid (PD machinery) assembles at each division site, the machinery constricts the division sites, and the divided mitochondria and plastids segregate into the daughter cells. MD and PD machineries form a complex structure before the constriction phase, which is called the MD/PD machinery complex (Yoshida et al. 2009). However, the linkage between the MD and PD machineries disappears during subsequent cell division processes, and the complex disassembles into MD and PD machineries (see Chap. 14). These spatiotemporal behaviors suggest that an undetermined molecular mechanism mediates constriction of the organellar division site.

Multi-omics analyses of the isolated MD/PD machinery complex identified a novel kinesin-like protein, which was named the three-organelle division-inducing protein (TOP, gene ID: CMR497C) (Yoshida et al. 2013). TOP has a kinesin motor domain in the carboxy-terminal region and is widely conserved in eukaryotes, especially in photosynthetic eukaryotes. *TOP* gene expression levels increase during S phase and early M phase (Fig. 16.2), along with subsequent increases in TOP protein levels. During S phase, TOP is localized on the MD/PD machinery complex and appears to mediate assembly of the two spindle poles during S-G2 phases. Subsequently, TOP accumulates on spindle poles and spindle microtubules. These observations suggest that TOP is involved in assembly of the spindle apparatus and nuclear division. The consequences of downregulation of *TOP* gene expression indicate that TOP is required for the progression of plastid and mitochondrial division in addition to nuclear division. Cytological analyses revealed that TOP-mediated organelle proliferation is regulated by phosphorylation. Aurora kinase (AUR), a serine/threonine kinase, colocalizes with TOP and appears near each mitochondrion before the constriction phase of mitochondrial and plastid division. Subsequently, AUR accumulates at each mitochondrial division site (Kato et al. 2011; Yoshida et al. 2013). Many components in the MD/PD machinery complex undergo protein phosphorylation during mitochondrial and plastid division, suggesting that AUR may regulate their activity by phosphorylation. Downregulation of TOP activity blocks mitochondrial and plastid division and disrupts AUR localization at the mitochondrial division site, resulting in AUR dispersion in the cytosol. These combined results indicate that the kinesin-like protein TOP is a crucial functional regulator of the MD/PD machinery complex dynamics, which appears to be mediated by Aurora kinase.

## 16.5 Actin and Myosin Families

The function of actomyosin in *C. merolae* remains to be elucidated. The *C. merolae* genome encodes one classical actin protein (CMM237C) and five actin-related proteins [CMS412C (51% identity with actin CMM237C at the amino acid level), CML073C (29% identity), CMS185C (26% identity), CMR270C (37% identity), and CMJ223C (24% identity)] (Matsuzaki et al. 2004). The genome does not encode any classical myosin proteins or dynein motor domains. However, the mRNA transcripts and proteins of CMM237C and CMS412C cannot be detected using conventional methods. Quantitative RT-PCR transcriptomics analysis has been successfully used to observe slight changes in actin gene expression patterns during the cell cycle (Moriyama et al. 2014), but further work is needed. Interestingly, actin expression is easily detected in the red alga *Cyanidium caldarium* RK-1, which is cytomorphologically analogous to *C. merolae*, and cell division in that species is mediated by an actin-based contractile ring that likely contains myosin (Takahashi et al. 1998). Cell division and intracellular transport in *C. merolae* cells appear to proceed via uncharacterized mechanisms that may be more ancient than the actomyosin system.

## 16.6 Elongation Factor-1α

*C. merolae* cell division is characterized by some unusual features compared with other eukaryotes. First, the cleavage furrow or the electron-dense structure representing the contractile ring is not observed during cytokinesis. Second, cytochalasin (which inhibits actin polymerization) does not affect cytokinesis. Third, neither the expression of *actin* nor actin-related protein can be detected. Fourth, the genome does not encode myosin. Together, these data suggest that cell division in *C. merolae* does not proceed via the conventional actomyosin system (Suzuki et al. 1995).

To elucidate the mechanism of *C. merolae* cytokinesis, immunohistochemistry analyses have been attempted to identify proteins localized in the cell division plane. To date, the eukaryotic elongation factor-1α (eEF-1α) is the only known protein localized in the cell division plane during *C. merolae* cytokinesis. Elongation factor-1α is a highly conserved member of the GTPase superfamily, which functions in a number of non-translational cellular processes such as interacting with cytoskeletal proteins (Durso and Cyr 1994; Shiina et al. 1994). In protozoa and animal cells, eEF-1α is localized in the contractile ring and functions in cytokinesis (Numata et al. 2000; Bunai et al. 2006; Fujimoto and Mabuchi 2010). The universal function of eEF-1α in cytokinesis, and the localization of eEF-1α in the *C. merolae* cell division plane, suggests that it may mediate cytokinesis in the absence of the actomyosin system in *C. merolae*. Immunohistochemical analyses using an anti-eEF-1α antibody (Imoto et al. 2011) have been conducted to determine the eEF-1α distribution throughout the cell cycle in *C. merolae*. The results showed that eEF-1α is continuously expressed during the cell cycle and its levels increase during the cytokinesis-G1 phase. Initially, eEF-1α is distributed throughout the cytoplasm during G1-early mitosis; then, it accumulates and localizes along the equatorial plane when the spindle forms between the spindle poles. Subsequently, eEF-1α levels increase along the equatorial plane at the initiation of cytokinesis. As cytokinesis proceeds, eEF-1α levels decline along the equatorial plane. The two daughter cells are connected by a narrow bridge at the final stage of cytokinesis, but eEF-1α is not detected in this structure. These observations strongly suggest that eEF-1α functions in cell membrane contraction, but not in membrane severance. These combined results suggest that *C. merolae* cytokinesis occurs in two steps: initial contraction of the cell membrane mediated by eEF-1α and severance of the narrow membranous bridge connecting the daughter cells mediated by an unidentified mechanism.

## 16.7 Conclusions and Perspectives

*C. merolae* has become an invaluable model cell to elucidate organelle division. Pioneering studies have demonstrated that organelle division and inheritance are tightly regulated and sequentially orchestrated in *C. merolae* cells. Spindle structures and kinesin superfamily proteins are crucial for organelle division, both directly and indirectly. Organelles are transported along a microtubule network, which appears to be driven by kinesin superfamily proteins. The C-kinesin TOP induces the accumulation of Aurora kinase onto the MD/PD machinery complex and activates organellar constriction by the complex. This suggests that the distribution of other mitotic kinases may be regulated by microtubule motors. However, the precise molecular mechanisms driving these processes are the subject of intensive and ongoing research. These studies particularly aim to elucidate how the cell division cycle and the organelle division cycle are synchronized. Transcriptomics analyses of synchronized *C. merolae* cultures throughout cell division and cytokinesis have identified several candidates during cytokinesis (Fujiwara et al. 2009). Future analyses of these genes will provide valuable insight into ancient molecular mechanisms of cytokinesis.

**Acknowledgments** This work was supported by a Human Frontier Science Program Long Term Fellowship (LT000356/2011-L to Y.Y.) and a grant from the Japan Society for the Promotion of Science Fellowship (no. 14J04556 to Y.I.).

## References

Bunai F, Ando K, Ueno H, Numata O (2006) Tetrahymena eukaryotic translation elongation factor 1A (eEF1A) bundles filamentous actin through dimer formation. J Biochem 140:393–399. doi.org/10.1093/jb/mvj169

Durso NA, Cyr RJ (1994) A calmodulin-sensitive interaction between microtubules and a higher plant homolog of elongation factor-1 alpha. Plant Cell 6:893–905. https://doi.org/10.1105/tpc.6.6.893

Fujimoto H, Mabuchi I (2010) Elongation factors are involved in cytokinesis of sea urchin eggs. Genes Cells 15:123–135. https://doi.org/10.1111/j.1365-2443.2009.01370.x

Fujiwara T, Misumi O, Tashiro K et al (2009) Periodic gene expression patterns during the highly synchronized cell nucleus and organelle division cycles in the unicellular red alga *Cyanidioschyzon merolae*. DNA Res 16:59–72. https://doi.org/10.1093/dnares/dsn032

Hirokawa N (1998) Kinesin and dynein superfamily proteins and the mechanism of organelle transport. Science 279:519–526. https://doi.org/10.1126/science.279.5350.519

Hirokawa N, Noda Y, Tanaka Y, Niwa S (2009) Kinesin superfamily motor proteins and intracellular transport. Nat Rev Mol Cell Biol 10:682–696. https://doi.org/10.1038/nrm2774

Imoto Y, Fujiwara T, Yoshida Y et al (2010) Division of cell nuclei, mitochondria, plastids, and microbodies mediated by mitotic spindle poles in the primitive red alga *Cyanidioschyzon merolae*. Protoplasma 241:63–74. https://doi.org/10.1007/s00709-010-0107-y

Imoto Y, Nishida K, Yagisawa F et al (2011) Involvement of elongation factor-1α in cytokinesis without actomyosin contractile ring in the primitive red alga *Cyanidioschyzon merolae*. Cytologia 76:431–437. https://doi.org/10.1508/cytologia.76.431

Kato S, Imoto Y, Ohnuma M et al (2011) Aurora kinase of the red alga *Cyanidioschyzon merolae* is related to both mitochondrial division and mitotic spindle formation. Cytologia 76:465–472

Kobayashi Y, Kanesaki Y, Tanaka A et al (2009) Tetrapyrrole signal as a cell-cycle coordinator from organelle to nuclear DNA replication in plant cells. Proc Natl Acad Sci U S A 106:803–807. https://doi.org/10.1073/pnas.0804270105

Kobayashi Y, Imamura S, Hanaoka M, Tanaka K (2011) A tetrapyrrole-regulated ubiquitin ligase controls algal nuclear DNA replication. Nat Cell Biol 13:483–487. https://doi.org/10.1038/ncb2203

Maruyama S, Kuroiwa H, Miyagishima S et al (2007) Centromere dynamics in the primitive red alga *Cyanidioschyzon merolae*. Plant J 49:1122–1129. https://doi.org/10.1111/j.1365-313X.2006.03024.x

Matsuzaki M, Misumi O, Shin-I T et al (2004) Genome sequence of the ultrasmall unicellular red alga *Cyanidioschyzon merolae* 10D. Nature 428:653–657. https://doi.org/10.1038/nature02398

Moriyama T, Tajima N, Sekine K, Sato N (2014) Localization and phylogenetic analysis of enzymes related to organellar genome replication in the unicellular rhodophyte *Cyanidioschyzon merolae*. Genome Biol Evol 6:228–237. https://doi.org/10.1093/gbe/evu009

Nishida K, Yagisawa F, Kuroiwa H et al (2005) Cell cycle-regulated, microtubule-independent organelle division in *Cyanidioschyzon merolae*. Mol Biol Cell 16:2493–2502. https://doi.org/10.1091/mbc.E05

Numata O, Kurasawa Y, Gonda K, Watanabe Y (2000) Tetrahymena elongation factor-1α is localized with calmodulin in the division furrow. J Biochem 127:51–56. https://doi.org/10.1093/oxfordjournals.jbchem.a022583

Shiina N, Gotoh Y, Kubomura N et al (1994) Microtubule severing by elongation factor 1. Science 266:282–285. https://doi.org/10.1126/science.7939665

Sumiya N, Fujiwara T, Era A, Miyagishima S (2016) Chloroplast division checkpoint in eukaryotic algae. Proc Natl Acad Sci 113:E7629–E7638. https://doi.org/10.1073/pnas.1612872113

Suzuki K, Kawazu T, Mita T, Takahashi H, Itoh R, Toda K, Kuroiwa T (1995) Cytokinesis by a contractile ring in the primitive red alga *Cyanidium caldarium* RK-1. Eur J Cell Biol 67:170–178

Takahashi H, Takano H, Kuroiwa H et al (1998) A possible role for actin dots in the formation of the contractile ring in the ultra-micro alga *Cyanidium caldarium* RK-1. Protoplasma 202:91–104. https://doi.org/10.1007/BF01280878

Yagisawa F, Fujiwara T, Kuroiwa H et al (2012) Mitotic inheritance of endoplasmic reticulum in the primitive red alga *Cyanidioschyzon merolae*. Protoplasma 249:1129–1135. https://doi.org/10.1007/s00709-011-0359-1

Yagisawa F, Fujiwara T, Ohnuma M et al (2013) Golgi inheritance in the primitive red alga, *Cyanidioschyzon merolae*. Protoplasma 250:943–948. https://doi.org/10.1007/s00709-012-0467-6

Yoshida Y, Kuroiwa H, Hirooka S et al (2009) The bacterial ZapA-like protein ZED is required for mitochondrial division. Curr Biol 19:1491–1497. https://doi.org/10.1016/j.cub.2009.07.035

Yoshida Y, Fujiwara T, Imoto Y et al (2013) The kinesin-like protein TOP promotes aurora localisation and induces mitochondrial, chloroplast and nuclear division. J Cell Sci 126:2392–2400. https://doi.org/10.1242/jcs.116798

# Part VI
# Metabolism

**Kan Tanaka and Naoki Sato**

Metabolism is a function that supplies energy and biomolecules for the cell, and has been usually considered well conserved among organisms as compared with other biological subjects. Thus, metabolic pathways and their regulatory aspects have been estimated based on information obtained from very limited number of model organisms. However, as genome information of divergent organisms, including prokaryotes and eukaryotes, are being made clear, it is now clear that the metabolic pathways and enzymes are far more diverse than hitherto anticipated and much remains to be investigated to understand the whole view.

Complete genome sequence of *Cyanidioschyzon merolae* was determined in 2004 and analyzed extensively thereafter. The revealed metabolic pathways indicated considerable difference from those found in other model organisms, and especially, it was found that the overall structure of metabolism was not so analogous to land plants that have been considered typical as the photosynthetic eukaryotes. These findings are significantly enriching our knowledge on the unexpected diversity of eukaryotes. In Chap. 17, updated comparative genomic data are used to characterize photosynthetic mechanisms in red algae in general, or in particular, *C. merolae*, with respect to green plants. Chapter 18 summarizes the simplified nitrogen assimilation system in this simplest autotrophic eukaryote, and identification of the fundamental transcriptional regulator, whose activation pathway appears to be connected with the general checkpoint kinase TOR as in other eukaryotes. Chapter 19 is a concise summary of various pathways of carbon metabolism in red algae, that are obviously similar to those in green plants in many aspects but distinct from the latter in some key enzymes, all of which makes up red algal metabolic originality, potentially useful in algal biotechnology.

# Chapter 17
# Photosynthesis

**Naoki Sato and Takashi Moriyama**

**Abstract** Photosynthesis is the essential biological process that ultimately drives all forms of life. The photosynthetic apparatus of red algae is distinct from that of green algae and plants (green plants) in that phycobilisomes transmit light energy to Photosystem II. There are two types of red algal phycobilisomes, one having phycoerythrins and the other having only phycocyanins, which are similar to those in cyanobacteria. Phycobilisomes also act as a nitrogen reserve. The structure of Photosystem I is essentially similar in red algae and green plants (with some differences in small subunits), whereas the extrinsic proteins of Photosystem II protecting the oxygen-evolving center differ in the two types of photosynthetic organisms. Another major difference is that more components of the two photosystems are encoded by the chloroplast genome in red algae than in green plants, in which most are encoded by the nuclear genome. *Cyanidioschyzon merolae* serves as a model system for the study of red algal photosynthesis.

**Keywords** Chloroplast genome • Comparative genomics • Photosystem I • Photosystem II • Phycobilisome

## 17.1 Introduction

Photosynthesis is the ultimate driving force of all living beings, exploiting sunlight to produce biochemical free energy mainly in the form of reduced carbon (carbohydrate and/or lipids) and oxidants (most commonly oxygen). Oxygenic photosynthesis, or photosynthesis producing oxygen as the oxidant, is common in algae and plants, and it has been considered a great heritage from cyanobacteria through endosymbiosis (see, e.g., Archibald 2014). Despite their common ancestry, the components and modes of functioning of individual photosynthetic mechanisms are not identical in chloroplasts and cyanobacteria.

N. Sato (✉) • T. Moriyama
Department of Life Sciences, Graduate School of Arts and Sciences, The University of Tokyo, Meguro-ku, Tokyo, Japan
e-mail: naokisat@bio.c.u-tokyo.ac.jp



T. Kuroiwa et al. (eds.), *Cyanidioschyzon merolae*,
https://doi.org/10.1007/978-981-10-6101-1_17

Marked differences are also found between red algae and "green plants" (also called Viridiplantae, namely, green algae and plants).

As shown by the absorption spectra in Fig. 17.1, *Cyanidioschyzon merolae* contains three types of pigment: chlorophyll *a*, phycobiliproteins, and carotenoids. Red algae do not contain chlorophyll *b*, which is an important accessory light-harvesting pigment in Viridiplantae and some cyanobacteria (e.g., *Prochlorococcus*). Chlorophyll *a* exists in light-harvesting chlorophyll-protein complexes, as well as in the reaction center complexes of Photosystem I (PS I) and Photosystem II (PS II).

*C. merolae* contains phycocyanin and allophycocyanin as major phycobiliproteins, whereas many marine red algae also contain phycoerythrin. The presence of phycocyanin gives *C. merolae* cells their blue-green appearance. The same is true for other red algae in Cyanidiales. As will be described, the two

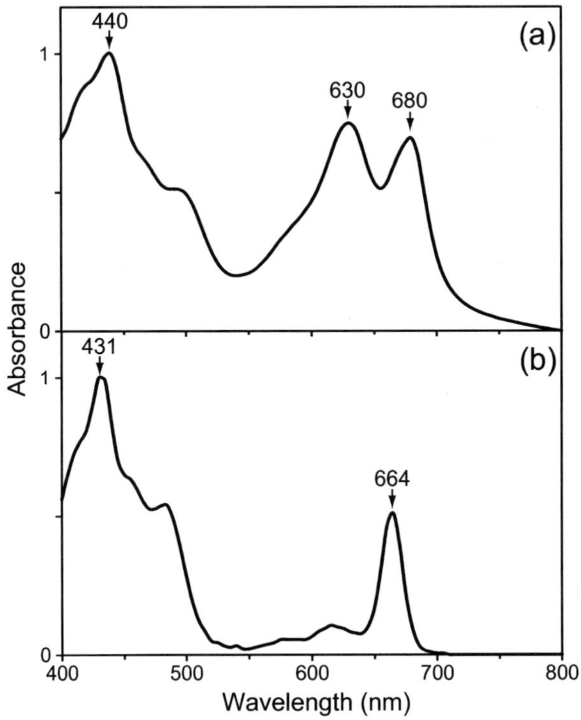


**Fig. 17.1** Absorption spectra of (**a**) intact cells of *Cyanidioschyzon merolae* and (**b**) extracted pigments. *C. merolae* cells contain chlorophyll *a*, phycobiliproteins, and carotenoids as major pigments. In intact cells, protein-associated chlorophyll *a* has absorption maxima at approximately 680 and 440 nm, whereas phycobiliproteins and carotenoids absorb light at about 630 nm and 450–480 nm, respectively. Acetone extract of the cells (b) contains chlorophyll *a* and carotenoids, because phycobiliproteins are not soluble in organic solvents. The absorption peaks of these pigments in extract were different from those in intact cells, in which the pigments exist as protein-pigment complexes

types of phycobiliprotein are not radically different at the level of amino acid sequences, and the presence of phycobiliproteins, whether red or blue, characterizes the photosynthetic properties of red algae. These pigment proteins are also potential resources as safe coloring agents for use in the food industry. Photosynthetic production of oil and polysaccharides from red algae has become more important in recent years.

The chloroplast (plastid) is the site of photosynthesis in red algae, as in many other photosynthetic eukaryotes (Fig. 17.2). The chloroplast is enclosed by two layers of the envelope membrane ("Env" in Fig. 17.2c). Thylakoid membranes ("T") are not stacked in red algae. Phycobilisomes are arranged on the thylakoid membranes as regular arrays, which can be observed in thin sections depending on the viewing angle ("PBS"). In the middle of the chloroplast are nucleoid regions with low electron densities ("Nuc"), which contain chloroplast DNA, occasionally visualized as thin fibers.

In this chapter, we describe the protein constituents of antenna systems, PS I, PS II, electron transport systems, adenosine triphosphate (ATP) synthase, and the carbon fixation cycle, essentially based on comparative genomic data for various red algae. We will also present some recent studies on the functional aspects of photosynthesis in red algae. We employ our original, comparative genomics data based on whole-genome clustering using the software Gclust

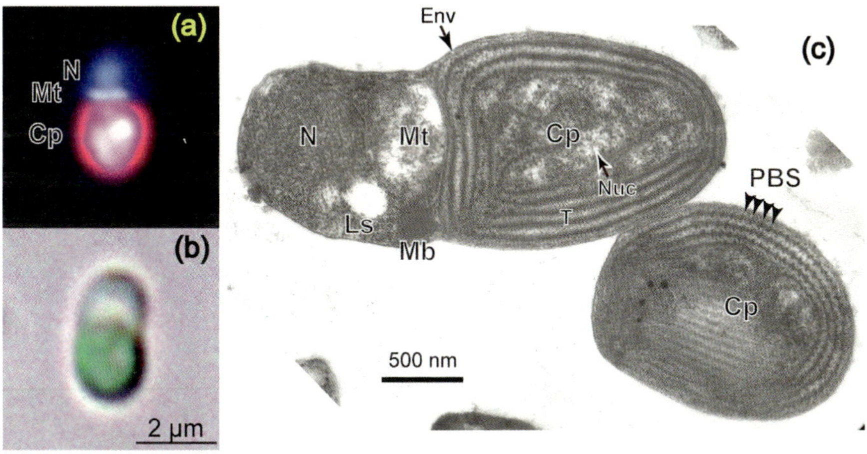


**Fig. 17.2** (**a**) Fluorescence micrograph, (**b**) differential interference image, and (**c**) electron micrograph of *Cyanidioschyzon merolae* cells. Note that fluorescence in the chloroplast was *deep red* (not *magenta* as seen here) in the actual view field, due to emission by phycobiliproteins rather than chlorophyll *a*. In the transmission electron microscopy (TEM) image of a thin section (c), two cells are shown: one in longitudinal section and the other in transverse section. Several layers of the thylakoid membrane are visible within the chloroplasts. No grana stacking was observed. In the *middle* of the chloroplast, nucleoid regions are visible at *lower* electron density. Depending on the angle, arrays of phycobilisomes (PBS) are clearly seen (*Cp* chloroplast, *Env* chloroplast envelope membranes, *Ls* lysosome, *Mb* microbody, *Mt* mitochondrion, *N* cell nucleus, *Nuc* chloroplast nucleoid region, *T* thylakoid membrane)

**Table 17.1** Sources of genomic data used for comparative genomics

| ID | Name | Chloroplast genome | Mitochondrial genome | Nuclear genome URL | Ref |
|---|---|---|---|---|---|
| CME | *Cyanidioschyzon merolae* | AB002583 | D89861 | Merolae server<br>http://merolae.s.u-tokyo.ac.jp/ | [1] |
| GaSU | *Galdieria sulphuraria* | KJ700459 | KJ700460 | EnsemblPlants: ASM34128v1<br>http://plants.ensembl.org/ | [2] |
| PoPU | *Porphyridium purpureum* | AP012987 | — | Porphyridium genome project<br>http://cyanophora.rutgers.edu/porphyridium/ | [3] |
| PyYE | *Pyropia yezoensis* | KC517072 | NC_017837 | NRIFS Susabi-nori genome<br>http://nrifs.fra.affrc.go.jp/ | [4] |
| ChCR | *Chondrus crispus* | HF562234 | NC_001677 | EnsemblPlants: ASM35022v2<br>http://plants.ensembl.org/ | [5] |
| ATH | *Arabidopsis thaliana* (angiosperm) | AP000423 | Y08501 | TAIR<br>http://www.arabidopsis.org/ | [6] |

For chloroplast and mitochondrial genomes, GenBank accession numbers are shown. "—" indicates that this genome data is unavailable

References for nuclear genomes: [1] Matsuzaki et al. (2004), Nozaki et al. (2007), [2] Schönknecht et al. (2013), [3] Bhattacharya et al. (2013), [4] Nakamura et al. (2013), [5] Collén et al. (2013), and [6] Lamesch et al. (2011)

(Sato 2009). Table 17.1 lists the organisms that we compared. The results are available as the dataset Gclust2016R ("R" for red algae) on the Gclust server (http://gclust.c.u-tokyo.ac.jp/).

## 17.2 Light-Harvesting Antenna Systems

In photosynthesis, sunlight is first absorbed by the pigments serving as (peripheral) "antennae" that harvest light quanta and supply their energy to the "core antenna" chlorophylls in the reaction center complexes, which finally transmit excitation energy to the reaction centers. This peculiar, funnel-like organization is important for the efficient use of incident light with minimal investment to cost-intensive reaction centers. It also permits adjustment of the photosynthetic capacity under varying light conditions (i.e., according to light quality and intensity). Light-harvesting chlorophyll is also present in the core complexes of PS I and PS II, as well as in antenna chlorophyll-binding proteins, but they have different absorption maxima to allow excitation energy to efficiently flow into the reaction centers. This variation in the absorption maxima of chlorophyll is realized by differences in the mode of binding to the proteins, but the mechanistic details remain to be fully elucidated. In this section, we describe the antenna systems distinct from the core complexes of photosystems. The present chapter provides an update on the protein components in light-harvesting antenna systems, as well as PS I and PS II, which have been previously published (Vanselow et al. 2009; Su et al. 2010).

Red algal antenna systems are different from those of green plants, mainly because chlorophyll *b* is absent and phycobilisomes are the light-harvesting system for PS II. Light-harvesting chlorophyll *a*/*b*-binding proteins, found abundantly in the chloroplasts of Viridiplantae, are not present in red algae. A particular type of light-harvesting chlorophyll protein, Lhcr, is present as light-harvesting chlorophyll protein I (LHC I) in red algae. Most of the phycobilisome components are encoded by the chloroplast genome (Table 17.2), whereas all Lhcr proteins are encoded by the nuclear genome. We explain these two types of light-harvesting systems in detail below.

Phycobilisome is a large protein complex (see, e.g., Liu et al. 2005; Watanabe and Ikeuchi 2013) composed of a core in contact with PS II and several rods (typically six rods per core) that radiate from the core. The core is a bundle of three core cylinders, each made up of stacked unit disks. Each disk is a double trimer (or hexamer) of heterodimers consisting of α- and β-allophycocyanins (encoded by *apcA* and *apcB* genes, respectively). The rod is also made up of stacks of disks, each of which is a double trimer of heterodimers consisting of α- and β-phycocyanins (encoded by *cpcA* and *cpcB* genes, respectively) or α- and β-phycoerythrins (encoded by *cpeA* and *cpeB* genes, respectively). Additionally, linker proteins connect disks within the core (ApcC) or within the rod (CpcC), connect the rod and the core (CpcG), connect the core and the membrane (ApcE), and connect the rod and the membrane (CpcL/CpcG2, acting in the binding of PS I and rods in cyanobacteria); some linker proteins are also located at the tip of the rod (CpcD, only found in cyanobacteria). Phycoerythrin-related proteins are named CpeX instead of CpcX (where X is a character in the gene name).

Traditionally, phycocyanins have been classified into two types. C-phycocyanin is mainly present in cyanobacteria and Cyanidiales. The α-subunit of C-phycocyanin binds a phycocyanobilin, whereas the β-subunit of C-phycocyanin binds two phycocyanobilins. R-phycocyanin, mainly present in red-colored red algae, binds both phycocyanobilin and phycoerythrobilin.

Phycobilisomes in *Porphyridium purpureum* were observed by atomic force microscopy (AFM) (Liu et al. 2008), and dramatic changes in the supramolecular architecture of thylakoid membranes have been shown to occur following nitrogen starvation (Zhao et al. 2016). The mobility of phycobilisome was demonstrated by

**Table 17.2** Components of phycobilisomes in red algae

| Component | CME | GaSU | PoPU | PyYE | ChCR |
|---|---|---|---|---|---|
| ApcA | Cp | Cp | Cp | Cp | Cp |
| ApcB | Cp | Cp | Cp | Cp | Cp |
| ApcC | CMO250C | EME31123 | 2113.4 | g9334 | CDF37252 |
| ApcD | Cp | Cp | Cp | Cp | Cp |
| ApcE | Cp | Cp | Cp | Cp | Cp |
| ApcF | Cp | Cp | Cp | Cp | Cp |
| CpcA | Cp | Cp | Cp | Cp | Cp |
| CpcB | Cp | Cp | Cp | Cp | Cp |
| CpcC or CpeC | CPM166C | EME29162 | 3399.1?<br>479.22 | g2303<br>g2306<br>g1935? | ? |
| CpcE | CMJ043C | EME31428 | 4473.7 | g7170<br>g9023 | CDF35399 |
| CpcF | CMJ044C | EME26156 | 2146.6 | ? | ? |
| CpcG | Cp | Cp | Cp | Cp | Cp |
| CpeA | None | None | Cp | Cp | Cp |
| CpeB | None | None | Cp | Cp | Cp |

"Cp" indicates that the gene is encoded by the chloroplast genome. All nuclear-encoded genes are indicated by their gene identifiers. The absence of a gene is indicated by "none." In PyYE, the gene number is indicated beginning from "g." Question marks indicate that the gene is not detected in the current genomic data but that the gene is likely to exist

the fluorescence recovery after photobleaching (FRAP) experiments in *Porphyridium cruentum*, but not in *Cyanidium caldarium* (Kaňa et al. 2014), and has been correlated with membrane fluidity. Mobility can account for the well-known light-induced increase in fluorescence, which is a result of state 2-to-state 1 transition, favoring light energy partition to PS II. Energy decoupling between phycoerythrin and phycocyanin has been shown to be an initial adaptation to high light in *P. cruentum* (Liu et al. 2008). Such dynamic changes in phycobilisomes have not been documented in *C. merolae*.

Lhcr proteins act as the light-harvesting system of red algal PS I. They have been characterized in *P. purpureum* (Tan et al. 1997), *G. sulphuraria* (Marquardt et al. 2001), and *C. merolae* (Busch et al. 2010) and exhibit a degree of homology to LHC I proteins of green plants (also called Lhca1–Lhca4) and diatom LHC I. Each Lhca protein in green plants is known to have three membrane-spanning α-helices and to bind eight to ten chlorophyll molecules. Comparative genomic analysis using Gclust identified five subgroups of red algal Lhcr proteins (Fig. 17.3). All three Lhcr proteins of *C. merolae* belong to the Lhcr2 subgroup. Lhcr5 is most closely related to the large family of fucoxanthin-binding chlorophyll *a*/*c*-binding proteins of diatoms that act as LHC I.

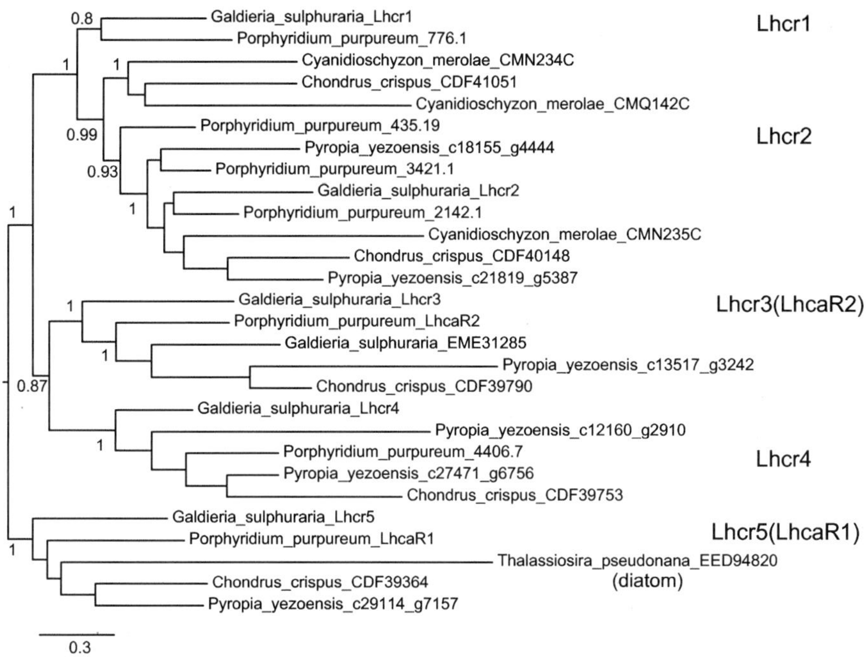


**Fig. 17.3** Phylogenetic tree of red algal Lhcr (LHC I) proteins that act as antennae for PS I. Lhcr protein sequences were obtained by whole-genome clustering. Known sequences were described in Busch et al. (2010), Su et al. (2010), and Thangaraj et al. (2011). This tree was calculated by the maximum likelihood method using PhyML software (Guindon et al. 2010) and drawn with FigTree software (http://tree.bio.ed.ac.uk/software/figtree/). Each value on the major branches indicates the confidence level of branching. Five major clades correspond to Lhcr1–5. Each clade contains proteins of *G. sulphuraria*, *P. purpureum*, *P. yezoensis*, and *C. crispus*, except Lhcr1. All three Lhcr proteins of *C. merolae* belong to the Lhcr2 clade. Lhcr5 is closest to the very large family of fucoxanthin-chlorophyll *a*/*c*-binding diatom proteins. In this figure, only one diatom protein is shown

## 17.3 Photosystems

### *17.3.1 Photosystem I*

In the noncyclic electron transport mode of photosynthesis, PS I uses excitation energy to oxidize PS II via the electron transport chain and to reduce $NADP^+$ to NADPH. The basic components and modes of action of PS I have been mainly characterized in green plants and cyanobacteria. Structural studies of red algal PS I have been published for *C. caldarium* (Gardian et al. 2007), *Galdieria sulphuraria* (Thangaraj et al. 2011), and *C. merolae* (Busch et al. 2010). These studies include the isolation of PS I particles, biochemical analysis of polypeptides, and electron microscopic single-particle analysis. The PS I particle is a trimer of PS I complexes in cyanobacteria; however, in red algae, as in green

**Table 17.3** Comparison of PS I components in red algae and *Arabidopsis thaliana*

| Component | CME | GaSU | PoPU | PyYE | ChCR | ATH |
|---|---|---|---|---|---|---|
| PsaA | Cp | Cp | Cp | Cp | Cp | Cp |
| PsaB | Cp | Cp | Cp | Cp | Cp | Cp |
| PsaC | Cp | Cp | Cp | Cp | Cp | Cp |
| PsaD | Cp | Cp | Cp | Cp | Cp | 1G03130 |
| | | | | | | 4G02770 |
| PsaE | Cp | Cp | Cp | Cp | Cp | 2G20260 |
| | | | | | | 4G28750 |
| PsaF | Cp | Cp | Cp | Cp | Cp | 1G31330 |
| PsaG | None | None | None | None | None | 1G55670 |
| PsaH | None | None | None | None | None | 3G16140 |
| | | | | | | 1G52230 |
| PsaI | Cp | Cp | Cp | Cp | Cp | Cp |
| PsaJ | Cp | Cp | Cp | Cp | Cp | Cp |
| PsaK | Cp | Cp | Cp | Cp | Cp | 1G30380 |
| PsaL | Cp | Cp | Cp | Cp | Cp | 4G12800 |
| PsaM | Cp | Cp | Cp | Cp | Cp | None |
| PsaN | None | None | None | None | None | 5G64040 |
| PsaO | CMP086C | EME28128 | 3695.1 | g722 | CDF32153 | 1G08380 |

"Cp" indicates that the gene is encoded by the chloroplast genome. All nuclear-encoded genes are indicated by their gene identifiers. The absence of a gene is indicated by "none." PsaX, present in certain species of cyanobacteria, is not found in any of the organisms listed in this table

plants, the PS I supercomplex consists of a monomer of PS I complex and Lhcr antenna (Gardian et al. 2007; Vanselow et al. 2009; Busch et al. 2010; Thangaraj et al. 2011).

Components of the PS I complex are listed in Table 17.3. The PS I complex is assembled around a core consisting of two large subunits, PsaA and PsaB, which form the P700 reaction center that contains chlorophyll *a* and *a*′. Each monomer of the cyanobacterial PS I complex contains 96 chlorophyll molecules, as well as 22 carotenoids, 2 phylloquinones, 3 $Fe_4S_4$ clusters, and 4 lipids (Fromme et al. 2001). The pea PS I complex, with four Lhca subunits, showed the plant-specific subunits PsaG and PsaH, as well as previously undetected PsaK (Amunts et al. 2010, Fig. 17.4). In contrast with green plants, most red algal PS I subunits are encoded by the chloroplast genome. The three genes *psaG*, *psaH*, and *psaN* are not present in red algae. Vanselow et al. (2009) reported that they found PsaN in *G. sulphuraria*; however, the indicated amino acid sequence was not truly homologous to known PsaN (we detected no significant homologous genomic or chloroplast sequences with a TBLASTN E-value $<0.1$ in any known red algal genomes).

The *C. merolae* PS I complex has been found to contain three Lhcr proteins, as well as the phycobilisome subunits CpcC and CpcG, but not ApcE (Busch et al. 2010). The authors suggested a PS I-linked phycobilisome sub-complex. Since red

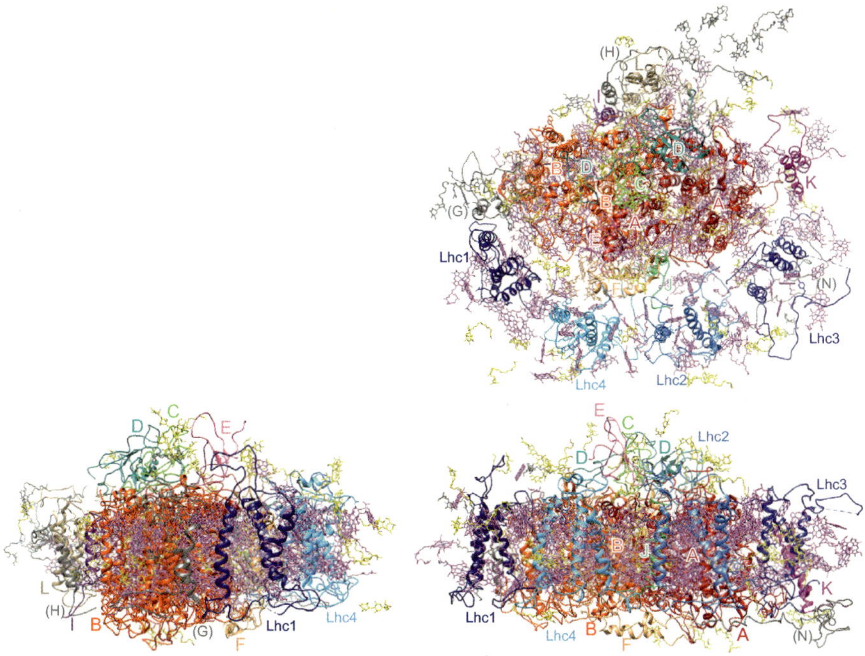


**Fig. 17.4** Model of Photosystem I (PS I) supercomplex in *green* plants, used as a reference to visualize the structure of the *red* algal PS I complex. This figure was constructed from a coordinate file (PDB code: 3LW5) using the UCSF Chimera software (Pettersen et al. 2004) showing three views of an identical complex: *left, top, and side (right bottom)*. In the side view, the stroma is located at the *top*. Chlorophyll is indicated in orchid, while other nonprotein compounds in gold and iron atoms in *black*. The PsaG, PsaH, and PsaN subunits that are present in plant PS I, but not in *red* algal PS I, are shown in *dark gray*, with annotations in parentheses. All of these structures are located at the periphery of the complex. Lhc proteins are shown in four different hues of *blue*. These are likely to be replaced by Lhcr proteins in red algae

algae do not have CpcG2/CpcL, which is supposed to anchor the phycobilisome rod directly to PS I in cyanobacteria (Watanabe and Ikeuchi 2013), another mechanism is likely to anchor the phycobilisome sub-complex to PS I. In a proteomic analysis by Busch et al. (2010), PsaM and PsaO were not detected in the PS I preparation. We can visualize the structure of *C. merolae* PS I based on the structure of plant PS I (Fig. 17.4). In this figure, plant-specific subunits (G, H, and N) are shown in dark gray with annotations in parentheses. These subunits are located in the periphery of the PS I complex. If we suppose that they are replaced by subunits specific to red algae, then they are expected to form an interaction interface with other complexes such as PS II and cytochrome $b_6f$ complexes. In addition, the Lhca1–Lhca4 plant subunits could be replaced by homologous Lhcr proteins with little alteration of the overall structure. Vanselow et al. (2009) suggested that two molecules of CMQ142C, and one each of CMN234C and CMN235C, may be associated with a

PS I complex. The details of the arrangement of Lhcr proteins will be elucidated by on-going structural analysis.

### *17.3.2 Photosystem II*

PS II uses light energy to oxidize water to produce oxygen and to reduce PS I via the electron transport chain. Red algal PS II receives light energy from phycobilisomes, as described above. PS II is assembled around a core consisting of PsbA (D1), PsbD (D2), PsbB (CP47), and PsbC (CP43). D1 and D2 proteins form the reaction center containing P680, a dimer of chlorophyll *a*. CP47 and CP43 are light-harvesting proteins transferring light energy to the D1/D2 core. Many components of PS II, such as PsbA, B, C, D, E, F, H, I, J, K, and L, are encoded by the chloroplast genome in both red algae and green plants. In red algae, however, many additional components are also encoded by the chloroplast genome (Table 17.4). The extrinsic PS II subunits (i.e., those facing the lumen or the internal space of the thylakoid membrane) are different in green plants, cyanobacteria, and red algae. PsbP and PsbQ bind to PsbO (the so-called 33-kDa protein) in green plants, whereas PsbU and PsbV (cytochrome c550) bind to PsbO in cyanobacteria. Extrinsic proteins of red algal PS II are similar to their cyanobacterial counterparts; however, the subunit PsbQ′, which is homologous to PsbQ, is also present in purified red algal PS II particles (Ohta et al. 2003; Enami et al. 2008; Ifuku 2015). A homolog of PsbP is also encoded in the genomes of cyanobacteria (called "cyano-P") and red algae ("PsbP-like"), but these homologs have not been detected in purified PS II particles and are not considered to be directly involved in the oxygen evolution during photosynthesis (Sato 2010; Ifuku 2015).

Single-particle analysis of PS II has been performed in *P. purpureum* (Bumba et al. 2004) and *C. caldarium* (Gardian et al. 2007). A PS II particle is a dimer of the PS II complex. In reconstructed images, extrinsic proteins were identified as distinct protrusions. Crystals of the *C. caldarium* PS II complex were prepared (Adachi et al. 2009), and the X-ray crystal structure of the PS II dimer was determined at 2.76 Å resolution (Ago et al. 2016). The overall structure (Fig. 17.5) was similar to the previously determined structures of cyanobacterial PS II (Guskov et al. 2009; Umena et al. 2011).

## 17.4 Electron Transport and Adenosine Triphosphate Synthesis

The cytochrome $b_6f$ complex accepts electrons from PS II and donates them to PS I. During electron transport within thylakoid membranes, protons are translocated through the membranes into the lumen to form a proton gradient or proton-motive

**Table 17.4** Small PS II subunits in red algae and *Arabidopsis thaliana*

| Component | CME | GaSU | PoPU | PyYE | ChCR | ATH |
|---|---|---|---|---|---|---|
| PsbM | CMT182C | EME25842 | 3802.1 | g6715<br>g6570 | ? | Cp |
| PsbN | Cp | Cp | Cp | Cp | Cp | Cp |
| PsbO | CMI290C | EME29918 | 2663.4 | g3681<br>g9974<br>g3682<br>g8790 | CDF33932 | 5G66570<br>3G50820<br>4G37230 |
| PsbP | None | None | None | None | None | 1G06680<br>2G30790 |
| PsbP-like (non-PS II component) | CMN290C | EME32636 | 4487.13 | g6492 | ? | 3G55330<br>2G39470 |
| PsbQ | None | None | None | None | None | 4G05180<br>4G21280<br>2G01918<br>3G01440<br>1G14150 |
| PsbQ′ | CMC133C | EME28144 | 469.1 | g2626<br>g8010<br>g9246 | CDF38494 | None |
| PsbR | None | None | None | None | None | 1G79040 |
| PsbS/NPQ24 | None | None | None | None | None | 1G44575 |
| PsbT | Cp | Cp | Cp | Cp | Cp | Cp |
| PsbU | CMI248C | EME29916 | 555.1 | g10039<br>g4763 | CDF33931 | None |
| PsbV/Cyt c550 | Cp | Cp | Cp | Cp | Cp | None |
| PsbW/Psb28 | Cp | Cp | Cp | Cp | Cp | 2G30570<br>4G28660 |
| PsbX | Cp | Cp | Cp | Cp | Cp | 2G06520 |
| PsbY/Ycf32 | Cp | Cp | Cp | Cp | Cp | 1G67740 |
| PsbZ | Cp | Cp | Cp | Cp | Cp | Cp |
| Psb27 | CMK176C | EME28982 | 4542.1 | g4989 | ? | 1G03600<br>1G05385 |
| Psb30/Ycf12 | Cp | Cp | Cp | Cp | Cp | None |

The subunits, PsbA, B, C, D, E, F, H, I, J, K, and L, are encoded by the chloroplast genome in all organisms analyzed and are not listed in this table. Alternative protein names are as follows: PsbA = D1, PsbB = CP47, PsbC = CP43, PsbD = D2

"Cp" indicates that the gene is encoded by the chloroplast genome. All nuclear-encoded genes are indicated by their gene identifiers. The absence of a gene is indicated by "none." Question marks indicate that the gene is not detected in the current genomic data but that the gene is likely to exist

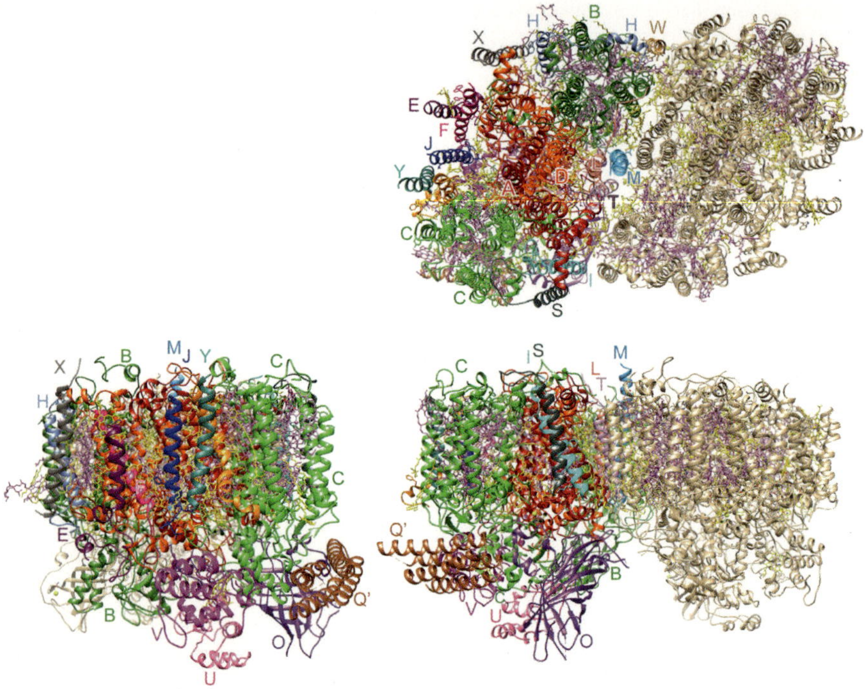


**Fig. 17.5** X-ray crystal structure of the Photosystem II (PS II) dimer in *Cyanidium caldarium*. This figure was created from a coordinate file (PDB code: 4YUU) using the UCSF Chimera software (Pettersen et al. 2004) and shows three views of an identical complex: *left*, *top*, *and side* (*right bottom*). The side view is oriented such that the stroma is facing on the *top*. The subunits of one of the dimers are labeled. Chlorophyll is shown in orchid, and other nonprotein compounds are shown in *gold*. Manganese atoms are presented in *black*

force including membrane potential. A special mechanism called the "Q cycle" is involved in translocating four protons for every two electrons transferred from plastoquinone to the cytochrome $b_6f$ complex (Cramer et al. 2006). The proton-motive force drives adenosine triphosphate (ATP) synthase to form ATP from adenosine diphosphate (ADP) and phosphate, which accompanies the translocation of protons back to the stroma. All subunits except AtpC are encoded by the chloroplast genome in both red algae and green plants (Table 17.5). Table 17.5 lists electron transport components in red algal chloroplasts. Many components are encoded by the chloroplast genome in both red algae and green plants.

In contrast with green plants, red algae lack plastocyanin, a copper-bearing protein. Cytochrome $c_6$ (encoded by the *petJ* gene) fulfils the role of plastocyanin in red algae, transferring electrons from the cytochrome $b_6f$ complex to PS I. In cyanobacteria and green algae, both plastocyanin and cytochrome $c_6$ are present,

**Table 17.5** Components of photosynthetic electron transport and ATP synthesis in red algae and *A. thaliana*

| Component | CME | GaSU | PoPU | PyYE | ChCR | ATH |
|---|---|---|---|---|---|---|
| Cytochrome $b_6f$ complex | | | | | | |
| PetA/cytochrome *f* | Cp | Cp | Cp | Cp | Cp | Cp |
| PetB/cytochrome $b_6$ | Cp | Cp | Cp | Cp | Cp | Cp |
| PetC/Rieske Fe-S protein | CMI281C | EME32261 | 529.1 | g3914 | CDF77454 | 4G03280 |
| | CMO066C | EME29050 | | g5986 | | |
| | CMQ165C | | | g8111 | | |
| PetD/subunit IV | Cp | Cp | Cp | Cp | Cp | Cp |
| PetG | Cp | Cp | Cp | Cp | Cp | Cp |
| PetL | Cp | Cp | Cp | Cp | Cp | Cp |
| PetM | Cp | Cp | Cp | Cp | Cp | 2G26500 |
| PetN | Cp | Cp | Cp | Cp | Cp | Cp |
| Oxidizing end of PS I | | | | | | |
| PetE/plastocyanin | None | None | None | None | None | 1G76100 |
| PetJ/cytochrome $c_6$ | Cp | Cp | Cp | Cp | Cp | 5G45040 |
| Reducing end of PS I | | | | | | |
| PetF/ferredoxin | Cp | Cp | Cp | Cp | Cp | 1G10960 |
| | | | | | | 1G60950 |
| | | | | | | 2G27510 |
| PetH/ferredoxin-NADP oxidoreductase | CMR052C | EME29997 | 2186.5 | g1708 | CDF39465 | 1G20020 |
| | | | 2104.21 | g4243 | CDF40763 | 1G30510 |
| | | | | | | 4G05390 |
| | | | | | | 5G66190 |
| Cyclic electron transport | | | | | | |
| PGR5 | CMT289C | EME26925 | ? | g4010 | ? | 2G05620 |
| | | EME26926 | | | | |
| PGRL1 | CMT288C | EME31250 | 3389.1 | g6544 | CDF39306 | 4G22890 |
| | | | | | | 4G11960 |
| | | | | | | 5G59400 |
| ATP synthase | | | | | | |
| AtpC | CMQ087C | EME31202 | 2104.20 | g9396 | CDF36141 | 4G04640 |
| | | | | g7529 | | 1G15700 |
| | | | | g614 | | |

All components of ATP synthase except AtpC (AtpA, B, D, E, F, G, H, and I) are encoded by the chloroplast genome in all of the organisms listed and are not listed here

"Cp" indicates that the gene is encoded by the chloroplast genome. All nuclear-encoded genes are indicated by their gene identifiers. The absence of a gene is indicated by "none." Question marks indicate that the gene is not detected in the current genomic data but that the gene is likely to exist

but the latter functions mainly under copper deficiency, where the synthesis of plastocyanin is repressed. Plant cytochrome $c_6$, having a lower redox potential, might have a different function. Although green plant homologs of *petJ* are encoded by the nuclear genome, red algal *petJ* is encoded by the chloroplast genome (Howe et al. 2006).

Homologs of proteins involved in the cyclic electron flow from PS I, such as PGR5 and PGRL1, which have been studied in plants and green algae, are also found in red algae (Table 17.5). Cyclic electron flow is, however, not well characterized in red algae.

## 17.5 Carbon Fixation

The components of the Calvin-Benson cycle in red algae are listed in Table 17.6. Many components are common in the oxidative pentose phosphate cycle (Sect. 19.3.1). As has been noted before (Matsuzaki et al. 2004), the red algal carbon fixation cycle is a hybrid of enzymes originating from endosymbiotic cyanobacteria (marked "C"), a eukaryotic host ("E"), and α-proteobacteria (the large and small subunits of Rubisco and Rubisco activase). Similar C and E patterns in red algae and green plants suggest that they share a common ancestor, following the exchange of cytosolic and chloroplast enzymes that occurred during establishment of the endosymbiont. Red algal Rubisco, having a high $CO_2/O_2$ specificity factor (Sugawara et al. 1999), could be utilized for constructing transgenic plants having a high $CO_2$ fixation capacity, keeping photorespiration to a minimum. However, there are two problems with this: the compatibility of Rubisco activase (see below) and a trade-off relationship between low oxygenase activity and the carboxylation rate (Tcherkez et al. 2006). This would make it difficult to increase photosynthesis efficiency substantially using red algal Rubisco.

A special type of Rubisco activase CbbX (also called CfxQ), distinct from green plant Rubisco activase, is necessary for remodeling red algal Rubisco into its active form. This CbbX protein is encoded by two different *cbbX* genes, one in the chloroplast genome and the other in the nuclear genome. Curiously, these two homologs form a heterooligomer that activates Rubisco (Loganathan et al. 2016).

## 17.6 Regulation of Photosynthesis

Regulation of photosynthesis is important not only for maximizing the efficiency of energy yield but also for protecting the photosystems from strong light. As described earlier (Sect. 17.2), initial steps in photosynthetic regulation are exerted by the energy transfer within phycobilisomes and the mobility of phycobilisomes that supply energy to either PS II or PS I.

**Table 17.6** Components of the carbon fixation cycle in red algae and *A. thaliana*

| Component | CME | GaSU | PoPU | PyYE | ChCR | ATH |
|---|---|---|---|---|---|---|
| Rubisco large (RbcL) subunit | Cp | Cp | Cp | Cp | Cp | Cp |
| Rubisco small (RbcS) subunit | Cp | Cp | Cp | Cp | Cp | 1G67090 |
| | | | | | | 5G38430 |
| | | | | | | 5G38420 |
| | | | | | | 5G38410 |
| Rubisco activase (CfxQ/CbbX) | Cp | Cp | Cp | Cp | Cp | [Unrelated activase] |
| | CMT172C | EME30696 | 2255.9 | g330-g329 | CDF40453 | |
| PGK (C) | CMJ305C | EME32655 | 3567.1 | g3725 | CDF35582 | 3G12780 |
| GAPDH (C) | CMJ042C | EME27669 | 4494.2 | g2079 | CDF35658 | 1G12900-C |
| | | | | | | 3G26650-C |
| | | | | | | 1G42970-C |
| | | | | | | 1G79530-E |
| | | | | | | 1G16300-E |
| TPI (E) | CMQ172C | EME29541 | 462.12 | ? | CDF39086 | 2G21170 |
| FBA (E) | CMI049C | EME25973 | 3490.7 | g2840 | CDF40481 | 2G21330 |
| | | EME27200 | 3540.3 | g2842 | | 4G38970 |
| | | | | g4300 | | 2G01140 |
| FBPase (E) | CMO245C | EME26967 | 3540.6 | g2976 | CDF33502 | 3G54050 |
| | | EME26922 | | g251 | | |
| SBPase (E) | CMI196C | EME27211 | 3690.1 | g1779 | CDF38025 | 3G55800 |
| TKT (C) | CMO121C | EME30775 | 4449.1? | g7461 | CDF36704 | 2G45290 |
| | CMO128C | | | g3879 | | 3G60750 |
| RPE (C) | CMT633C | EME30801 | 617.2 | g232 | CDF34278 | 5G61410 |
| RPI (E) | CMO291C | EME27197? | 592.4 | g3910 | CDF35864? | 3G04790 |
| | | EME28712? | | | | 5G44520 |
| PRK (C) | CMF117C | EME30780 | 3598.1 | g7274 | CDF36858 | 1G32060 |
| CP12 (C) | CMR315C | EME30187 | 2201.4 | g1897 | CDF32098 | 2G47400 |
| | | | | | | 3G62410 |
| | | | | | | 1G76560 |

*Rubisco* ribulose-1,5-bisphosphate carboxylase/oxygenase, *PGK* phosphoglycerate kinase, *GAPDH* glyceraldehyde 3-dehydrogenase, *TPI* triose-phosphate isomerase, *FBA* fructose 1,6-bisphosphate aldolase (class I), *FBPase* fructose 1,6-bisphosphatase, *SBPase* sedoheptulose 1,7-bisphosphatase, *TKT* transketolase, *RPE* ribulose-5-phosphate epimerase, *RPI* ribose 5-phosphate isomerase, *PRK* phosphoribulokinase. CfxQ of PyYE appears to be split into two neighboring contigs in the draft genome sequences. "C" or "E" after the enzyme name indicates probable origin (C = cyanobacterial origin and E = eukaryotic origin). The two ATH GAPDH entries marked "E" are of eukaryotic origin but localize in the chloroplast. Red algal Rubisco is of α-proteobacterial origin

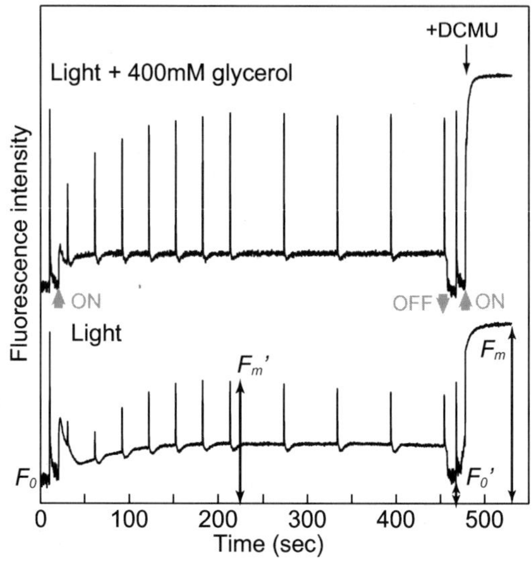


**Fig. 17.6** Kinetics of pulse-amplitude modulated (PAM) fluorometry in dark-adapted *C. merolae* cells. The cells were grown photoautotrophically (lower trace) or mixotrophically in the presence of 400 mM glycerol (upper trace). $F_v'/F_m'$ values (note that $F_v' = F_m' - F_0'$) were higher in the presence of glycerol, indicating higher photosynthetic activity. DCMU (3-(3,4-dichlorophenyl)-1,1-dimethylurea) was used to measure the maximal fluorescence level ($F_m$) representing full PS II activity

Figure 17.6 shows fluorescence induction kinetics measured by pulse-amplitude modulated (PAM) fluorometry in dark-adapted *C. merolae* cells. The basal level of fluorescence under actinic light decreased initially in the control. This does not occur in land plants and cyanobacteria. The high level of variable fluorescence ($F_v' = F_m' - F_0'$) was noted in the culture with glycerol, suggesting a higher photochemical efficiency in the presence of glycerol. This result is in contrast with those in *G. sulphuraria* (Oesterhelt et al. 2007a), where the addition of glucose to the medium decreased $F_v/F_m$ values, with a concomitant decrease in the $O_2$ evolution rate and Rubisco level. *G. sulphuraria* enters heterotrophic growth in the presence of glucose, whereas *C. merolae* maximizes its photosynthetic activity, possibly using glycerol as a metabolic intermediate (see Chap. 19).

The diverse origin of Calvin-Benson cycle enzymes (Sect. 17.5) has an interesting implication for the regulation of carbon fixation. In angiosperms, the activation of glyceraldehyde 3-phosphate dehydrogenase (GAPDH), fructose 1,6-bisphosphatase (FBPase), sedoheptulose 1,7-bisphosphatase (SBPase), and phosphoribulokinase (PRK) is achieved through the reduction of two regulatory cysteine residues mediated by the ferredoxin-thioredoxin system. In red algae, virtually no redox regulation was found in PRK activity, whereas GAPDH was regulated by redox (Maberly et al. 2010). PRK and GAPDH form a ternary complex with an 8.5-kDa protein called CP12 in chloroplasts and cyanobacteria (Graciet et al. 2004; see also Table 17.6). In *Chlamydomonas*, CP12 has been shown to inhibit the activity of PRK and GAPDH, whereas the effect was small in red algae (Maberly et al. 2010). CP12 was also found to form a high molecular weight complex with PRK and GAPDH in *G. sulphuraria* (Oesterhelt et al. 2007b). Interestingly, all red algal CP12 lack the two conserved cysteine residues in the *N*-terminal domain, whereas the two other cysteine residues in the *C*-terminal domain are conserved in plants, algae, and cyanobacteria (see Gclust2016R dataset, Cluster 1402).

FBPase of *G. sulphuraria*, having the conserved cysteine residues for thioredoxin regulation, responds only partially to redox regulation; this is in contrast with spinach enzyme, which is completely inactivated in its oxidized form (Reichert et al. 2003). Redox regulation of Calvin-Benson cycle enzymes has not been studied in *C. merolae*.

The transcriptional regulation of photosynthesis has been documented in *C. merolae*. The roles of sigma factors in activation of chloroplast transcription, and hence photosynthesis, have been determined (Fujii et al. 2013, 2015). An interesting characteristic of red algal chloroplasts is the retention of various prokaryotic response regulators (Ycf27 and Ycf29), a homolog of nitrogen regulator NtcA (Ycf28), and a homolog of CbbR (Ycf30), which are encoded by the chloroplast genome (Minoda et al. 2010). Minoda et al. (2010) showed that the chloroplast-encoded Rubisco operon is activated by CbbR. Red algal chloroplasts are, therefore, partially self-autonomous in their gene expression. Further studies are needed to fully elucidate the transcriptional regulation of the red algal chloroplast genome.

## References

Adachi H, Umena Y, Enami I et al (2009) Towards structural elucidation of eukaryotic photosystem II: purification, crystallization and preliminary X-ray diffraction analysis of photosystem II from a red alga. Biochim Biophys Acta 1787:121–128

Ago H, Adachi H, Umena Y et al (2016) Novel features of eukaryotic photosystem II revealed by its crystal structure analysis from a red alga. J Biol Chem 291:5676–5687

Amunts A, Toporik H, Borovikova A, Nelson N (2010) Structure determination and improved model of plant photosystem I. J Biol Chem 285:3478–3486

Archibald J (2014) One plus one equals one. Symbiosis and the evolution of complex life. Oxford University Press, Oxford

Bhattacharya D, Price DC, Chan CX et al (2013) Genome of the red alga *Porphyridium purpureum*. Nat Commun 4:1941

Bumba L, Havelková-Doušová H, Hušák M, Vácha F (2004) Structural characterization of photosystem II complex from red alga *Porphyridium cruentum* retaining extrinsic subunits of the oxygen-evolving complex. Eur J Biochem 271:2967–2975

Busch A, Nield J, Hippler M (2010) The composition and structure of photosystem I-associated antenna from *Cyanidioschyzon merolae*. Plant J 62:886–897

Collén J, Porcel B, Carré W et al (2013) Genome structure and metabolic features in the red seaweed *Chondrus crispus* shed light on evolution of the Archaeplastida. Proc Natl Acad Sci U S A 110:5247–5252

Cramer WA, Zhang H, Yan J et al (2006) Transmembrane traffic in the cytochrome $b_6f$ complex. Annu Rev Biochem 75:769–790

Enami I, Okumura A, Nagao R et al (2008) Structures and functions of the extrinsic proteins of photosystem II from different species. Photosynth Res 98:349–363

Fromme P, Jordan P, Krauss N (2001) Structure of photosystem I. Biochim Biophys Acta 1507:5–31

Fujii G, Imamura S, Hanaoka M, Tanaka K (2013) Nuclear-encoded chloroplast RNA polymerase sigma factor SIG2 activates chloroplast-encoded phycobilisome genes in a red alga, *Cyanidioschyzon merolae*. FEBS Lett 587:3354–3359

Fujii G, Imamura S, Era A et al (2015) The nuclear-encoded sigma factor SIG4 directly activates transcription of chloroplast *psbA* and *ycf17* genes in the unicellular red alga *Cyanidioschyzon merolae*. FEMS Microbiol Lett 362:fnv063

Gardian Z, Bumba L, Schrofel A et al (2007) Organisation of photosystem I and photosystem II in red alga *Cyanidium caldarium*: encounter of cyanobacterial and higher plant concepts. Biochim Biophys Acta 1767:725–731

Graciet E, Lebreton S, Gontero B (2004) Emergence of new regulatory mechanisms in the Benson-Calvin pathway via protein-protein interactions: a glyceraldehyde-3-phosphate dehydrogenase/CP12/phosphoribulokinase complex. J Exp Bot 55:1245–1254

Guindon S, Dufayard JF, Lefort V et al (2010) New algorithms and methods to estimate maximum-likelihood phylogenies: assessing the performance of PhyML 3.0. Syst Biol 59:307–321

Guskov A, Kern J, Gabdulkhakov A et al (2009) Cyanobacterial photosystem II at 2.0-Å resolution and the role of quinones, lipids, channels and chloride. Nat Struct Mol Biol 16:334–342

Howe CJ, Schlarb-Ridley BG, Wastl J et al (2006) The novel cytochrome $c_6$ of chloroplasts: a case of evolutionary *bricolage*? J Exp Bot 57:13–22

Ifuku K (2015) Localization and functional characterization of the extrinsic subunits of photosystem II: an update. Biosci Biotechnol Biochem 79:1223–1231

Kaňa R, Kotabová E, Lukeš M et al (2014) Phycobilisome motility and its role in the regulation of light harvesting in red algae. Plant Physiol 165:1618–1631

Lamesch P, Berardini TZ, Li D et al (2011) The arabidopsis information resource (TAIR): improved gene annotation and new tools. Nucleic Acids Res 40:D1202–D1210

Liu L-N, Chen X-L, Zhang Y-Z, Zhou B-C (2005) Characterization, structure and function of linker polypeptides in phycobilisomes of cyanobacteria and red algae: an overview. Biochim Biophys Acta 1708:133–142

Liu L-N, Aartsma TJ, Thomas J-C et al (2008) Watching the native supramolecular architecture of photosynthetic membrane in red algae. Topography of phycobilisomes and their crowding, diverse distribution patterns. J Biol Chem 283:34946–34953

Loganathan N, Tsai YCC, Mueller-Cajar O (2016) Characterization of the heterooligomeric red-type rubisco activase from red algae. Proc Natl Acad Sci U S A 113:14019–14024

Maberly SC, Courcelle C, Groben R, Gontero B (2010) Phylogenetically-based variation in the regulation of the Calvin cycle enzymes, phosphoribulokinase and glyceraldehyde-3-phosphate dehydrogenase, in algae. J Exp Bot 61:735–745

Marquardt J, Lutz B, Wans S et al (2001) The gene family coding for light-harvesting polypeptides of photosystem I of the red alga *Galdieria sulphuraria*. Photosynth Res 68:121–130

Matsuzaki M, Misumi O, Shin-I T et al (2004) Genome sequence of the ultrasmall unicellular red alga *Cyanidioschyzon merolae* 10D. Nature 428:653–657

Minoda A, Weber APM, Tanaka K, Miyagishima S (2010) Nucleus-independent control of the Rubisco operon by the plastid-encoded transcription factor Ycf30 in the red alga *Cyanidioschyzon merolae*. Plant Physiol 154:1532–1540

Nakamura Y, Sasaki N, Kobayashi M et al (2013) The first symbiont-free genome sequence of marine red alga, Susabi-nori (*Pyropia yezoensis*). PLoS One 8:e57122

Nozaki H, Takano H, Misumi O et al (2007) A 100%-complete sequence reveals unusually simple genomic features in the hot-spring red alga *Cyanidioschyzon merolae*. BMC Biol 5:28

Oesterhelt C, Schmälzlin E, Schmitt JM, Lokstein H (2007a) Regulation of photosynthesis in the unicellular acidophilic red alga *Galdieria sulphuraria*. Plant J 51:500–511

Oesterhelt C, Klocke S, Holtgrefe S et al (2007b) Redox regulation of chloroplast enzymes in *Galdieria sulphuraria* in view of eukaryotic evolution. Plant Cell Physiol 48:1359–1373

Ohta H, Suzuki T, Ueno M et al (2003) Extrinsic proteins of photosystem II. An intermediate member of the PsbQ protein family in red algal PS II. Eur J Biochem 270:4156–4163

Pettersen EF, Goddard TD, Huang CC et al (2004) UCSF chimera—a visualization system for exploratory research and analysis. J Comput Chem 25:1605–1612

Reichert A, Dennes A, Vetter S, Scheibe R (2003) Chloroplast fructose 1,6-bisphosphatase with changed redox modulation: comparison of the *Galdieria* enzyme with cysteine mutants from spinach. Biochim Biophys Acta 1645:212–217

Sato N (2009) Gclust: *trans*-kingdom classification of proteins using automatic individual threshold setting. Bioinformatics 25:599–605

Sato N (2010) Phylogenomic and structural modeling analyses of the PsbP superfamily reveal multiple small segment additions in the evolution of photosystem II-associated PsbP protein in green plants. Mol Phylogenet Evol 56:176–186

Schönknecht G, Chen W-H, Ternes CM et al (2013) Gene transfer from bacteria and archaea facilitated evolution of an extremophilic eukaryote. Science 339:1207–1210

Su N-N, Xie B-B, Zhang X-Y et al (2010) The supramolecular architecture, function, and regulation of thylakoid membranes in red algae: an overview. Photosynth Res 106:73–87

Sugawara H, Yamamoto H, Shibata N et al (1999) Crystal structure of carboxylase reaction-oriented ribulose 1,5-bisphosphate carboxylase/oxygenase from a thermophilic red alga, *Galdieria partita*. J Biol Chem 274:15655–15661

Tan S, Ducret A, Aebersold R, Gantt E (1997) Red algal LHC I genes have similarities with both Chl *a*/*b*- and *a*/*c*-binding proteins: a 21 kDa polypeptide encoded by *LhcaR2* is one of the six LHC I polypeptides. Photosynth Res 53:129–140

Tcherkez GGB, Farquhar GD, Andrews TJ (2006) Despite slow catalysis and confused substrate specificity, all ribulose bisphosphate carboxylases may be nearly perfectly optimized. Proc Natl Acad Sci U S A 103:7246–7251

Thangaraj B, Jolley CC, Sarrou I et al (2011) Efficient light harvesting in a dark, hot, acidic environment: the structure and function of PSI-LHCI from *Galdieria sulphuraria*. Biophys J 100:135–143

Umena Y, Kawakami K, Shen JR, Kamiya N (2011) Crystal structure of oxygen-evolving photosystem II at a resolution of 1.9Å. Nature 473:55–60

Vanselow C, Weber APM, Krause K, Fromme P (2009) Genetic analysis of the photosystem I subunits from the red alga, *Galdieria sulphuraria*. Biochim Biophys Acta 1787:46–59

Watanabe M, Ikeuchi M (2013) Phycobilisome: architecture of a light-harvesting supercomplex. Photosynth Res 116:265–276

Zhao L-S, Su H-N, Li K et al (2016) Supramolecular architecture of photosynthetic membrane in red algae in response to nitrogen starvation. Biochim Biophys Acta 1857:1751–1758

# Chapter 18
# Nitrogen Metabolism

**Sousuke Imamura and Kan Tanaka**

**Abstract** *Cyanidioschyzon merolae* is able to grow on ammonium, as well as nitrate, as the sole nitrogen source; however, this alga does not have a canonical nitrite reductase (NiR). Interestingly, a sulfite reductase-like protein (SiRB) functions as the NiR in this alga. Unlike the genomes of higher plants, the *C. merolae* genome encodes only one glutamine synthetase (GS), which is localized in the cytosol. Thus, the nitrogen assimilation pathway in *C. merolae* is complex and unique, since the intermediate metabolites of nitrogen assimilation must translocate back and forth between the chloroplast and the cytosol during the nitrogen assimilation process. Using genetic and molecular biology approaches, we found that the MYB-type transcription factor, MYB1, is the central regulator of nitrogen assimilation. Nitrogen depletion-induced transcripts of nitrogen assimilation genes were completely absent from the MYB1-knockout strain. Furthermore, direct interactions between MYB1 and the promoter regions of nitrogen assimilation genes were observed in vivo and in vitro. Although MYB1-dependent transcripts of nitrogen assimilation genes were detected in nitrate-grown cells, they disappeared upon the addition of a preferred nitrogen source such as ammonium or glutamine. These findings suggested that MYB1 mediates the nitrogen catabolite repression-sensitive transcription of nitrogen assimilation genes in response to nitrogen status in *C. merolae*.

**Keywords** Ammonium assimilation • MYB-type transcription factor • Nitrate assimilation • Nitrogen assimilation • Nitrogen catabolite repression • Sulfite reductase • Target of rapamycin

## 18.1 Introduction

All living organisms are composed of biomolecules, such as amino acids, nucleic acids, carbohydrates, and lipids, that mainly consist of carbon, nitrogen, sulfur, and phosphorus. Among these essential elements, nitrogen is a major component of

---

S. Imamura (✉) • K. Tanaka (✉)
Laboratory for Chemistry and Life Science, Institute of Innovative Research, Tokyo Institute of Technology, R1-29, 4259 Nagatsuta, Midori-ku, Yokohama 226-8503, Japan
e-mail: simamura@res.titech.ac.jp; kntanaka@res.titech.ac.jp

 
T. Kuroiwa et al. (eds.), *Cyanidioschyzon merolae*,
https://doi.org/10.1007/978-981-10-6101-1_18

amino acids and nucleic acids, and the nitrogen requirement is second only to the carbon requirement for the maintenance and growth of organisms.

Nitrogen is a more important element to autotrophic organisms such as plants and algae than to heterotrophic organisms, since they can obtain carbon from the ambient air by photosynthesis. However, the molecular mechanism underlying nitrogen assimilation and its control in plant lineages are poorly understood. Thus, we analyzed the nitrogen assimilation pathway and its regulation in the red alga *Cyanidioschyzon merolae*.

In this chapter, we summarize and discuss the nitrogen assimilation pathway of ammonium and nitrate in *C. merolae*, and the regulation of genes involved in nitrogen assimilation in this alga.

## 18.2 Nitrogen Assimilation Genes and Pathway

The main nitrogen sources of *C. merolae* under normal growth conditions are ammonium ($NH_4^+$) and nitrate ($NO_3^-$). Figure 18.1 illustrates a possible nitrogen assimilation pathway in *C. merolae* based on the results of previous studies (Terashita et al. 2006; Imamura et al. 2010). The genes related to nitrogen assimilation are summarized in Table 18.1. We describe the ammonium and nitrate assimilation pathways separately below.

### *18.2.1 Ammonium Assimilation*

Ammonium taken up from the environment into cells is via a specific ammonium transporter (AMT) and is then assimilated into glutamine by glutamine synthetase (GS). It is predicted that AMT is encoded by two genes, *CMT526C* and *CMK126C* (*AMT1* and *AMT2* in this chapter, respectively), in *C. merolae*. The transcription of these genes is induced under nitrogen depletion conditions (Imamura et al. 2009, 2015), suggesting that these two proteins are high-affinity ammonium transporters. However, it is necessary to analyze the biochemical properties of these proteins, such as their *Km* values, to further characterize their role in ammonium assimilation.

The plants and eukaryotic algae studied so far have two GS isoforms; GS1, which is cytosolic, and GS2, which is localized in chloroplasts (Lam et al. 1996). GS1 is the main form in roots but is also found in the leaves of vascular plants. Therefore, it is thought to be primarily responsible for the assimilation of ammonium from the environment. The main form in leaves is GS2, which is thought to be involved in recycling ammonium generated by photorespiration (Linka and Weber 2005). A single nuclear gene encodes GS2, whereas multiple genes encode GS1.

Interestingly, the *C. merolae* genome has only gene encoding GS, raising the possibility that GS localizes only in the cytoplasm or chloroplast or in both.

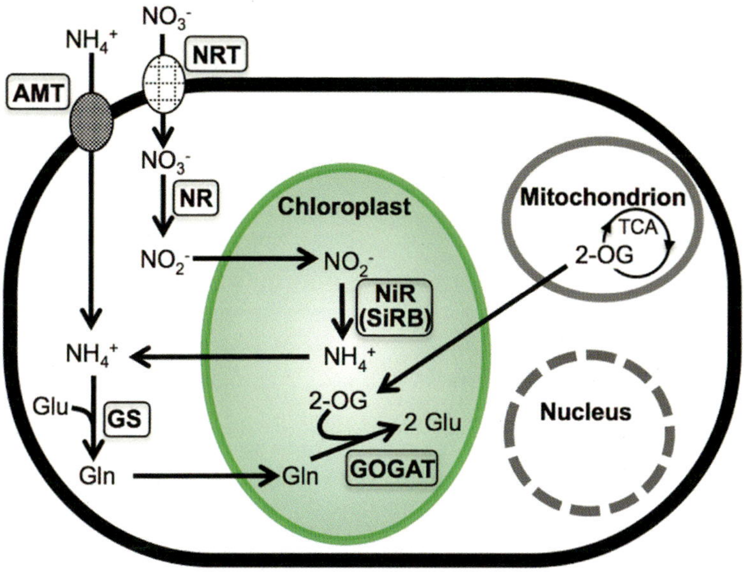


**Fig. 18.1** Scheme of nitrogen assimilation pathway in *Cyanidioschyzon merolae*. *AMT* ammonium transporter; *GS* glutamine synthetase; *NRT* nitrate transporter; *NR* nitrate reductase; *NiR* (*SirB*) nitrite reductase; *GOGAT* glutamate synthase; *Gln* glutamine; *Glu* glutamate; *2-OG* 2-oxoglutarate; *TCA* tricarboxylic acid cycle

**Table 18.1** List of genes related to nitrogen assimilation in *C. merolae*

| Gene number | Function (abbreviation) |
|---|---|
| CMG018C | Nitrate transporter (NRT) |
| CMG019C | Nitrate reductase (NR) |
| CMG021C | Nitrite reductase (NiR, SiRB) |
| CMI233C | Glutamine synthetase (GS) |
| CMJ282C | Nitrogen transcription factor (MYB1) |
| CMK126C | Ammonium transporter (AMT2) |
| CMT526C | Ammonium transporter (AMT1) |
| CMV060C[a] | Glutamate synthase (GOGAT) |

[a]Denote gene encoded on the chloroplast genome. Others are encoded on the nuclear genome

Therefore, we experimentally verified the intracellular localization of GS to characterize the nitrogen assimilation pathway in *C. merolae*. The immunofluorescence microscopy using antibodies against rice GS1 (Ishiyama et al. 2004), which specifically recognize the *C. merolae* GS, showed a cytoplasmic localization (Fig. 18.2a). Furthermore, after disruption of *C. merolae* cells and fractionation of the cellular compartments by density gradient centrifugation, the GS protein cofractionated with cytosolic α-tubulin but not chloroplast-localized RbcL or mitochondrion-localized EF-Tu (Fig. 18.2b). These results demonstrated that the *C. merolae* GS localizes only in the cytoplasm (Terashita et al. 2006).

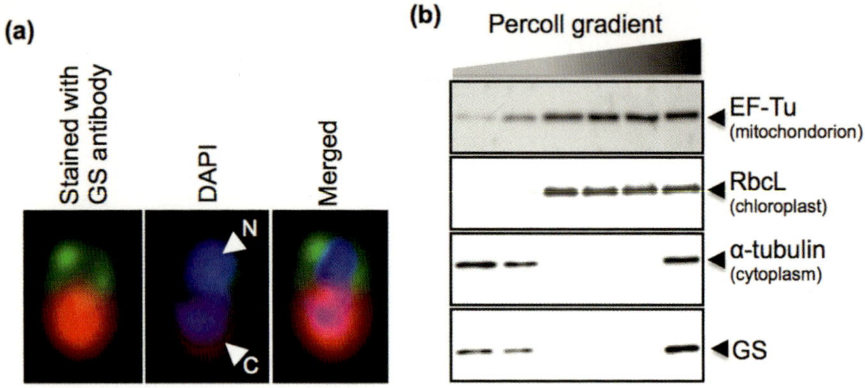


**Fig. 18.2** Intracellular localization of glutamine synthetase (GS) in *Cyanidioschyzon merolae*. (**a**) Immunostaining with anti-GS antibody. Green signal indicates GS localization (*left*). N and C indicate nucleus and chloroplast, respectively. DAPI staining of cells (DAPI) and merged image of immunostaining and DNA fluorescence (Merged) are shown. (**b**) Immunoblot analysis of GS after cell fractionation. After cell disruption, organelles were fractionated by percoll density gradient centrifugation and were used for immunoblot analyses with indicated antibodies (Figure modified from Terashita et al. 2006)

Ammonium is assimilated into glutamine in the cytoplasm, and then glutamate synthase (GOGAT) catalyzes the transfer of the amide group from glutamine to 2-oxoglutarate (2-OG) to yield two molecules of glutamate (Merrick and Edwards 1995). Since GOGAT is encoded by the chloroplast genome, it was thought to be localized in the chloroplast. Therefore, for the GS-GOGAT cycle to operate, glutamine must be transported from the cytoplasm into the chloroplast. In addition, 2-OG, which is a substrate of GOGAT along with glutamine, is generated from the TCA cycle in the mitochondrion and transported to the chloroplast after exiting the cytoplasm or being directly transported to the chloroplast (Fig. 18.1).

Our unpublished data showed that transcripts of photorespiration-related genes are induced under low-$CO_2$ conditions in this alga, indicating that the photorespiration pathway operates in this alga and that ammonium generated from this pathway must be recycled. It is interesting to uncover the role of the cytoplasmic GS in ammonium recycling.

## *18.2.2 Nitrate Assimilation*

### 18.2.2.1 Genes Involved in Nitrate Assimilation

In general, incorporated nitrate in the cell is first reduced to nitrite ($NO_2^-$) in the cytosol by nitrate reductase (NR) using NADH or NADPH (Merrick and Edwards 1995). Nitrite is then reduced to ammonium in the chloroplast by the ferredoxin-dependent nitrite reductase (NiR). In *C. merolae*, the genes encoding nitrate

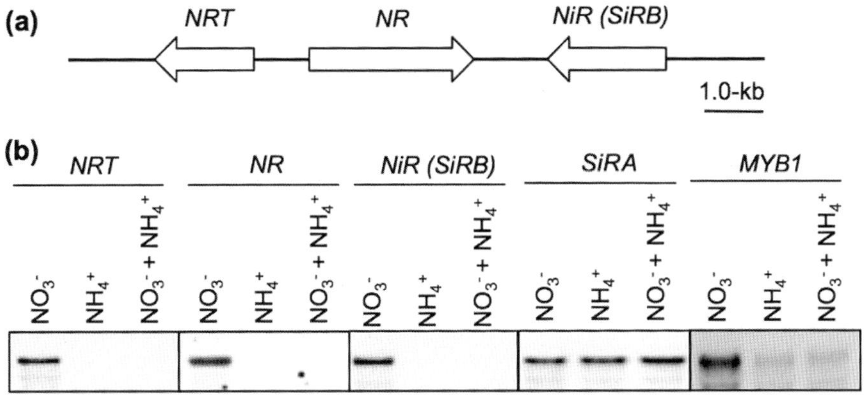


**Fig. 18.3** Nitrate assimilation genes and their expression in *Cyanidioschyzon merolae*. (**a**) Schematic structure of nitrate assimilation gene cluster in nuclear genome. (**b**) Transcript levels of *NRT*, *NR*, *NiR* (*SiRB*), *SiRA*, and *MYB1*. *C. merolae* cells were grown in media containing indicated nitrogen sources. Gene transcripts were detected by northern blotting (Figure modified from Imamura et al. 2010)

reductase (NR, CMG019C) and the MFS (major facilitator superfamily)-type nitrate (and/or nitrite) transporter (NRT, CMG018C) are adjacent in the genome (Table 18.1, Fig. 18.3a). Surprisingly, no candidate genes encoding NiR have been found in the *C. merolae* genome by BLAST searches using other organisms' NiR sequences as queries, even though *C. merolae* can grow with nitrate as the sole nitrogen source. Therefore, we searched for similar proteins and found two candidate genes encoding putative sulfite reductases (SiRs) (SiRA, CMJ117C; SiRB, CMG021C). It has been reported that SiR proteins have a structure closely related to that of NiR (Nakayama et al. 2000). One of these genes, SiRB, is located next to the gene encoding NR in the *C. merolae* genome (Fig. 18.3a). Thus, we hypothesized that SiRB could function as an NiR in *C. merolae*.

To test this possibility, we first quantified the transcript levels of *SiRB* in both ammonium- and nitrate-grown cells (Imamura et al. 2009, 2010). Transcripts of *SiRB* were detectable in nitrate-grown cells but not in ammonium-grown ones (Fig. 18.3b). The transcription patterns of *NRT* and *NR* were the same as that of *SiRB*. On the other hand, *SiRA* was constitutively expressed irrespective of the nitrogen source. The protein expression patterns of SiRA and SiRB were well correlated with their transcription patterns. That is, SiRB was detected in nitrate-grown cells but not in ammonium-grown cells, whereas SiRA was present under both nitrogen conditions (Sekine et al. 2009, 2017). These observations supported the hypothesis that SiRB functions as an NiR in the cell.

Next, we examined the function of SiRB by genetic approaches. We constructed an SiRB-knockout strain, named SI021, and examined its growth with nitrate as the nitrogen source. The growth of SI021 was comparable to that of its parental strain, M4, when ammonia was the nitrogen source, but its growth was clearly inhibited when nitrate was the sole nitrogen source (Imamura et al. 2010). We expressed the

SiRB protein in an NiR-deficient mutant, PN1, (Suzuki et al. 1995), of the cyanobacterium *Leptolyngbya boryana* and monitored cell growth with nitrate as the nitrogen source. The SiRB-expressing *L. boryana* cells could grow after changing the nitrogen source from ammonia to nitrate in the growth medium (Imamura et al. 2010). Thus, the results of these genetic analyses strongly indicated that SiRB physiologically functions as an NiR in *C. merolae*.

In in vitro biochemical assays using recombinant SiRA and SiRB proteins, SiRA had high *Kcat* values for both sulfite and nitrite, but SiRB had virtually only nitrite reduction activity (Sekine et al. 2009, 2017). These results were consistent with the in vivo data and demonstrated that SirB functions as an NiR in the cell. Thus, we renamed SirB as NiR based on its function.

The mutation S217C in the active site of NiR (SiRB), a residue that is specific to NiRs among various SiR-like enzymes, partially recovered sulfite reduction activity, suggesting that this residue is a major determinant of substrate specificity (Sekine et al. 2017). This finding also suggested that mutation(s) in the ancestral SiR protein conferred nitrite reductase activity to the protein during evolution.

#### 18.2.2.2 Nitrate Assimilation Pathway

It had been predicted that NR localizes and generates nitrite in the cytoplasm. Immunofluorescence microscopy analyses confirmed that NiR (SiRB) localizes in the chloroplast (Imamura et al. 2010). Therefore, it is conceivable that nitrite is transferred to the chloroplast and then reduced to ammonium, which returns to the cytoplasm to be reduced into glutamine. After that, the glutamine is transferred to the chloroplast and is used as the substrate for glutamate synthesis in the GS-GOGAT cycle (Fig. 18.1). The transporters that shuttle nitrite, ammonium or glutamine across the chloroplast membrane have not yet been identified.

## 18.3 Regulation of Expression of Nitrogen Assimilation Genes

### *18.3.1 Role of GATA-Type Transcription Factors in Nitrogen Regulation*

Next, we tried to identify a transcription factor (TF) controlling the expression of nitrogen assimilation-related genes under nitrogen depletion conditions. We used information on the nitrogen regulation system in *Saccharomyces cerevisiae* as a reference, since the TF(s) contributing to nitrogen assimilation are poorly understood in plant lineages. In *S. cerevisiae*, the GATA-type TF, Gln3, is the central TF regulating nitrogen assimilation (Cooper 2002). The activity and intracellular localization of Gln3 are controlled at the posttranslational level, especially by its phosphorylation status (Rai et al. 2015). Gln3 is phosphorylated in the presence of preferred nitrogen sources, such as glutamine and ammonia, and is retained in the cytoplasm by the Gln3-binding protein Ure2 that recognizes the phosphorylation status (Beck and Hall 1999). The kinase TOR (target of rapamycin) phosphorylates Gln3. TOR is a conserved serine/threonine protein kinase among eukaryotes, and it plays critical roles in growth regulation in response to several environmental and physiological conditions (Virgilio and Loewith 2006; Wullschleger et al. 2006). After TOR inhibition under nitrogen depletion, Gln3 is dephosphorylated by the PP2A phosphatase and enters the nucleus. Consequently, expression of nitrogen assimilation-related genes is induced in response to nitrogen depletion conditions.

We assumed that the same type of TF as that in *S. cerevisiae* would control the transcription of nitrogen assimilation genes in *C. merolae*. Therefore, we investigated whether four GATA-type TFs encoded in the nuclear genome of *C. merolae* (CMB029C, CMD180C, CME102C, and CMR493C) control the transcription of nitrogen-related genes. However, there was no positive evidence that these four TFs were involved in such regulation. The results of northern blot and immunoblot analyses indicated that the transcript and protein levels of the genes encoding GATA-type TFs did not change after nitrogen depletion. Also, there were no band shifts in an electrophoresis mobility shift assay (EMSA) with purified recombinant GATA-type proteins and the promoter regions of *NRT*, *NiR* (*SiRB*), and *NR* (Imamura et al. 2009).

*Arabidopsis* GATA-type TFs have been studied using a reverse genetics approach (Bi et al. 2005). In that study, the authors obtained and analyzed homozygous T-DNA insertion lines for 23 of the 30 *Arabidopsis* GATA-type TFs. They found that the transcription of *GNC* was inducible by nitrate. However, transcript profiling experiments showed that a considerable proportion of the downregulated genes in the loss-of-function mutants were involved in carbon metabolism, but not nitrogen metabolism. These findings together with our results suggested that GATA-type TFs do not regulate the expression of nitrogen assimilation genes in plant cells.

### 18.3.2 MYB-Type Transcription Factor, MYB1, Is a Central Nitrogen Regulator

Next, we conducted a DNA microarray analysis using RNA prepared before and 2 h after nitrogen depletion and searched for TFs whose transcript levels were increased. Only CMJ282C, which encodes an R2R3 MYB-type TF, showed upregulated transcript levels after nitrogen depletion. We designated this TF as MYB1 (Imamura et al. 2009).

The MYB1 protein was not detectable under nitrogen-replete conditions, but had appeared by 2 h after nitrogen depletion and peaked at 4 h. Its presence was well correlated with the expression of nitrogen assimilation genes, *NRT*, *NR*, *NiR (SiRB)*, and *AMT1*, and GS (Fig. 18.4a, b). However, *MYB1* transcripts were detected irrespective of nitrogen status, implying that there is posttranscriptional regulation of MYB1 expression (Imamura et al. 2009).

Nitrogen depletion-responsive transcripts were absent from the MYB1-knockout strain SI282, indicating that MYB1 activates the transcription of nitrogen assimilation genes in response to nitrogen depletion (Fig. 18.4c). Further in vivo and in vitro experiments supported this conclusion. For example, chromatin immunoprecipitation and EMSA analyses demonstrated a specific interaction between

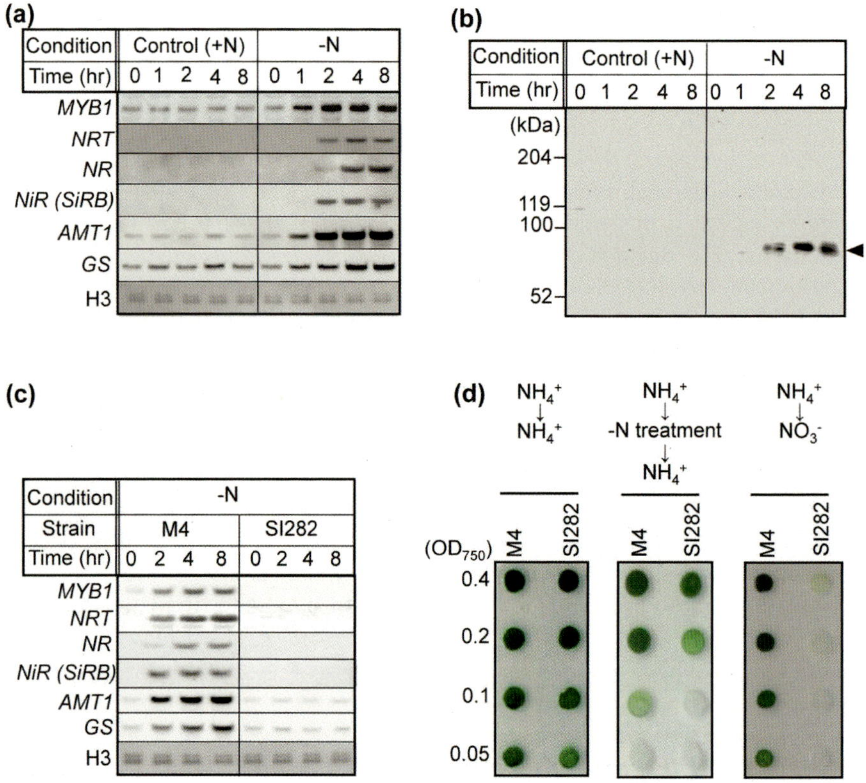


**Fig. 18.4** Expression of *MYB1* and nitrogen assimilatory genes in *Cyanidioschyzon merolae* under nitrogen depletion conditions. (**a**) Transcript levels of *MYB1* and nitrogen assimilation genes under nitrogen depletion (−N) conditions in *C. merolae* wild type. H3 (histone H3 gene, *CMN176C)* is shown as loading control. (**b**) Protein level of MYB1 under nitrogen depletion (−N) conditions. Arrowhead indicates position of endogenous MYB1. (**c**) Transcript levels of *MYB1* and nitrogen assimilation genes under nitrogen depletion (−N) conditions in M4 and SI282 (MYB1-knockout strain) cells. (**d**) Viability of M4 and SI282 cells. M4 and SI282 cells were exposed to nitrogen depletion (*middle*) or control (*left*) conditions for 8 h, then diluted to $OD_{750}$ of 0.4, 0.2, 0.1, or 0.05 with MA2 medium, and spotted onto a plate containing $NH_4^+$ as the sole nitrogen source. Cells of M4 and SI282 in logarithmic growth phase were diluted with medium containing $NO_3^-$ and spotted onto a plate containing nitrate as the sole nitrogen source, as described above (*right*) (Figure modified from Imamura et al. 2009)

MYB1 and the promoter regions of nitrogen assimilation genes (Imamura et al. 2009). Compared with the parental strain, SI282 showed decreased cell viability after exposure to nitrogen-depleted conditions for 8 h (Fig. 18.4d). These results clearly demonstrated that MYB1 plays a central role to directly activate nitrogen assimilation genes in response to nitrogen status (Fig. 18.6) (Imamura et al. 2009). It should be noted that it remains unclear whether MYB1 regulates *AMT2* transcription, as it does *AMT1*.

### *18.3.3 Nitrogen Catabolite Repression of Nitrogen Assimilation Genes*

The nitrogen assimilation genes, *AMT1*, *GS*, *NRT*, *NR*, and *NiR (SiRB)*, as well as the regulator, *MYB1*, were constantly expressed in cells grown with nitrate as the sole nitrogen source but repressed in those grown with ammonium (Fig. 18.3b). This raised the possibility of a nitrate-specific regulatory pathway. To test this possibility, *C. merolae* cells were subjected to nitrogen depletion conditions for 4 h and then supplied with nitrate. Total RNAs were sequentially isolated from the cells and analyzed. If a nitrate-specific regulatory pathway was present, then nitrogen depletion-induced increases in the transcript levels of these genes should have been further enhanced. However, there was no transcriptional induction of those genes after addition of nitrate (Fig. 18.5).

On the other hand, transcripts of nitrogen assimilation genes completely disappeared or decreased to the basal level after adding ammonium to nitrogen-depleted *C. merolae* cells (Fig. 18.5). The transcript levels of these genes similarly decreased after adding glutamine, but not after adding glutamate. These results suggested that *C. merolae* cells sense ammonium and/or glutamine, but not nitrate, as the nitrogen signal (Fig. 18.6). Cells of *C. merolae* wild type grew slower in medium containing nitrate than in medium containing ammonia (Imamura et al. 2010), indicating that nitrate is not the preferred nitrogen source for *C. merolae*. Thus, the expression of nitrogen assimilation genes seems to be regulated via nitrogen catabolite repression. This hypothesis is well supported by the observation that the transcript levels of nitrogen assimilation genes in cells grown with nitrate and ammonium were almost the same as those in cells grown with ammonium (Fig. 18.3b).

In the case of MYB1, its transcription pattern was the same as that of nitrogen assimilation genes, and the amount of nitrogen depletion-responsive MYB1 protein decreased to zero after adding ammonia to the cells (Imamura et al. 2009). These findings indicated that MYB1 mediates the NCR-sensitive transcription of nitrogen assimilation genes in response to nitrogen status in *C. merolae* (Fig. 18.6).

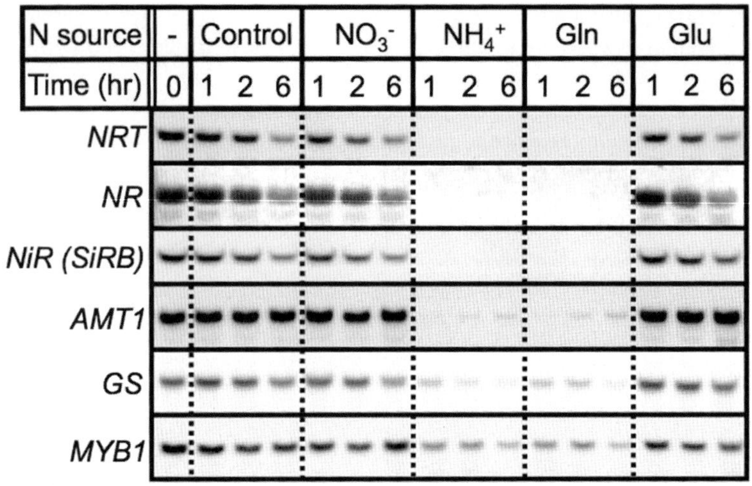


**Fig. 18.5** Nitrogen catabolite repression-sensitive transcription of nitrogen assimilation genes in *Cyanidioschyzon merolae*. *C. merolae* cells were exposed to nitrogen depletion conditions for 4 h (set time 0); then fed ammonium ($NH_4^+$), nitrate ($NO_3^-$), glutamine (Gln), glutamate (Glu), or water (control) at a final concentration of 5 mM; and sequentially harvested at indicated times. After isolating RNAs from cells, northern blot analyses were conducted with indicated gene-specific probes (Figure modified from Imamura et al. 2009)

### *18.3.4 Regulation of Nitrate-Responsive Expression of Nitrate Assimilation Genes*

Nitrate-responsive transcription of *NRT*, *NR*, and *NiR (SiRB)* was not detected in the MYB1-knockout strain, SI282 (Imamura et al. 2010). Furthermore, SI282 could not grow in medium containing nitrate (Fig. 18.4d), indicating that MYB1 is the indispensable TF to control the expression of nitrate assimilation genes.

In *Arabidopsis thaliana*, NIN-like protein (NLP) family proteins specifically bind to the promoter regions of nitrate-responsive genes and are activated posttranslationally in a process modulated by nitrate (Yanagisawa 2014). Thus, NILP is the central regulator of nitrate signaling in *A. thaliana*. NLP family proteins contain the RWP-RK domain, which may function in dimerization and DNA binding. In *C. merolae*, CMN310C, which is annotated as a hypothetical protein, contains the RWP-RK domain. The role of CMN310C in nitrate regulation is unknown, but its transcript levels in *C. merolae* cells were comparable in medium containing nitrate and in medium containing ammonia (Imamura et al. unpublished data). These results suggested that CMN310C is involved in types of regulation other than nitrate regulation and/or is regulated posttranslationally in *C. merolae*, similar to *A. thaliana* NLP proteins.

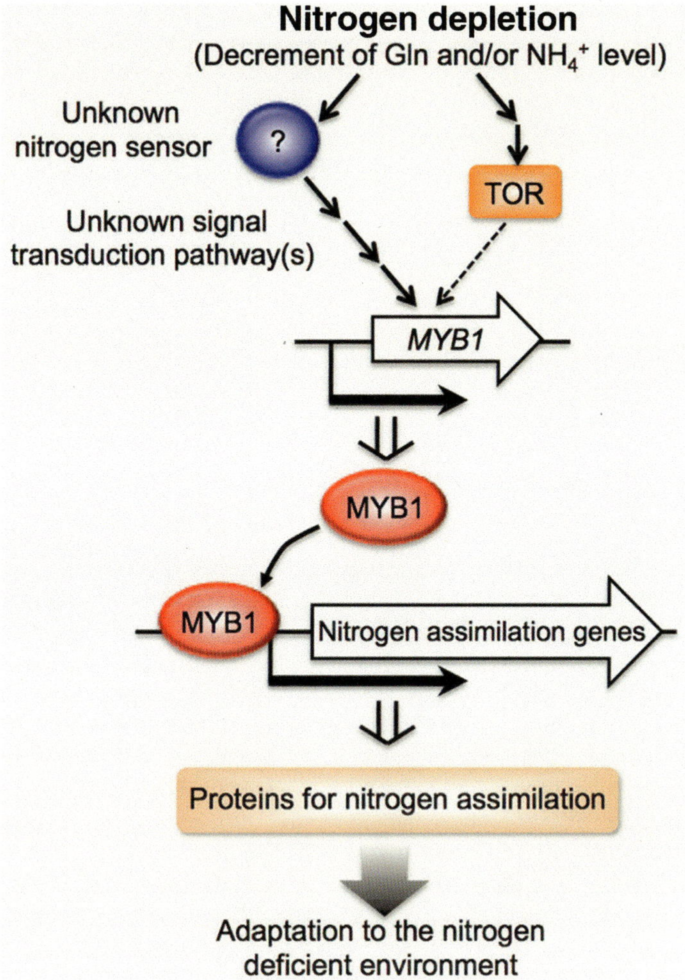


**Fig. 18.6** A possible model of MYB1-mediated nitrogen-responsive transcription (See text for details)

### *18.3.5 TOR Signaling Partially Regulates MYB1-Dependent Transcription*

As mentioned above, Gln3 is regulated by TOR and expression of MYB1 is controlled at the posttranscriptional level. It is well known that TOR senses the status of nitrogen-containing nutrients and transmits the signal to downstream regulators (Shimobayashi and Hall 2016). In eukaryotes, TOR inactivation by rapamycin mimics cellular responses to nitrogen depletion (Beck and Hall 1999; Cooper 2002; Crespo et al. 2005). Thus, we hypothesized that TOR is also involved in MYB1-dependent transcription. To test this possibility, we performed qRT-PCR analyses with RNA isolated from *C. merolae* strain F12 (Imamura et al. 2013), which has a rapamycin-sensitive phenotype, after rapamycin treatment. Transcripts of nitrogen assimilation-related genes, *AMT1*, *GS*, *NRT*, *NR*, and *NiR (SirB)* and the regulator *MYB1*, were increased after rapamycin treatment, even in the presence of ammonium, suggesting that TOR is involved in MYB1-mediated transcription. However, northern blot analyses showed that the increase in the transcript levels of these genes after rapamycin treatment was smaller than that under nitrogen depletion, indicating that TOR signaling is not the main regulatory pathway of MYB1-dependent transcription in *C. merolae* (Fig. 18.6).

## 18.4 Future Perspectives

Previous studies using bacteria have indicated that they do not sense the extracellular concentration of the nitrogen source, but sense intracellular nitrogen sources, anabolic products, and metabolites that fluctuate with anabolic reactions in the cells. For example, in *E. coli*, glutamine, which is the initial nitrogen assimilation product, and 2-OG, which supplies the carbon skeleton for nitrogen assimilation, are monitored by GlnD and PII, respectively (Merrick and Edwards 1995; Arcondeguy et al. 2001). The signal induces nitrogen depletion-responsive transcription via a two-component regulatory system. In cyanobacteria, the 2-OG level is monitored by PII and NtcA and regulates nitrogen-responsive transcription (Osanai and Tanaka 2007; Ohashi et al. 2012). Therefore, in bacteria, the pathways from metabolites that fluctuate in response to nitrogen conditions to activate transcription are relatively well-characterized. In contrast, such knowledge is very limited for most eukaryotes, especially plants and algae.

In *C. merolae*, TOR signaling is partially involved in MYB1-dependent transcription; however, it is unclear what kind of signal or signal transduction pathway (s) is the central regulator for such transcriptional activation (Fig. 18.6). Nitrogen assimilation in *C. merolae* cells is complex, because nitrogen metabolites, such as nitrite, ammonium, and glutamine, translocate across organelles (Fig. 18.1). It is still unknown which metabolite(s) in which intracellular location(s) function as the nitrogen signal (Figs. 18.1 and 18.6). Further comprehensive analyses of nitrogen signaling in the cell using new approaches, such as forward genetics and biochemical approaches, are necessary to address these questions. *C. merolae* is a suitable organism for such analyses because of its many good features, simple cell architecture and genome, low gene redundancy, weak cell wall, and capacity for genetic manipulation by homologous recombination.

## References

Arcondeguy T, Jack R, Merrick M (2001) PII signal transduction proteins, pivotal players in microbial nitrogen control. Microbiol Mol Biol Rev 65:80–105

Beck T, Hall MN (1999) The TOR signalling pathway controls nuclear localization of nutrient-regulated transcription factors. Nature 402:689–692

Bi YM, Zhang Y, Signorelli T, Zhao R, Zhu T, Rothstein S (2005) Genetic analysis of *Arabidopsis* GATA transcription factor gene family reveals a nitrate-inducible member important for chlorophyll synthesis and glucose sensitivity. Plant J 44:680–692

Cooper TG (2002) Transmitting the signal of excess nitrogen in *Saccharomyces cerevisiae* from the tor proteins to the GATA factors: connecting the dots. FEMS Microbiol Rev 26:223–238

Crespo JL, Díaz-Troya S, Florencio FJ (2005) Inhibition of target of rapamycin signaling by rapamycin in the unicellular green alga *Chlamydomonas reinhardtii*. Plant Physiol 139:1736–1749

Imamura S, Kanesaki Y, Ohnuma M, Inouye T, Sekine Y, Fujiwara T, Kuroiwa T, Tanaka K (2009) R2R3-type MYB transcription factor, CmMYB1, is a central nitrogen assimilation regulator in *Cyanidioschyzon merolae*. Proc Natl Acad Sci U S A 106:12548–12553

Imamura S, Ishiwata A, Watanabe S, Yoshikawa H, Tanaka K (2013) Expression of budding yeast FKBP12 confers rapamycin susceptibility to the unicellular red alga *Cyanidioschyzon merolae*. Biochem Biophys Res Commun 439:264–269

Imamura S, Terashita M, Ohnuma M, Maruyama S, Minoda A, Weber AP, Inouye T, Sekine Y, Fujita Y, Omata T, Tanaka K (2010) Nitrate assimilatory genes and their transcriptional regulation in a unicellular red alga *Cyanidioschyzon merolae*: genetic evidence for nitrite reduction by a sulfite reductase-like enzyme. Plant Cell Physiol 51:707–717

Imamura S, Kawase Y, Kobayashi I, Sone T, Era A, Miyagishima SY, Shimojima M, Ohta H, Tanaka K (2015) Target of rapamycin (TOR) plays a critical role in triacylglycerol accumulation in microalgae. Plant Mol Biol 89:309–318

Ishiyama K, Inoue E, Watanabe-Takahashi A, Obara M, Yamaya T, Takahashi H (2004) Kinetic properties and ammonium-dependent regulation of cytosolic isoenzymes of glutamine synthetase in *Arabidopsis*. J Biol Chem 279:16598–16605

Lam HM, Coschigano KT, Oliveira IC, Melo-Oliveira R, Coruzzi GM (1996) The molecular-genetics of nitrogen assimilation into amino acids in higher plants. Annu Rev Plant Physiol Plant Mol Biol 47:569–593

Linka M, Weber AP (2005) Shuffling ammonia between mitochondria and plastids during photorespiration. Trends Plant Sci 10:461–465

Merrick MJ, Edwards RA (1995) Nitrogen control in bacteria. Microbiol Rev 59:604–622

Nakayama M, Akashi T, Hase T (2000) Plant sulfite reductase: molecular structure, catalytic function and interaction with ferredoxin. J Inorg Biochem 82:27–32

Ohashi Y, Shi W, Takatani N, Aichi M, Maeda S, Watanabe S, Yoshikawa H, Omata T (2012) Regulation of nitrate assimilation in cyanobacteria. J Exp Bot 62:1411–1424

Osanai T, Tanaka K (2007) Keeping in touch with PII: PII-interacting proteins in unicellular cyanobacteria. Plant Cell Physiol 48:908–914

Rai R, Tate JJ, Shanmuganatham K, Howe MM, Nelson D, Cooper TG (2015) Nuclear Gln3 import is regulated by nitrogen catabolite repression whereas export is specifically regulated by glutamine. Genetics 201:989–1016

Sekine K, Sakakibara Y, Hase T, Sato N (2009) A novel variant of ferredoxin-dependent sulfite reductase having preferred substrate specificity for nitrite in the unicellular red alga *Cyanidioschyzon merolae*. Biochem J 423:91–98

Sekine K, Moriyama T, Kim J, Hase T, Sato N (2017) Characterization of two ferredoxin-dependent sulfite reductases having different substrate specificity in the red alga *Cyanidioschyzon merolae*. J Biochem 162:37–43

Shimobayashi M, Hall MN (2016) Multiple amino acid sensing inputs to mTORC1. Cell Res 26:7–20

Suzuki I, Kikuchi H, Nakanishi S, Fujita Y, Sugiyama T, Omata T (1995) A novel nitrite reductase gene from the cyanobacterium *Plectonema boryanum*. J Bacteriol 177:6137–6143

Terashita M, Maruyama S, Tanaka K (2006) Cytoplasmic localization of the single glutamine synthetase in a unicellular red alga, *Cyanidioschyzon merolae* 10D. Biosci Biotechnol Biochem 70:2313–2315

Virgilio CD, Loewith R (2006) The TOR signalling network from yeast to man. Int J Biochem Cell Biol 38:1476–1481

Wullschleger S, Loewith R, Hall MN (2006) TOR signaling in growth and metabolism. Cell 124:471–484

Yanagisawa S (2014) Transcription factors involved in controlling the expression of nitrate reductase genes in higher plants. Plant Sci 229:167–171

# Chapter 19
# Carbon Metabolism

**Takashi Moriyama, Natsumi Mori, and Naoki Sato**

**Abstract** Photosynthetic eukaryotes synthesize organic compounds through photosynthesis. The compounds are then consumed for energy production or converted into other compounds through various metabolic pathways in organelles. Recently, studies on carbohydrate and lipid metabolism were performed on the red alga *Cyanidioschyzon merolae*. These studies were possible due to advancements in genomics and transformation techniques. The *C. merolae* genome encodes enzymes related to basic carbohydrate and lipid metabolism, and the subcellular distribution of metabolic pathways of *C. merolae* is basically identical to that of land plants. However, some metabolic enzymes, such as NAD-dependent malic enzyme, glyoxylate cycle enzymes, some mitochondrial translocators, and several fatty acid desaturases, are not found in *C. merolae*, indicating the metabolic pathways of *C. merolae* are simpler than those of green plants. In this chapter, we will summarize our current understanding of carbon metabolism and describe unique features of *C. merolae* metabolism by comparing metabolic enzymes among different species of sequenced red algae.

**Keywords** Carbohydrate • Desaturase • Fatty acid • Floridean starch • Lipid • Translocator

## 19.1 Introduction

Plants and algae synthesize triose phosphates, such as glyceraldehyde 3-phosphate (GAP) and dihydroxyacetone phosphate (DHAP), from carbon dioxide ($CO_2$) through photosynthesis. The synthesized triose phosphates are used for energy production during respiration and are a source of organic compounds. Various

T. Moriyama • N. Mori • N. Sato (✉)
Department of Life Sciences, Graduate School of Arts and Sciences, The University of Tokyo, Meguro-ku, Tokyo, Japan
e-mail: moripoly@bio.c.u-tokyo.ac.jp; moppu@bio.c.u-tokyo.ac.jp; naokisat@bio.c.u-tokyo.ac.jp



T. Kuroiwa et al. (eds.), *Cyanidioschyzon merolae*,
https://doi.org/10.1007/978-981-10-6101-1_19

organelles are involved in carbon metabolism. In land plants, glycolytic pathways are present in both the cytosol and plastids (Dennis and Miernyk 1982). Organic acids such as pyruvate, produced from glycolysis, are further oxidized by the tricarboxylic acid (TCA) cycle in mitochondria, similar to non-photosynthetic eukaryotes, such as mammals and yeasts. As storage carbohydrates, starch and sucrose are synthesized in plastids and the cytosol, respectively. Photosynthesis also provides materials for lipid synthesis. The glycerol backbone (glycerol 3-phosphate) is produced from triose phosphate in the cytosol and plastids. For fatty acid synthesis, triose phosphate synthesized through photosynthesis is transported to the cytosol and then converted to phospho*enol*pyruvate (PEP) by the glycolytic pathway. PEP is further transported to plastids and converted into acetyl-CoA. Fatty acids are synthesized through the condensation of malonyl-CoA derived from acetyl-CoA in plastids. Lipids are synthesized from glycerol 3-phosphate and fatty acids in plastids (Joyard et al. 2010) or the endoplasmic reticulum (ER) (Li-Beisson et al. 2013). In a minor pathway, fatty acids are also synthesized in mitochondria (Wada et al. 1997).

The basic metabolic pathways described above are also present in the unicellular red alga *Cyanidioschyzon merolae*. Exploiting the fact that the *C. merolae* genome contains a limited number of duplicated genes, exhaustive subcellular localization analyses of carbon metabolic enzymes were performed by observing the localization of GFP-fused proteins in transiently transformed *C. merolae* (Moriyama et al. 2014; Mori et al. 2016). Studies revealed that the subcellular distribution of *C. merolae* metabolic pathway proteins is nearly identical to that of land plants (Fig. 19.1). One difference is that starch is synthesized and accumulated in the cytosol, as in other red algae (Viola et al. 2001). In addition to localization analyses, metabolic studies, in particular on lipids, were intensively performed in recent years, resulting in detailed characterization of the metabolic processes in *C. merolae*.

*C. merolae* has enzymes related to basic carbohydrate and lipid metabolism; however, compared with land plants, metabolic enzymes, including transaldolase (in the oxidative pentose phosphate pathway (oxPPP)), pyruvate decarboxylase (in the fermentation pathway), and several fatty acid desaturases, are not found in *C. merolae*. Therefore, the metabolic pathways of *C. merolae* are simpler than those of land plants and green algae. This makes *C. merolae* ideal for investigating the fundamentals of carbon metabolic pathways conserved in photosynthetic eukaryotes. In this chapter, we summarize studies on carbohydrate and lipid metabolism in *C. merolae* and focus on the unique features of carbon metabolism in *C. merolae*, which were revealed by comparing the sequences of metabolic enzymes from different species of red algae.

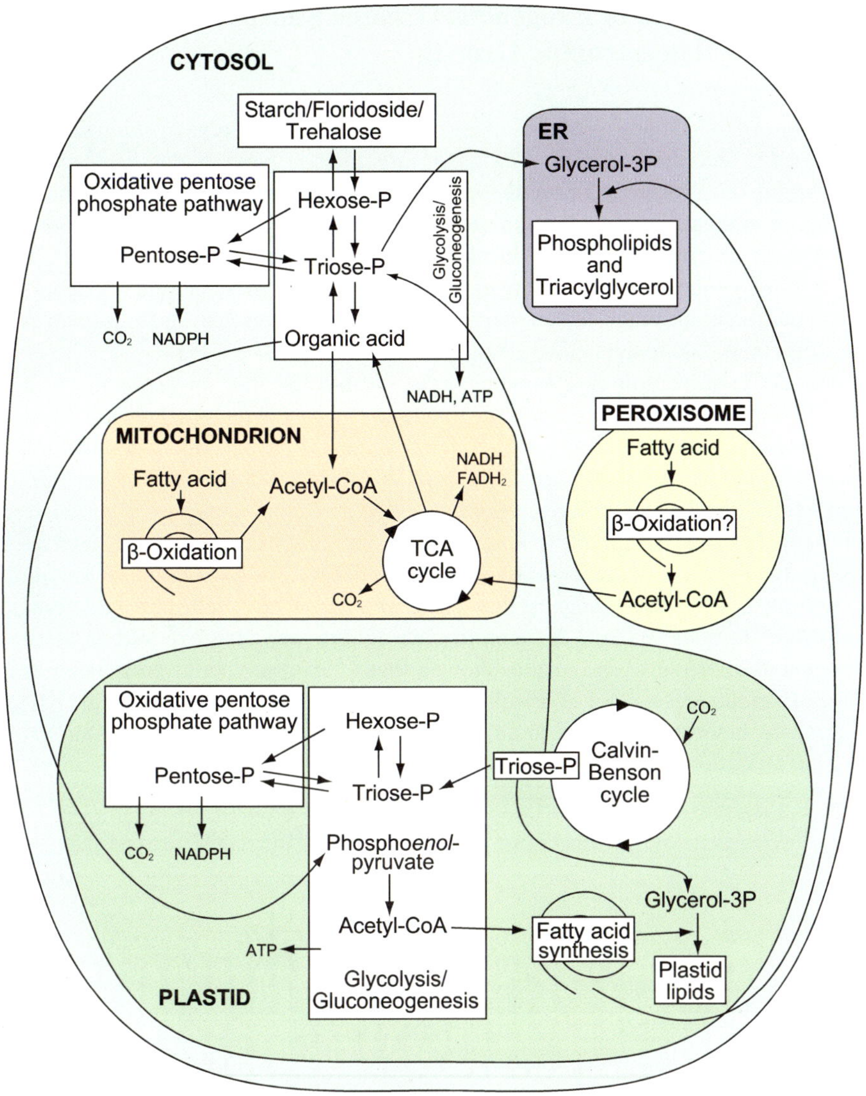


**Fig. 19.1** Overview of carbon metabolism pathways in *C. merolae*. Triose phosphates (Triose-P), including glyceraldehyde 3-phosphate (GAP) and dihydroxyacetone phosphate (DHAP), are synthesized by photosynthesis in the plastid and then metabolized in various organelles. Starch is synthesized in the cytosol in red algae. In *C. merolae*, it is believed that the fatty acid degradation pathway, β-oxidation, occurs in the mitochondrion and peroxisome. This model was based on localization studies of enzymes involved in central carbon metabolism (Moriyama et al. 2014) and lipid and fatty acid metabolism (Mori et al. 2016)

## 19.2 Utilization of Exogenous Organic Substrates and Heterotrophic Growth

It is known that many types of algae, such as the green alga *Chlamydomonas reinhardtii* and the red alga *Galdieria sulphuraria,* grow under heterotrophic conditions (Gross and Schnarrenberger 1995). However, *C. merolae* was thought to be an obligate photoautotroph (Raven et al. 1990); therefore, utilization of exogenous organic substrates in this species remained poorly studied. The *C. merolae* genome is estimated to encode a minimal set of metabolic transporters, although there are much fewer transporters in *C. merolae* than in *G. sulphuraria* (Barbier et al. 2005; Tyra et al. 2007).

Recently, utilization of exogenous organic substrates in *C. merolae* was evaluated by measuring the respiratory rate in nutrient-deprived cells that contained very little starch (Moriyama et al. 2015). Most of the exogenous substrates including sugars, alcohols, carboxylic acids, and amino acids increased the respiratory rate 3- to 19-fold (Fig. 19.2). However, the effect of the exogenous substrates was small in nutrient-rich cells. In mixotrophic conditions, the exogenous substrates increased the growth rate, but the increase was limited to cells grown under low-light conditions. These results suggest that *C. merolae* can utilize environmental organic substrates to some extent. A heterotrophic culture was also exposed to several exogenous substrates. The results showed that *C. merolae* could grow heterotrophically in the presence of glycerol at high concentrations (Moriyama et al. 2015). Curiously, heterotrophic growth did not continue infinitely but stopped after six or

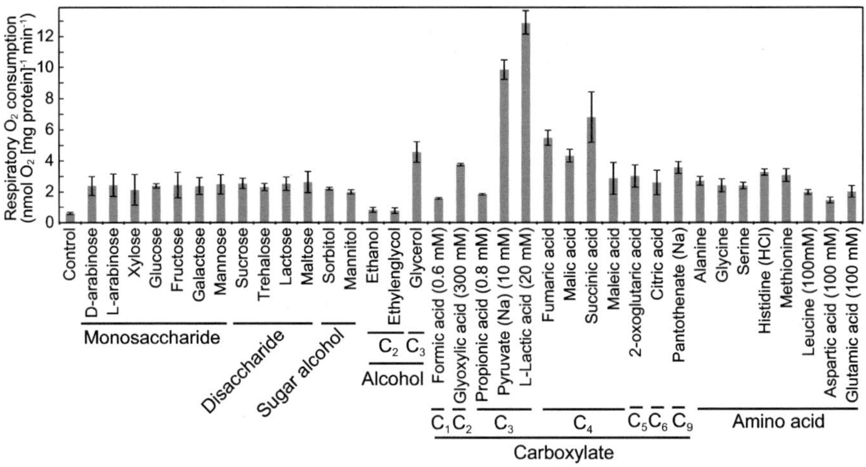


**Fig. 19.2** Effect of exogenous organic substrates on *C. merolae* cell respiration. The respiratory rate was determined after 200 mM of the indicated organic substrates were added to the culture. The concentrations of formic acid, glyoxylic acid, propionic acid, pyruvate, lactic acid, leucine, aspartic acid, and glutamic acid are shown in parentheses (Modified from Moriyama et al. 2015)

seven cell divisions (Moriyama et al. unpublished data). Therefore, we still need to find a better way to prolong heterotrophic growth.

## 19.3 Carbohydrate Metabolism

### *19.3.1 Glycolysis/Gluconeogenesis and the Oxidative Pentose Phosphate Pathway*

Through glycolysis, glucose and its derivatives are used to make organic acids, such as pyruvate. Glucose is also oxidized in the oxPPP, which functions to generate NADPH and pentose sugars, rather than ATP. In *Arabidopsis thaliana*, glycolysis (Embden-Meyerhof pathway, the most common glycolytic pathway) and oxPPP occur in both the cytosol and plastids (Dennis and Miernyk 1982). The localization of oxPPP enzymes in the cytosol and plastids is not the same, namely, all of the oxPPP enzymes are found in plastids, whereas only the first five enzymes of oxPPP are found in the cytosol (Fig. 19.3). Differences in the subcellular distribution of glycolysis and oxPPP enzymes were also reported in photosynthetic eukaryotes, such as *Chlamydomonas reinhardtii* (Johnson and Alric 2013) and *Phaeodactylum tricornutum,* a diatom (Gruber et al. 2009).

Subcellular localization studies of enzymes related to central carbon metabolism revealed that glycolysis and oxPPP proteins have similar localization in both *C. merolae* and green tissue cells from *A. thaliana* (Moriyama et al. 2014, Fig. 19.3). However, the gene encoding the oxPPP enzyme transaldolase was not found in the *C. merolae* genome; this finding will be discussed in more detail later. Plant-specific glycolytic enzymes, such as pyrophosphate (PPi)-dependent phosphofructokinase (PFP) and NADP-dependent non-phosphorylating GAP dehydrogenase (GAPN), are conserved in red algae and are localized to the cytosol in both *C. merolae* and land plants. Carbohydrate metabolism-related enzymes that are unique to *C. merolae* or red algae are summarized in Table 19.1 and are described below.

In glycolysis, glucose is first phosphorylated by glucokinase or hexokinase to glucose 6-phosphate (G6P). Generally speaking, glucokinase and hexokinase are conserved in prokaryotes and eukaryotes, respectively. Red algae, however, have glucokinase(s), but not hexokinase in common. Among green plants, hexokinases are found in land plants and the filamentous terrestrial charophyte *Klebsormidium flaccidum*. In contrast, glucokinase in general, is found in green algae, but the exact distribution of hexokinase is species-dependent. Red algal glucokinases have an extension sequence comprised of approximately 70 amino acids at the *C*-terminus (Weber et al. 2004). It is thought that this region forms a transmembrane helix that is embedded in the plasma membrane. In contrast, bacterial and green algal glucokinases do not have this *C*-terminal extension region. The hexokinases from

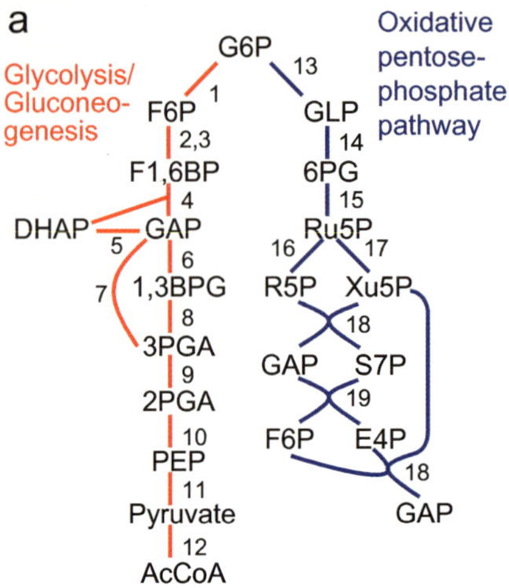


| ID | Enzyme |
|---|---|
| 1 | Phosphoglucose isomerase |
| 2 | Phosphofructokinase |
| 3 | Fructose-1,6-bisphosphatase |
| 4 | Fructose-1,6-bisphosphate aldolase |
| 5 | Triosephosphate isomerase |
| 6 | GAP dehydrogenase (phosphorylating) |
| 7 | GAP dehydrogenase (non phosphorylating) |
| 8 | Phosphoglycerate kinase |
| 9 | Phosphoglycerate mutase |
| 10 | Enolase |
| 11 | Pyruvate kinase |
| 12 | Pyruvate dehydrogenase |
| 13 | G6P dehydrogenase |
| 14 | 6-phosphogluconolactonase |
| 15 | 6PG dehydrogenase |
| 16 | dehydrogenase |
| 17 | Ribulose-5-phosphate-3-epimerase |
| 18 | Transketolase |
| 19 | Transaldolase |

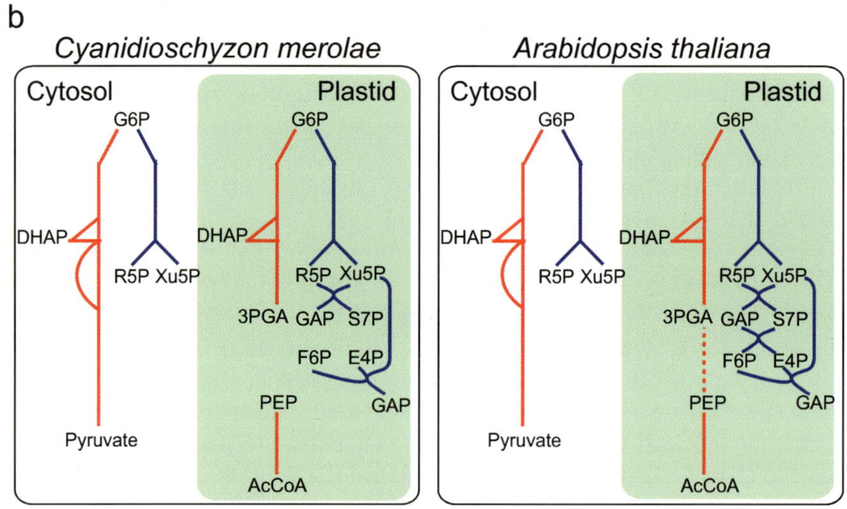


**Fig. 19.3** Subcellular localization of glycolysis and the oxidative pentose phosphate pathway (oxPPP) in *C. merolae* and land plants. (**a**) General glycolytic pathway (*red line*) and oxPPP (*blue line*). Numbers between intermediates indicate the reaction process ID, and the corresponding enzymes are shown in the table on the right. Abbreviations of intermediates: *1,3BPG* 1,3-bisphoglycerate, *2PGA* 2-phosphoglycerate, *3PGA* 3-phosphoglycerate, *6PG* 6-phosphogluconate, *AcCoA* acetyl-CoA, *DHAP* dihydroxyacetone phosphate, *E4P* erythrose 4-phosphate, *F1,6BP* fructose 1,6-bisphosphate, *F6P* fructose 6-phosphate, *G6P* glucose 6-phosphate, *GAP* glyceraldehyde 3-phosphate, *GLP* glucono 1,5-lactone 6-phosphate, *PEP* phosphoenolpyruvate, *R5P* ribose 5-phosphate, *Ru5P* ribulose 5-phosphate, *S7P* sedoheptulose 7-phosphate, *Xu5P* xylulose 5-phosphate. (**b**) Glycolysis and oxPPP in *C. merolae* and *A. thaliana*. Phosphoglycerate mutase and enolase have not been reported in the chloroplasts of green tissues in *A. thaliana*. The *dashed line* indicates a reaction step that occurs only in non-photosynthetic tissues. The *C. merolae* genome does not contain a gene-encoding transaldolase (Modified from Moriyama et al. 2014)

**Table 19.1** List of carbohydrate metabolism-related enzymes unique to *C. merolae* and red algae

| Enzyme | *Cme* | *Gsu* | *Ppu* | *Ccr* | *Pye* |
|---|---|---|---|---|---|
| *Glycolysis/gluconeogenesis* | | | | | |
| Glucokinase | 1 | 1 | 2 | 1 | 1 |
| Fusion protein of inorganic pyrophosphatase and fructose-1,6-bisphosphatase | 1 | 1 | 1 | 1 | 1 |
| Phosphoglycerate mutase | 1 | 2 | 1 | 4, 1 (Pt) | 2, 1 (Pt) |
| Enolase | 1 | 1 | 2 | 5 | 4 |
| *Metabolism of pyruvate* | | | | | |
| Malic enzyme, $NADP^+$-dependent | 1 | – | 2 | 1 | 1 |
| Malic enzyme, $NAD^+$-dependent | – | 1 | 1 | 1 | 1 |
| Lactate dehydrogenase | 5 | – | – | – | – |
| Pyruvate decarboxylase | – | – | 1 | – | – |
| Phosphate acetyltransferase | – | – | 2 | – | – |
| Acetate kinase | – | – | 1 | – | – |
| *Oxidative pentose phosphate pathway* | | | | | |
| Transaldolase | – | 1 | – | 1 | 1 |
| *Glyoxylate cycle* | | | | | |
| Isocitrate lyase and malate synthase | – | – | – | – | – |

The number of enzymes encoded by each red algal genome is shown. Enzymes were identified by BLAST searches. Genome sequences were obtained from the *Cyanidioschyzon merolae* (*Cme*) genome project (http://merolae.biol.s.u-tokyo.ac.jp/; Matsuzaki et al. 2004; Nozaki et al. 2007), *Porphyridium purpureum* (*Ppu*) genome project (http://cyanophora.rutgers.edu/porphyridium/; Bhattacharya et al. 2013), National Research Institute of Fisheries Science (http://nrifs.fra.affrc.go.jp/ResearchCenter/5_AG/genomes/nori/index.html) for *Pyropia yezoensis* (*Pye*, Nakamura et al. 2013) and GenomeNet Database (http://www.genome.jp/) for *Galdieria sulphuraria* (*Gsu*, Weber et al. 2004; Schönknecht et al. 2013), and *Chondrus crispus* (*Ccr*, Collén et al. 2013). Abbreviations: Pt, encoded by the plastid genome; –, not found

some land plants bind to the plasma membrane though an *N*-terminal extension sequence rich in hydrophobic amino acids (Wiese et al. 1999).

The plastid in *C. merolae* does not have phosphoglycerate mutase (PGAM) or enolase (ENO); therefore glycolysis is discontinuous between 3-phosphoglycerate (3PGA) and PEP with respect to metabolic intermediates, similar to that in green tissue cells of *A. thaliana* (Fig. 19.3). However, among red algae, the *PGAM* gene was found in the plastid genomes of the multicellular red algae *Chondrus crispus* and *Pyropia yezoensis* (Table 19.1). Additionally, these nuclear genomes contained four or five *ENO* genes, one of which encoded an ENO with an *N*-terminal extension sequence expected as a transit peptide. Therefore, the full glycolytic pathway may occur in the plastids of these red algae.

There is another unique glycolysis gene in red algae. The genomes of red algae contain a gene encoding a pyrophosphatase/fructose-1,6-bisphosphatase fusion protein (PPase/FBP). This enzyme is believed to be localized to the cytosol.

Because PPase degrades pyrophosphate to phosphate, this fusion enzyme may be involved in the regulation of glycolysis/gluconeogenesis flux via regulation of PPi-dependent PFK activity.

The gene for transaldolase, an oxPPP enzyme, is conserved in almost all organisms but is not found in the genome of *C. merolae*. Consistently, protein extracts from *C. merolae* showed no detectable transaldolase activity (Moriyama et al. 2014). The *TAL* gene was also not found in the genome of *Porphyridium purpureum* (Table 19.1). However, the *TAL* gene was found in the genomes of the red algae, *G. sulphuraria*, *C. crispus,* and *P. yezoensis*. These results indicate that the presence of the *TAL* gene varies in red algae. Results from a study on a *tal* mutant of *Escherichia coli* (Nakahigashi et al. 2009) showed that two *C. merolae* glycolytic enzymes, ATP-dependent PFK and fructose-1,6-bisphosphate aldolase (FBA), found in the plastid, may compensate for the lack of the oxPPP enzyme transaldolase (Moriyama et al. 2014).

## *19.3.2 Fermentation Pathway*

Under anaerobic conditions, plants regenerate $NAD^+$ from NADH by lactate or alcohol fermentation to drive glycolysis. Lactate is synthesized from pyruvate by lactate dehydrogenase (LDH). During alcohol fermentation, pyruvate is converted to acetaldehyde by pyruvate decarboxylase (PDC), which is then reduced by alcohol dehydrogenase (AlcDH) to produce ethanol. Compared with land plants, the genome of *C. reinhardtii* has additional genes related to fermentation, such as pyruvate formate lyase (PFL), pyruvate:ferredoxin oxidoreductase (PFOR), bifunctional alcohol/acetaldehyde dehydrogenase, phosphate acetyltransferase (PAT), and acetate kinase (ACK) (Yang et al. 2014; Burgess et al. 2016). *C. reinhardtii* can anaerobically produce ethanol, acetate, and formate using these enzymes.

The *C. merolae* genome contains genes that encode AlcDH and aldehyde dehydrogenase (AldDH), but not PFL, PFOR, PAT, ACK, or PDC. PDC is believed to be essential for alcohol and acetate fermentation; however, *C. merolae* may lack these two types of fermentation. Although AlcDH and AldDH generally function to produce ethanol and acetate, these enzymes may be involved in other pathways in *C. merolae*. The red algae, *G. sulphuraria*, *C. crispus*, and *P. yezoensis* are similar to *C. merolae* with respect to these genes; however, the *P. purpureum* genome has additional genes encoding PAT, ACK, and PDC (Table 19.1). Among red algae, only *C. merolae* has cytosolic LDH (Moriyama et al. 2015). In an assay to examine the effect of exogenous substrates on the respiratory rate in *C. merolae*, exogenous lactate dramatically increased the respiration rate, whereas ethanol had only a slight effect (Fig. 19.2, Moriyama et al. 2015). Taken together with the genomic data, this suggests that *C. merolae* does not metabolize ethanol and produces lactate under anaerobic conditions.

### 19.3.3 Tricarboxylic Acid Cycle

Pyruvate produced by glycolysis is transported to mitochondria where it is further oxidized by the TCA cycle. As in other eukaryotes, the *C. merolae* mitochondrion contains a complete set of TCA cycle enzymes (Moriyama et al. 2014).

Land plants have alternative pathways for PEP metabolism in the cytosol (Fig. 19.4c). Oxaloacetate (OAA) and malate are found in the cytosol and mitochondria and are produced through the activities of the PEP metabolic enzymes, PEP carboxylase (PEPC), PEP carboxykinase (PEPCK), malate dehydrogenase (MDH), and NADP-dependent malic enzyme (NADP-ME). These PEP metabolic enzymes are also found in *C. merolae* in the cytosol (Moriyama et al. 2014). In land plants, malate that is transported from the cytosol to mitochondria can also act as a substrate in the TCA cycle. Specifically, malate is converted to pyruvate through the activity of the NAD-dependent malic enzyme (NAD-ME) (Tronconi et al. 2008). NAD-ME is conserved in plants and algae, but its function is not essential for normal growth (Jenner et al. 2001; Tronconi et al. 2008). On the other hand, *C. merolae* lacks NAD-ME; therefore, the transport of pyruvate to the mitochondrion is essential for respiration (Fig. 19.4a).

When OAA, the receptor for acetyl-CoA in the TCA cycle, is regenerated from malate, carbon can flow in the TCA cycle in a cyclical manner. However, in nitrogen assimilation, 2-oxoglutarate (2OG) is removed from mitochondria by a 2OG/malate translocator, which exchanges cytosolic malate for mitochondrial 2OG (Figs. 19.4b, d, and 19.6). Therefore, lost intermediates must be supplemented. In animal mitochondria, OAA is regenerated from pyruvate and $CO_2$ by pyruvate carboxylase (PYC). Similar to land plants, red algae lack PYC, whereas green algae possess PYC. Because land plants contain both a dicarboxylate translocator that exchanges cytosolic malate for mitochondrial inorganic phosphate (Fig. 19.6) and NAD-ME, they can maintain the nitrogen assimilation mode by recruiting both malate and pyruvate or only malate from the cytosol (Fig. 19.4d). In contrast, because *C. merolae* lacks NADP-ME, PYC, and dicarboxylate translocator (Fig. 19.6), both pyruvate and malate must be transported into the mitochondrion for nitrogen assimilation (Fig. 19.4b). In other words, in *C. merolae*, when malate is transported into the mitochondrion, nitrogen assimilation becomes active. This hypothesis is supported by the results of a study that used an assay to determine the effect of various exogenous substrates on the respiratory rate (Fig. 19.2). The results showed that pyruvate and lactate were more suitable substrates for respiration than intermediates of the TCA cycle, such as malate (Moriyama et al. 2015).

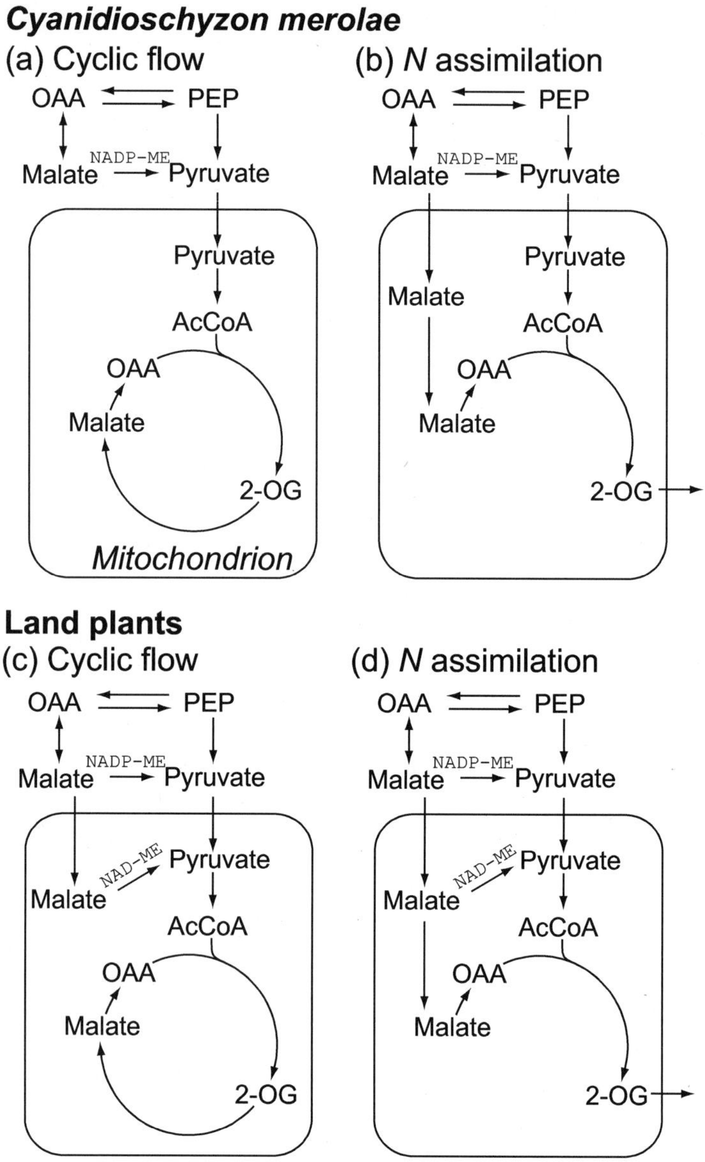


**Fig. 19.4** Flux modes of the tricarboxylic acid (TCA) cycle in *C. merolae* and land plants. *C. merolae* does not have mitochondrial NAD-dependent malic enzyme (NAD-ME). Therefore, pyruvate, not malate, must be transported to the mitochondrion for respiration (**a**) unlike land plants (**c**). In nitrogen assimilation in *C. merolae*, cytosolic malate is transported to the mitochondrion in exchange for mitochondrial 2-oxoglutarate (2OG) by the 2OG translocator (**b**). By recruiting both pyruvate and malate to the mitochondrion, nitrogen assimilation is maintained. In land plants (**d**), cytosolic malate is transported to mitochondria by a 2OG translocator and/or dicarboxylate translocator. Nitrogen assimilation is maintained by the recruitment of pyruvate and/or malate in land plants

### 19.3.4 Metabolism of Acetyl-CoA Derived from β-Oxidation

Land plants have two pathways for the metabolism of acetyl-CoA derived from fatty acid degradation, namely, β-oxidation. One is the glyoxylate cycle, which synthesizes sucrose (Fig. 19.5a), and the other is respiratory oxidation (Fig. 19.5b).

During germination in land plants that produce oil seeds, the glyoxylate cycle uses accumulated triacylglycerol (TAG) to produce sucrose, which is used as a carbon source in meristems. The glyoxylate cycle consists of five enzymes: citrate synthase (CS), aconitase (ACO), isocitrate lyase (ICL), malate synthase (MLS), and MDH (Graham 2008). Of these enzymes, ACO is located in the cytosol, and the remaining enzymes are located in glyoxysomes (specialized peroxisomes). Acetyl-CoA is incorporated into the glyoxylate cycle through CS and MLS (Fig. 19.5a). Then, succinate is produced by ICL and transported to mitochondria, where it is converted to malate by two enzymes in the TCA cycle. In the cytosol, malate is converted to sucrose through gluconeogenesis. In this pathway, because the two TCA cycle enzymes, isocitrate dehydrogenase (IDH) and 2OGDH, are bypassed, malate is synthesized without decarboxylation. Green plants and algae have this cycle, and they can produce sugar from accumulated oil (Plancke et al. 2014). The genomes of red algae do not contain *ICL* and *MLS* genes. In addition, peroxisomal MDH and cytosolic ACO are not found in *C. merolae* (Moriyama et al. 2014). These findings indicate that red algae do not have the glyoxylate cycle.

The pathway for respiratory oxidation has been reported in *A. thaliana* (Fig. 19.5b, Pracharoenwattana et al. 2005). Peroxisomal citrate, which is synthesized from acetyl-CoA and OAA, is transported to mitochondria. Conversely, mitochondrial OAA, which is produced from the TCA cycle, is transported to peroxisomes. That is, the TCA cycle occurs between mitochondria and peroxisomes, and acetyl-CoA derived from β-oxidation is used for respiration. *C. merolae* has three peroxisomal CSs and may have this pathway (Moriyama et al. 2014). However, the *C. merolae* genome does not contain a citrate translocator gene (Fig. 19.6). Additionally, β-oxidation enzymes have not been found in the peroxisome, although β-oxidation enzymes have been found to be localized to the mitochondrion (Mori et al. 2016). It has been reported that *C. merolae* accumulates TAG under nitrogen-depleted and normal culture conditions (Sumiya et al. 2015; Imamura et al. 2015; Toyoshima et al. 2016; Takusagawa et al. 2016); however, further studies are needed to clarify the pathway for utilization of accumulated TAG.

### 19.3.5 Organellar Translocator

In plants, G6P, triose phosphates such as GAP and DHAP, and xylulose-5-phosphate (Xu5P) produced by photosynthesis in plastids are transported to the cytosol and exchanged for inorganic phosphate (P*i*) by sugar phosphate translocators.

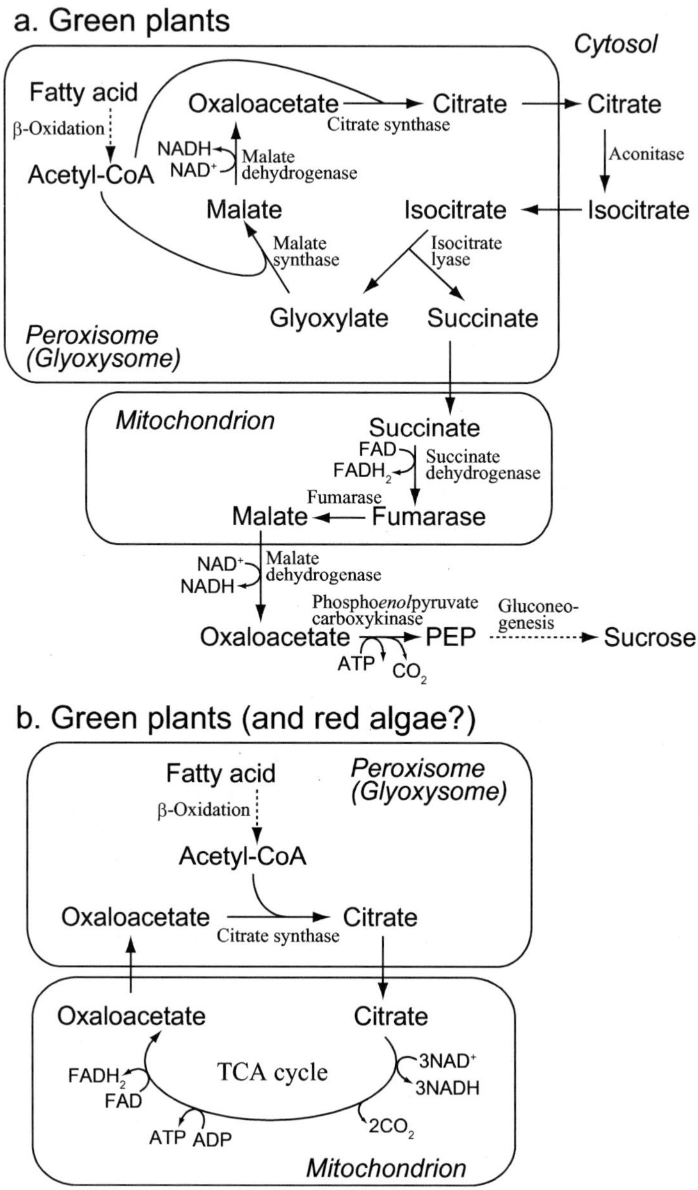


**Fig. 19.5** Metabolic pathways of acetyl-CoA derived from β-oxidation. (**a**) Glyoxylate cycle. Red algae do not have isocitrate lyase or malate synthase, which indicates the lack of a glyoxylate cycle. (**b**) Respiratory oxidation pathway. Acetyl-CoA is metabolized by the TCA cycle, which occurs across peroxisome and mitochondria

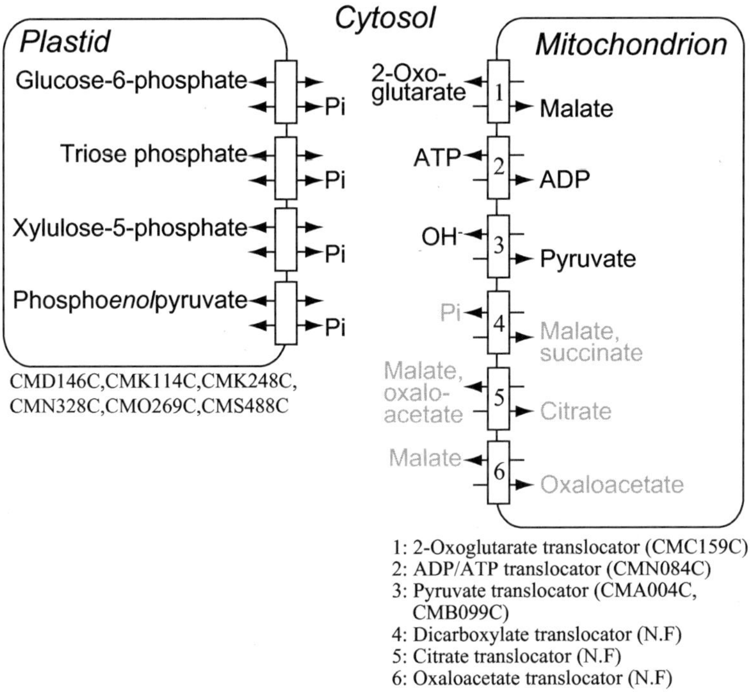


**Fig. 19.6** Plastid and mitochondrial translocators in *C. merolae*. Protein IDs of putative plastid translocators in *C. merolae* are shown below the plastid box. Land plants contain mitochondrial translocators 1–6, whereas *C. merolae* lacks translocators 4–6 (substrates in *gray*). Enzyme names and protein IDs corresponding to the number are shown below the mitochondria box. Abbreviations: *N.F* not found in the *C. merolae* genome

Conversely, PEP, synthesized by glycolysis in the cytosol, is transported to plastids through a PEP translocator. PEP is then used for the synthesis of fatty acids and aromatic amino acids by the shikimate pathway. Six sugar-phosphate translocator genes are found in the genome of *C. merolae*; however, the substrate specificity of each homologue still needs to be verified (Fig. 19.6, Barbier et al. 2005).

Among translocators known in plant mitochondria, the *C. merolae* genome contains genes for 2OG and ADP/ATP pyruvate translocators (Fig. 19.6, Barbier et al. 2005). The pyruvate translocator is necessary to supply pyruvate to the TCA cycle and was recently identified in animals, yeast (Bricker et al. 2012; Herzig et al. 2012), and *A. thaliana* (Li et al. 2014). Two pyruvate translocator genes are found in the *C. merolae* genome. On the other hand, dicarboxylate, citrate, and OAA translocator genes are not found in the *C. merolae* genome. The genomic data suggests that subcellular transport of metabolic intermediates in *C. merolae* is much simpler than in land plants.

## 19.4 Metabolism of Storage Sugars

Red algae accumulate photosynthetic products as soluble or insoluble carbohydrate in the cytosol (Viola et al. 2001). Floridean starch and (iso)floridoside were detected in *C. merolae* (De Luca and Moretti 1983; Hirabaru et al. 2010). Light microscopy after staining with Lugol's solution (Toyoshima et al. 2016; Takusagawa et al. 2016) and electron microscopy (Toyoshima et al. 2016) showed starch accumulation in the cytosol. The pathways for floridean starch, (iso) floridoside, trehalose, and matrix polysaccharide are described by Barbier et al. (2005) and are presented in Fig. 19.7 in a modified form.

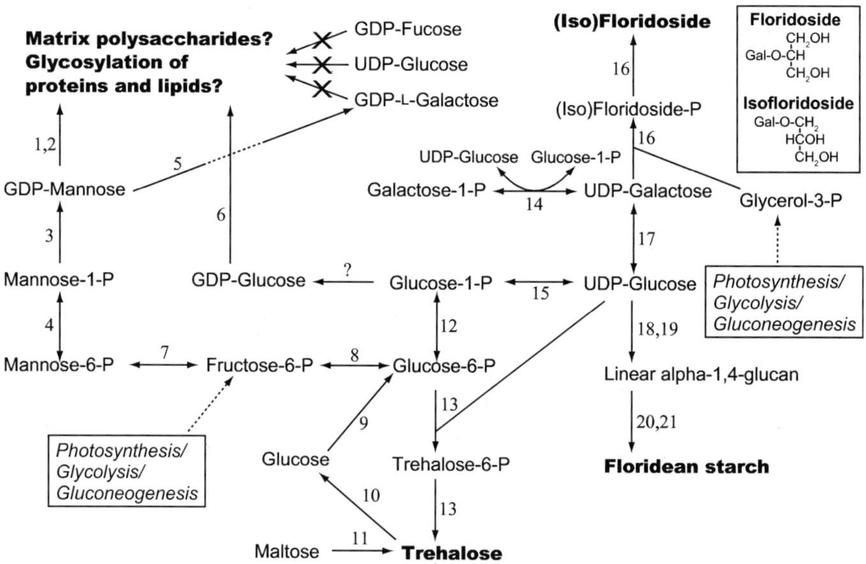


**Fig. 19.7** Metabolic pathways for storage and matrix polysaccharides in *C. merolae*. The number between the pair of intermediates refers to the enzyme that mediates the reaction: *1*, α-1,6-mannosyltransferase (CMT636C); *2*, β-1,4-mannosyltransferase (CMR050C); *3*, mannose-1-phosphate guanylyltransferase (CML145C, CMO003C); *4*, phosphomannomutase (CMQ067C, CMT314C); *5*, GDP-mannose 3′, 5′-epimerase (CMS308C); *6*, α-1,3-glucosyltransferase (CMH028C); *7*, mannose-6-phosphate isomerase (CMQ359C); *8*, phosphoglucose isomerase (CMT497C); *9*, glucokinase (CMO276C); *10*, trehalase (CMH100C, CMN097C); *11*, trehalose synthase (CMB158C); *12*, phosphoglucomutase (CMT285C); *13*, bifunctional trehalose phosphate synthase/phosphatase (CMO053C); *14*, galactose-1-phosphate uridylyltransferase (CMQ030C); *15*, UTP:glucose-1-phosphate uridylyltransferase (CMS159C); *16*, (Iso)floridoside phosphate synthase/phosphatase (CMI293C); *17*, UDP-glucose 4-epimerase/UDP-galactose 4-epimerase (CMA041C); *18*, glycogenin (CMK020C); *19*, Starch/glycogen synthase (CMM317C); *20*, 1,4-α-glucan branching enzyme (CMH144C); and *21*, 4-α-glucanotransferase (CMP352C, CMT204C). The *C. merolae* genome does not contain genes for fucosyltransferase, cellulose synthase, or galactosyltransferase. The reactions catalyzed by these enzymes are indicated by crosses. The chemical structures of floridoside and isofloridoside are shown in the *upper right* corner (This figure is a summary of the data in Barbier et al. (2005) and Pade et al. (2015))

### 19.4.1 Floridean Starch

The nature of floridean starch was studied in detail by Hirabaru et al. (2010). *Cyanidium caldarium* and *G. sulphuraria*, from the order Cyanidiales, both contain a glycogen-type starch, whereas *C. merolae* contains a semiamylopectin-type starch, similar to that found in other red algae such as *P. purpureum*. The starch granules of *C. merolae* are uniform spheres with a diameter of 0.2 μm and consist of only semiamylopectin. In contrast, starch from other red algae, such as *G. sulphuraria*, *C. caldarium*, and *P. purpureum*, consist of both semiamylopectin and amylose (Shimonaga et al. 2007; Shimonaga et al. 2008). Granule-bound starch synthase (GBSS) is found in the starch granules of cyanobacteria, green plants, and red algae including *P. purpureum* (Shimonaga et al. 2007; Plancke et al. 2008; Deschamps et al. 2008), whereas starch phosphorylase, not GBSS, is found in the starch granules of *C. merolae* (Hirabaru et al. 2010).

### 19.4.2 (Iso)Floridoside and Trehalose

Floridoside is accumulated as a photosynthetic product and an osmoticum. De Luca and Moretti (1983) found that floridoside and isofloridoside levels in *C. merolae* were much lower than in *C. caldarium* and *G. sulphuraria*. The authors washed the red algal cells (including *C. merolae*) several times with water. It is likely that this washing step caused the *C. merolae* cells, which lack a rigid cell wall, to burst. This may explain why the floridoside levels were lower in *C. merolae* than the other red algae. It is also possible that the floridoside content changed depending on the culture conditions (De Luca and Moretti 1983); therefore, it is possible that only *C. merolae* were cultured in conditions that did not favor floridoside accumulation.

Pade et al. (2015) recently discovered a floridoside synthesis pathway in *G. sulphuraria* that involves a bifunctional floridoside phosphate synthase/phosphatase. In the study, two enzymes with homology to trehalose-6-phosphate synthase/phosphatase had floridoside phosphate synthase/phosphatase activity in vitro and in heterologous expression assays in a cyanobacterium. Results from a phylogenetic analysis suggest that *C. merolae* also has a floridoside phosphate synthase/phosphatase (Fig. 19.7, enzyme ID 16) and a trehalose-6-phosphate synthase/phosphatase (Fig. 19.7, enzyme ID 13). In *C. merolae*, (iso)floridoside is expected to be synthesized from UDP-galactose and glycerol 3-phosphate by floridoside phosphate synthase/phosphatase, as in *G. sulphuraria*.

It is unknown if trehalose is found in *C. merolae*, but genes for trehalose synthase (Fig. 19.7, enzyme ID 11) and trehalose-6-phosphate synthase/phosphatase were found in the *C. merolae* genome. Therefore, it is very likely that *C. merolae* has the ability to metabolize trehalose (Barbier et al. 2005).

### 19.4.3 Putative Enzymes Involved in the Synthesis of a Cell Wall-Like Structure

*C. merolae* lacks a rigid cell wall, but it is thought that this alga is enclosed by a less rigid cell wall-like structure. As this cell wall-like structure is poorly understood, we examined the *C. merolae* genome for the presence of genes encoding enzymes likely related to cell wall synthesis.

The cell wall of land plants are composed mainly of cellulose. Cellulose synthase (CES) is conserved in land plants, but not in the charophyte *K. flaccidum* or the chlorophyte *C. reinhardtii*. The cell wall of *C. reinhardtii* is composed mainly of hydroxyproline-rich glycoprotein (Catt et al. 1978). Many types of macrophytic red algae contain cellulose (Gretz et al. 1980; Mukai et al. 1981; Siddhanta et al. 2013), and the cellulose synthase *A* gene (*CESA*) was isolated from *P. yezoensis* (Roberts and Roberts 2009). The genome of *C. crispus* also contains *CESA* and *CESA-like* genes. *CESA* and *CESA-like* genes were found in the genomes of multicellular red algae, but not unicellular red algae including *C. merolae*, *G. sulphuraria*, and *P. purpureum*.

In red algae, various matrix polysaccharides, such as (1 → 4)-β-D-mannan, (1 → 4)-β-D-xylan, (1 → 3)-β-D-xylan, glucomannan, sulfated mixed-linkage glucan, and (1 → 3/1 → 4)-β-D-xylan were found (Popper et al. 2011). The *C. merolae* genome contains genes that encode two mannosyltransferases and a glucosyltransferase (Fig. 19.7, enzyme ID 1, 2, and 6), but not fucosyltransferase or galactosyltransferase (Fig. 19.7, checked with cross marks, Barbier et al. 2005). If the genes encode mannosyltransferases and glucosyltransferase that are related to the synthesis of matrix polysaccharides in *C. merolae*, a cell wall-like structure may contain mannan, glucomannan, or cellulose. However, it is also possible that these transferases are involved in the glycosylation of proteins and lipids. In *C. merolae*, the possibility of a cell wall-like structure consisting of glycoproteins was described by Misumi et al. (2005).

## 19.5 Lipid Metabolism

### 19.5.1 Composition of Lipids and Fatty Acids

The composition of lipids and fatty acids have been determined in red algae including *C. merolae* (Sato and Moriyama 2007; Toyoshima et al. 2016), *G. sulphuraria* (Vítová et al. 2016; Sakurai et al. 2016), *P. purpureum* (Nichols and Appleby 1969; Khozin-Goldberg et al. 2000; Naumann et al. 2007; Oh et al. 2009), *C. crispus* (Pettitt et al. 1989; Melo et al. 2015), and *P. yezoensis* (Araki et al. 1986; Araki et al. 1987; Araki et al. 1989). The results are summarized in Fig. 19.8.

It has been reported that *C. merolae* contains the following lipid classes: monogalactosyldiacylglycerol (MGDG), digalactosyldiacylglycerol (DGDG),

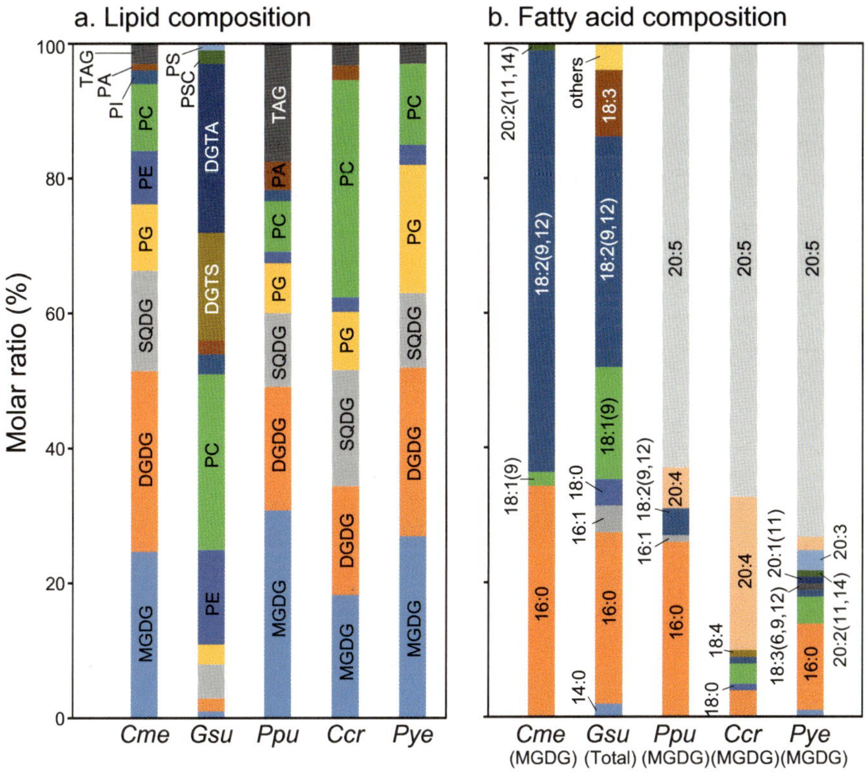


**Fig. 19.8** Composition of lipid classes and fatty acids in red algae. Reported composition of lipid classes (**a**) and fatty acids (**b**) in *C. merolae* (*Cme*), *G. sulphuraria* (*Gsu*), *P. purpureum* (*Ppu*), *C. crispus* (*Ccr*), and *P. yezoensis* (*Pye*) are shown. Fatty acids are represented by the number of carbon atoms [X] and the number of double bonds [Y], such as X:Y. Data were obtained from Toyoshima et al. (2016) for *C. merolae*, Vítová et al. (2016) for *G. sulphuraria*, Khozin-Goldberg et al. (2000) for *P. purpureum*, Pettitt et al. (1989) for *C. crispus*, and Araki et al. (1986) for *P. yezoensis*. In the lipid composition of *C. crispus*, values indicate weight percentage

sulfoquinovosyldiacylglycerol (SQDG), phosphatidylglycerol (PG), phosphatidylethanolamine (PE), phosphatidylcholine (PC), phosphatidylinositol (PI), phosphatidic acid (PA), and TAG. Phosphatidylserine (PS) and cardiolipin (CL) were not detected in *C. merolae* (Sato and Moriyama 2007). The *C. merolae* genome contains a gene with homology to CL synthase (CMN196C, Katayama et al. 2004). However, the protein product of this gene did not localize to the mitochondrion (Mori et al. 2016); thus its function is unclear. Under nitrogen-depleted conditions, the TAG content dramatically increased in *C. merolae* (Toyoshima et al. 2016; Takusagawa et al. 2016) like other algae such as *C. reinhardtii* (Merchant et al. 2012). The lipid compositions of the red algae, *P. purpureum*, *C. crispus*, and *P. yezoensis* were similar to that of *C. merolae* (Fig. 19.8a). In *G. sulphuraria*, two betaine lipids, diacylglyceryl-*N,N,N*-trimethylhomoserine

(DGTS) and diacylglyceryl-*O*-2′-(hydroxymethyl)-(*N*,*N*,*N*-trimethyl)-β-alanine (DGTA), and phosphosulfocholine (PSC) were also detected (Vítová et al. 2016). These two classes of lipids have a high proportion of lipids (Fig. 19.8a). DGTS was also found in *C. crispus* (Melo et al. 2015). However, it is unclear how betaine lipids are synthesized in *G. sulphuraria* and *C. crispus* because the genomes from these two species of red algae do not contain *BTA*, the only gene known to be involved in betaine lipid synthesis (Sato et al. 2017).

The fatty acid composition of *C. merolae* is simple compared with that of other red algae (Fig. 19.8b). In *C. merolae*, the major fatty acids are palmitic (16:0), stearic (18:0), oleic (18:1), and linoleic (18:2) acid. *C. merolae* does not contain unsaturated fatty acids with three or more double bonds, such as 18:3, 20:3, 20:4, and 20:5, which are found in other red algae (Fig. 19.8b). *C. merolae* has small amounts of C20 fatty acids, which are considered to be products of fatty acid elongation (FAE) in the ER (Mori et al. 2016).

### 19.5.2 Lipid and Fatty Acid Metabolic Pathways

Lipid metabolism genes, such as those involved in β-oxidation, the synthesis/elongation/desaturation of fatty acids, and the synthesis of glycolipids, phospholipids, and TAG, were identified in the genome of *C. merolae* (Sato and Moriyama 2007; Mori et al. 2016). Our group detected lipid metabolism genes in other red algae such as *G. sulphuraria*, *P. purpureum*, *C. crispus*, and *P. yezoensis* (Sato et al. 2017). Selected enzymes involved in red algal lipid synthesis are listed in Table 19.2. A comprehensive subcellular localization analysis on lipid metabolism enzymes was performed using GFP-tagged proteins in *C. merolae*, and a detailed metabolic map for lipid biosynthesis was constructed (Fig. 19.9a, Mori et al. 2016).

In land plants, fatty acids and plastid lipids, namely, MGDG, DGDG, SQDG, and PG, are synthesized in plastids, whereas TAG and phospholipids, such as PG, PS, PI, PE, and PC, are synthesized in the ER (Ohlrogge and Browse 1995; Mongrand et al. 1998). β-Oxidation for fatty acid degradation occurs in peroxisomes. In green plants, plastids contain stearoyl acyl-carrier-protein (ACP) desaturase and primarily synthesize 16:0 and 18:1 fatty acids. Fatty acids are also desaturated on complex lipids by Δ12 and Δ15 desaturases (the Δ + number shows the position of desaturation) in plastids and the ER.

In *C. merolae*, fatty acids are also synthesized in the plastid. However, this red alga has only a single type of 3-ketoacyl-ACP synthase (KAS), which has homology to KAS I and II (Table 19.2). Other red algae also have this type of KAS. KAS is a condensing enzyme in fatty acid synthesis. In bacteria and green plants, there are usually three types of KAS, namely, KAS I, KAS II, and KAS III, which show different substrate specificity based on the number of carbon atoms on the fatty acid. Because the amino acid sequence of KAS I and KAS II has high similarity to each other, it is difficult to distinguish between these KAS enzymes. Our phylogenetic analysis of KAS I and II showed that red algal KAS enzymes are a

**Table 19.2** List of lipid metabolism-related enzymes unique to *C. merolae* and red algae

| Enzyme | *Cme* | *Gsu* | *Ppu* | *Ccr* | *Pye* |
|---|---|---|---|---|---|
| *Fatty acid synthesis* | | | | | |
| 3-Ketoacyl-ACP synthase I/II (KAS I/II) | 1 | 1 | 1 | 1 | 1 |
| 3-Ketoacyl-ACP synthase III (KAS III) | – | – | 1 (Pt) | 1 (Pt) | 1 (Pt) |
| *Fatty acid desaturation* | | | | | |
| Stearoyl-ACP desaturase | – | – | – | – | – |
| Stearoyl-CoA desaturase (Cyt*b*5-fusion) | 1 | 1 | 1 | 1 | 1 |
| Fatty acid delta12-desaturase | 1 | 1 | 1 | 1 | 1 |
| Fatty acid delta15-desaturase | – | 1 | 1 | 1 | 1 |
| Fatty acid delta-5 desaturase (Cyt*b*5-fusion) | – | – | 1 | 1 | 2 |
| Fatty acid delta-6 desaturase (Cyt*b*5-fusion) | – | – | 1 | 1 | 1 |
| *Lipid synthesis* | | | | | |
| GlcDG synthase (MgdA) | 1 | – | – | – | 2 |
| Glucolipid epimerase (MgdE) | – | – | – | – | – |
| MGDG synthase (MGD) | 1 | 1 | 4 | 3 | 5 |
| DGDG synthase (DgdA/DGD) | 1 (Pt) | 1 (Pt) | 2 | 1 | 2 |
| PS synthase (base-exchange enzyme) | – | 1 | 1 | – | – |

The number of enzymes encoded by each red algal genome is shown. Enzymes were estimated by BLAST searches. Genome sequences were obtained from the same websites indicated in Table 19.1. Abbreviations: *Cme*, *C. merolae*; *Gsu*, *G. sulphuraria*; *Ppu*, *P. purpureum*; *Ccr*, *C. crispus*; *Pye*, *P. yezoensis*. Pt, encoded by the plastid genome; –, not found

monophyletic group and are located at an intermediate position between KAS I and KAS II of green plants (Sato et al. unpublished data); therefore, red algal KAS is indicated as KAS I/II in this chapter. Among red algae, the plastid genomes of *P. purpureum*, *C. crispus*, and *P. yezoensis* belonging to the Rhodophytina have the *fabH* gene-encoding KAS III (Table 19.2).

Another important finding in *C. merolae* fatty acid synthesis is that *C. merolae* lacks plastidial desaturases: stearoyl-ACP desaturase and Δ12 desaturase (Table 19.2). However, *C. merolae* does have FAD4 (Gao et al. 2009), which is a PG-specific desaturase that produces a minor fatty acid, 16:1(3-*trans*). Therefore, the primary products of fatty acid synthesis are saturated fatty acids, 16:0 and 18:0, which are both produced within the plastid (Fig. 19.9a). The synthesized fatty acids are transported to the ER and used in acyl lipid synthesis and/or subjected to FAE. In the ER, 18:0 fatty acids are desaturated by stearoyl-CoA desaturase (SCD) or Δ9 desaturase to produce 18:1, which is further desaturated by Δ12 desaturase to produce 18:2 (Fig. 19.9a, b). Thus, plastid lipid synthesis requires both unsaturated fatty acids (18:2) derived from ER lipids, such as PC, and saturated fatty acids (16:0) synthesized within plastids. This is called a coupled pathway (Fig. 19.9a) (Sato and Moriyama 2007).

The genomes of many species of red algae have genes encoding Δ5, Δ6, and Δ15 desaturases, which are required to synthesize polyunsaturated fatty acids (Table 19.2 and Fig. 19.9b). In fact, *P. purpureum*, *C. crispus*, and *P. yezoensis*

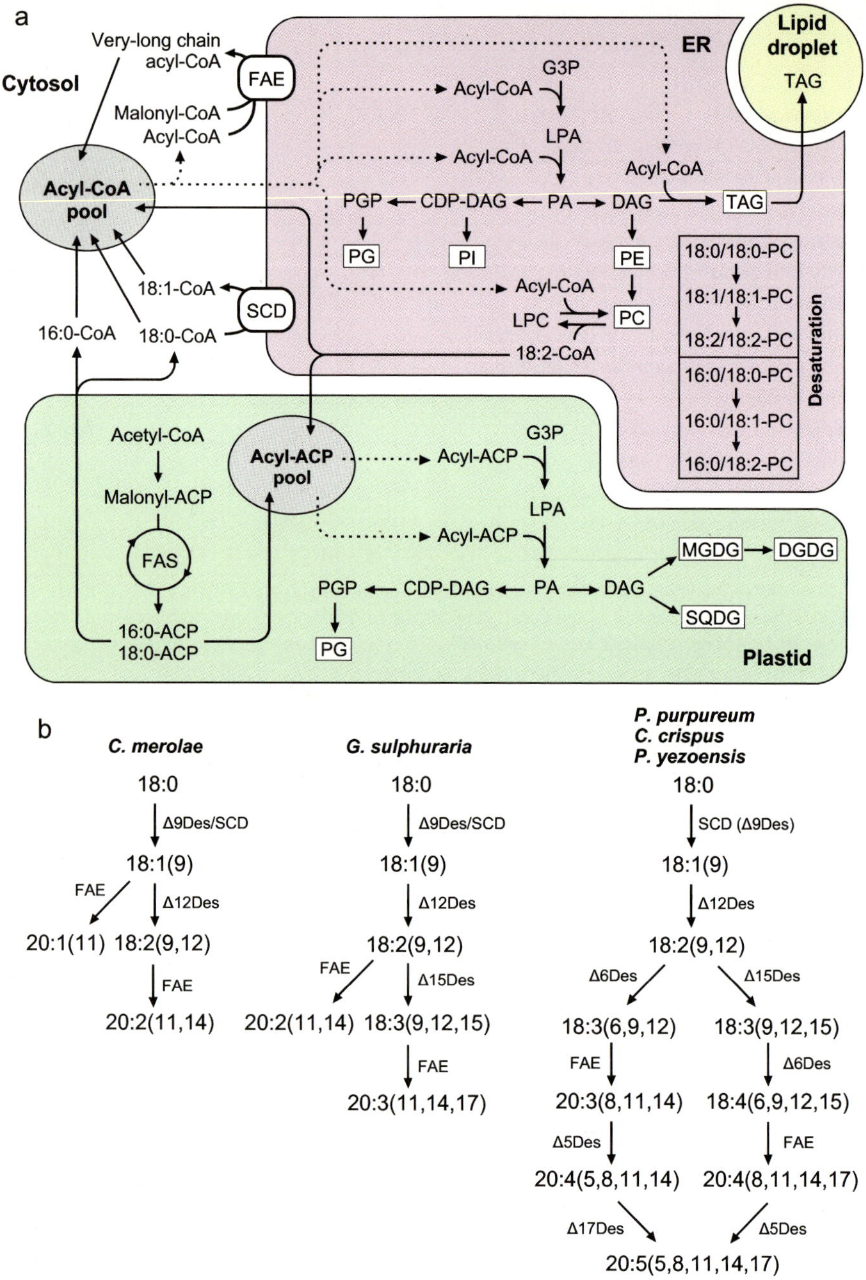


**Fig. 19.9** Simplified pathways of lipid metabolism in *C. merolae*. (**a**) Lipid synthesis in *C. merolae* (this model was based on Mori et al. (2016)). (**b**) Fatty acid elongation (FAE) and desaturation (Des) pathways in the red alga species, *C. merolae*, *G. sulphuraria*, *P. purpureum*, *C. crispus*, and *P. yezoensis*. The Δ + number indicates the position of desaturation. A gene for Δ9 desaturase was not found in the genomes of *C. crispus* and *P. yezoensis*. Abbreviations: *ACP* acyl

contain a large amount of arachidonic acid [20:4(5,8,11,14)] and eicosapentaenoic [20:5(5,8,11,14,17)] acid (Fig. 19.8b). On the other hand, these desaturases as well as 18:3 and C20 polyunsaturated fatty acids are not found in *C. merolae*.

In plants, MGDG is synthesized from diacylglycerol (DAG) and UDP-galactose by MGDG synthase (MGD1,2 or 3). However, in cyanobacteria, DAG is first converted to monoglucosyldiacylglycerol (GlcDG) through the action of GlcDG synthase (MgdA). GlcDG is then epimerized to produce MGDG by glucolipid epimerase (MgdE) (Awai et al. 2014). The *C. merolae* genome has a homologue of cyanobacterial *MgdA*, but not *MgdE* (Table 19.2). Because GlcDG was not detected in labeling studies using *C. merolae* cells (Sato and Moriyama 2007), we believe that eukaryotic MGD1 is the only enzyme involved in MGDG synthesis in *C. merolae*. For DGDG synthesis, *C. merolae* and *G. sulphuraria* have the cyanobacterial *dgdA* gene (*ycf82*) in the plastid genome, whereas other red algae contain genes encoding eukaryotic DGD1/2 (Table 19.2). *C. merolae* has two SQDG synthesis-related enzymes, UDP-sulfoquinovose synthase (SQD1) and sulfolipid synthase (SQD2). The activities of these enzymes were confirmed by complementation assays in cyanobacterial mutants to the corresponding genes (Sato et al. 2016a, *NB*: this paper is a publication from a different group of ours).

It is likely that, similar to land plants, *C. merolae* contains the same pathways for the synthesis of various phospholipids including PG, PI, and PE (Mori et al. 2016). There are two major pathways for PC synthesis in land plants, although complicated pathways that cross between the two pathways actually exist (Gibellini and Smith 2010; Cole et al. 2012). One is a pathway for PE methylation, and the other is a pathway for phosphocholine transfer from CDP-choline to DAG. The *C. merolae* genome only has genes related to the PE methylation pathway, including the PE methyltransferase (*PEMT*) gene and the phospholipid methyltransferase (*PLMT*) gene. The existence of the pathway was confirmed by labeling studies in intact cells (Sato et al. 2016b).

The structure of TAG in *C. merolae* was studied in detail using MALDI-TOF MS and lipase treatment studies. The results showed that the major molecular species of TAG are 16:0/18:2/18:0, 18:0/18:2/18:2, and 18:1/18:2/18:1. In other words, TAG contains mainly 18:2(9,12) at the C-2 position (Toyoshima et al. 2016). It was also reported that PC mainly contained 16:0 at the C-1 position and 18:2(9,12) at the C-2 position (Sato and Moriyama 2007). These results suggest that PC supplies the DAG moiety of TAG. However, the *C. merolae* genome does not contain genes for phosphatidylcholine:diacylglycerol cholinephosphotransferase (PDCT) and phospholipase C (PLC), which both produce DAG from PC (Toyoshima et al. 2016). Further studies are needed to elucidate the complete TAG synthesis pathway in *C. merolae*.

**Fig. 19.9** (continued) carrier protein, *G3P* glycerol 3-phosphate, *CDP* cytidine diphosphate, *DAG* diacylglycerol, *PGP* phosphatidylglycerol phosphate, *LPA* lysophosphatidic acid, *LPC* lysophosphatidylcholine, *SCD* stearoyl-CoA Δ9-desaturase (Modified from Sato et al. 2017)

## References

Araki S, Sakurai T, Omata T et al (1986) Lipid and fatty acid composition in the red alga *Porphyra yezoensis*. Jpn J Phycol 34:94–100

Araki S, Sakurai T, Kawaguchi A, Murata N (1987) Positional distribution of fatty acids in glycerolipids of the marine red alga, *Porphyra yezoensis*. Plant Cell Physiol 28:761–766

Araki S, Sakurai T, Oohusa T et al (1989) Characterization of sulfoquinovosyl diacylglycerol from marine red algae. Plant Cell Physiol 30:775–781

Awai K, Ohta H, Sato N (2014) Oxygenic photosynthesis without galactolipids. Proc Natl Acad Sci U S A 111:13571–13575

Barbier G, Oesterhelt C, Larson MD et al (2005) Comparative genomics of two closely related unicellular thermo-acidophilic red algae, *Galdieria sulphuraria* and *Cyanidioschyzon merolae*, reveals the molecular basis of the metabolic flexibility of *Galdieria sulphuraria* and significant. Plant Physiol 137:460–474

Bhattacharya D, Price DC, Chan CX et al (2013) Genome of the red alga *Porphyridium purpureum*. Nat Commun 4:1941

Bricker DK, Taylor EB, Schell JC et al (2012) A mitochondrial pyruvate carrier required for pyruvate uptake in yeast, *Drosophila*, and humans. Science 337:96–100

Burgess SJ, Taha H, Yeoman JA et al (2016) Identification of the elusive pyruvate reductase of *Chlamydomonas reinhardtii* chloroplasts. Plant Cell Physiol 57:82–94

Catt JW, Hills GJ, Roberts K (1978) Cell wall glycoproteins from *Chlamydomonas reinhardii*, and their self-assembly. Planta 138:91–98

Cole LK, Vance JE, Vance DE (2012) Phosphatidylcholine biosynthesis and lipoprotein metabolism. Biochim Biophys Acta 1821:754–761

Collén J, Porcel B, Carré W et al (2013) Genome structure and metabolic features in the red seaweed *Chondrus crispus* shed light on evolution of the Archaeplastida. Proc Natl Acad Sci U S A 110:5247–5252

De Luca P, Moretti A (1983) Floridosides in *Cyanidium caldarium*, *Cyanidioschyzon merolae* and *Galdieria sulphuraria* (Rhodophyta, Cyanidiophyceae). J Phycol 19:368–369

Dennis D, Miernyk J (1982) Compartmentation of nonphotosynthetic carbohydrate metabolism. Annu Rev Plant Physiol 33:27–50

Deschamps P, Colleoni C, Nakamura Y et al (2008) Metabolic symbiosis and the birth of the plant kingdom. Mol Biol Evol 25:536–548

Gao J, Ajjawi I, Manoli A et al (2009) Fatty acid desaturase4 of Arabidopsis encodes a protein distinct from characterized fatty acid desaturases. Plant J 60:832–839

Gibellini F, Smith TK (2010) The Kennedy pathway-*De novo* synthesis of phosphatidylethanolamine and phosphatidylcholine. IUBMB Life 62:414–428

Graham IA (2008) Seed storage oil mobilization. Annu Rev Plant Biol 59:115–142

Gretz MR, Aronson JM, Sommerfeld MR (1980) Cellulose in the cell walls of the bangiophyceae (rhodophyta). Science 207:779–781

Gross W, Schnarrenberger C (1995) Heterotrophic growth of two strains of the acido-thermophilic red alga *Galdieria sulphuraria*. Plant Cell Physiol 36:633–638

Gruber A, Weber T, Bártulos CR et al (2009) Intracellular distribution of the reductive and oxidative pentose phosphate pathways in two diatoms. J Basic Microbiol 49:58–72

Herzig S, Raemy E, Montessuit S et al (2012) Identification and functional expression of the mitochondrial pyruvate carrier. Science 337:93–96

Hirabaru C, Izumo A, Fujiwara S et al (2010) The primitive rhodophyte *Cyanidioschyzon merolae* contains a semiamylopectin-type, but not an amylose-type, α-glucan. Plant Cell Physiol 51:682–693

Imamura S, Kawase Y, Kobayashi I et al (2015) Target of rapamycin (TOR) plays a critical role in triacylglycerol accumulation in microalgae. Plant Mol Biol 89:309–318

Jenner HL, Winning BM, Millar AH et al (2001) NAD malic enzyme and the control of carbohydrate metabolism in potato tubers. Plant Physiol 126:1139–1149

Johnson X, Alric J (2013) Central carbon metabolism and electron transport in *Chlamydomonas reinhardtii*: metabolic constraints for carbon partitioning between oil and starch. Eukaryot Cell 12:776–793

Joyard J, Ferro M, Masselon C et al (2010) Chloroplast proteomics highlights the subcellular compartmentation of lipid metabolism. Prog Lipid Res 49:128–158

Katayama K, Sakurai I, Wada H (2004) Identification of an *Arabidopsis thaliana* gene for cardiolipin synthase located in mitochondria. FEBS Lett 577:193–198

Khozin-Goldberg I, HZ Y, Adlerstein D et al (2000) Triacylglycerols of the red microalga *Porphyridium cruentum* can contribute to the biosynthesis of eukaryotic galactolipids. Lipids 35:881–889

Li C-L, Wang M, Ma X-Y, Zhang W (2014) NRGA1, a putative mitochondrial pyruvate carrier, mediates ABA regulation of guard cell ion channels and drought stress responses in Arabidopsis. Mol Plant 7:1508–1521

Li-Beisson Y, Shorrosh B, Beisson F et al (2013) Acyl-lipid metabolism. Arab B 11:e0161

Matsuzaki M, Misumi O, Shin-I T et al (2004) Genome sequence of the ultrasmall unicellular red alga *Cyanidioschyzon merolae* 10D. Nature 428:653–657

Melo T, Alves E, Azevedo V et al (2015) Lipidomics as a new approach for the bioprospecting of marine macroalgae – unraveling the polar lipid and fatty acid composition of *Chondrus crispus*. Algal Res 8:181–191

Merchant SS, Kropat J, Liu B et al (2012) TAG, You're it! Chlamydomonas as a reference organism for understanding algal triacylglycerol accumulation. Curr Opin Biotechnol 23:352–363

Misumi O, Matsuzaki M, Nozaki H et al (2005) *Cyanidioschyzon merolae* genome. A tool for facilitating comparable studies on organelle biogenesis in photosynthetic eukaryotes. Plant Physiol 137:567–585

Mongrand S, Bessoule J-J, Cabantous F, Cassagne C (1998) The C16:3/C18:3 fatty acid balance in photosynthetic tissues from 468 plant species. Phytochemistry 49:1049–1064

Mori N, Moriyama T, Toyoshima M, Sato N (2016) Construction of global acyl lipid metabolic map by comparative genomics and subcellular localization analysis in the red alga *Cyanidioschyzon merolae*. Front Plant Sci 7:958

Moriyama T, Sakurai K, Sekine K, Sato N (2014) Subcellular distribution of central carbohydrate metabolism pathways in the red alga *Cyanidioschyzon merolae*. Planta 240:585–598

Moriyama T, Mori N, Sato N (2015) Activation of oxidative carbon metabolism by nutritional enrichment by photosynthesis and exogenous organic compounds in the red alga *Cyanidioschyzon merolae*: evidence for heterotrophic growth. Springer Plus 4:559

Mukai LS, Craigie JS, Brown RG (1981) Chemical composition and structure of the cell walls of the Conchocelis and thallus phases of *Porphyra tenera* (Rhodophyceae). J Phycol 17:192–198

Nakahigashi K, Toya Y, Ishii N et al (2009) Systematic phenome analysis of *Escherichia coli* multiple-knockout mutants reveals hidden reactions in central carbon metabolism. Mol Syst Biol 5:306

Nakamura Y, Sasaki N, Kobayashi M et al (2013) The first symbiont-free genome sequence of marine red alga, Susabi-nori (*Pyropia yezoensis*). PLoS One 8:e57122

Naumann I, Darsow KH, Walter C et al (2007) Identification of sulfoglycolipids from the alga *Porphyridium purpureum* by matrix-assisted laser desorption/ionisation quadrupole ion trap time-of-flight mass spectrometry. Rapid Commun Mass Spectrom 21:3185–3192

Nichols B, Appleby R (1969) The distribution and biosynthesis of arachidonic acid in algae. Phytochemistry 8:1907–1915

Nozaki H, Takano H, Misumi O et al (2007) A 100%-complete sequence reveals unusually simple genomic features in the hot-spring red alga *Cyanidioschyzon merolae*. BMC Biol 5:28

Oh SH, Han JG, Kim Y et al (2009) Lipid production in *Porphyridium cruentum* grown under different culture conditions. J Biosci Bioeng 108:429–434

Ohlrogge J, Browse J (1995) Lipid biosynthesis. Plant Cell 7:957–970
Pade N, Linka N, Ruth W et al (2015) Floridoside and isofloridoside are synthesized by trehalose 6-phosphate synthase-like enzymes in the red alga *Galdieria sulphuraria*. New Phytol 205:1227–1238
Pettitt T, Jones A, Harwood J (1989) Lipids of the marine red algae, *Chondrus crispus* and *Polysiphonia lanosa*. Phytochemistry 28:399–405
Plancke C, Colleoni C, Deschamps P et al (2008) Pathway of cytosolic starch synthesis in the model glaucophyte *Cyanophora paradoxa*. Eukaryot Cell 7:247–257
Plancke C, Vigeolas H, Höhner R et al (2014) Lack of isocitrate lyase in *Chlamydomonas* leads to changes in carbon metabolism and in the response to oxidative stress under mixotrophic growth. Plant J 77:404–417
Popper ZA, Michel G, Hervé C et al (2011) Evolution and diversity of plant cell walls: from algae to flowering plants. Annu Rev Plant Biol 62:567–590
Pracharoenwattana I, Cornah JE, Smith SM (2005) Arabidopsis peroxisomal citrate synthase is required for fatty acid respiration and seed germination. Plant Cell 17:2037–2048
Raven J, Johnston A, MacFarlane J (1990) Carbon metabolism. In: Cole KM, Sheath RG (eds) Biology of the red algae. Cambridge University Press, Cambridge, UK
Roberts E, Roberts AW (2009) A cellulose synthase (*CESA*) gene from the red alga *Porphyra yezoensis* (Rhodophyta). J Phycol 45:203–212
Sakurai T, Aoki M, Ju X et al (2016) Profiling of lipid and glycogen accumulations under different growth conditions in the sulfothermophilic red alga *Galdieria sulphuraria*. Bioresour Technol 200:861–866
Sato N, Moriyama T (2007) Genomic and biochemical analysis of lipid biosynthesis in the unicellular rhodophyte *Cyanidioschyzon merolae*: lack of a plastidic desaturation pathway results in the coupled pathway of galactolipid synthesis. Eukaryot Cell 6:1006–1017
Sato N, Kobayashi S, Aoki M et al (2016a) Identification of genes for sulfolipid synthesis in primitive red alga *Cyanidioschyzon merolae*. Biochem Biophys Res Commun 470:123–129
Sato N, Mori N, Hirashima T, Moriyama T (2016b) Diverse pathways of phosphatidylcholine biosynthesis in algae as estimated by labeling studies and genomic sequence analysis. Plant J 87:281–292
Sato N, Moriyama T, Mori N, Toyoshima M (2017) Lipid metabolism and potentials of biofuel and high added-value oil production in red algae. World J Microbiol Biotechnol 33:74
Schönknecht G, Chen W-H, Ternes CM et al (2013) Gene transfer from bacteria and archaea facilitated evolution of an extremophilic eukaryote. Science 339:1207–1210
Shimonaga T, Fujiwara S, Kaneko M et al (2007) Variation in storage alpha-polyglucans of red algae: amylose and semi-amylopectin types in *Porphyridium* and glycogen type in *Cyanidium*. Mar Biotechnol (NY) 9:192–202
Shimonaga T, Konishi M, Oyama Y et al (2008) Variation in storage alpha-glucans of the Porphyridiales (Rhodophyta). Plant Cell Physiol 49:103–116
Siddhanta AK, Kumar S, Mehta GK et al (2013) Cellulose contents of some abundant Indian seaweed species. Nat Prod Commun 8:497–500
Sumiya N, Kawase Y, Hayakawa J et al (2015) Expression of cyanobacterial acyl-ACP reductase elevates the triacylglycerol level in the red alga *Cyanidioschyzon merolae*. Plant Cell Physiol 56:1962–1980
Takusagawa M, Nakajima Y, Saito T, Misumi O (2016) Primitive red alga *Cyanidioschyzon merolae* accumulates storage glucan and triacylglycerol under nitrogen depletion. J Gen Appl Microbiol 62:111–117
Toyoshima M, Mori N, Moriyama T et al (2016) Analysis of triacylglycerol accumulation under nitrogen deprivation in the red alga *Cyanidioschyzon merolae*. Microbiology 162:803–812
Tronconi MA, Fahnenstich H, Gerrard Weehler MC et al (2008) Arabidopsis NAD-malic enzyme functions as a homodimer and heterodimer and has a major impact on nocturnal metabolism. Plant Physiol 146:1540–1552

Tyra HM, Linka M, Weber APM, Bhattacharya D (2007) Host origin of plastid solute transporters in the first photosynthetic eukaryotes. Genome Biol 8:R212

Viola R, Nyvall P, Pedersén M (2001) The unique features of starch metabolism in red algae. Proc Biol Sci 268:1417–1422

Vítová M, Goecke F, Sigler K, Řezanka T (2016) Lipidomic analysis of the extremophilic red alga *Galdieria sulphuraria* in response to changes in pH. Algal Res 13:218–226

Wada H, Shintani D, Ohlrogge J (1997) Why do mitochondria synthesize fatty acids? Evidence for involvement in lipoic acid production. Proc Natl Acad Sci U S A 94:1591–1596

Weber APM, Oesterhelt C, Gross W et al (2004) EST-analysis of the thermo-acidophilic red microalga *Galdieria sulphuraria* reveals potential for lipid A biosynthesis and unveils the pathway of carbon export from rhodoplasts. Plant Mol Biol 55:17–32

Wiese A, Gröner F, Sonnewald U et al (1999) Spinach hexokinase I is located in the outer envelope membrane of plastids. FEBS Lett 461:13–18

Yang W, Catalanotti C, D'Adamo S et al (2014) Alternative acetate production pathways in *Chlamydomonas reinhardtii* during dark anoxia and the dominant role of chloroplasts in fermentative acetate production. Plant Cell 26:4499–4518

# Part VII
# Application

**Osami Misumi and Tsuneyoshi Kuroiwa**

In the previous chapters, unique features about the cell structure, genome composition, metabolic system, etc. of *C. merolae* were introduced. The mechanisms of proliferation of mitochondria and chloroplast were elucidated by taking advantage of these characteristics of *C. merolae*. A lot of new findings of basic biology were brought by the alga. At the same time, research techniques and resources on *C. merolae* have also been developed. Next, we tried to combine the characteristics of *C. merolae* and the new experimental method for application research.

Because it has ability to withstand a very stressful environment, *C. merolae* can survive in the condition of high temperature and strong acidity. We expected that the *C. merolae* has various stress tolerance genes. A number of stress-responsive genes were identified by EST and microarray analysis. In Chap. 20, the effects of the stress-tolerance gene of *C. merolae* on improvement of stress tolerance of land plants are introduced. By introducing the gene of *C. merolae*, transgenic plants improved the ability to resist acid and high temperature. In Chap. 21, the experimental results on biomass production using *C. merolae* are reported. First, the culture conditions of biomass (especially triacylglycerol) synthesis and the fundamental mechanism of metabolic regulation in cells were studied (See Chaps. 17, 18 and 19). Based on these findings, we tried metabolic engineering with cyanobacterial acyl-ACP reductase by using newly developed transformation system of *C. merola*. This experimental result suggests that *C. merola* becomes a useful biomass producing cell by metabolic engineering. Regarding the induction method of biomass synthesis, modification of the medium composition and light condition using LED have also been developed. The possibilities and perspectives of *C. merola* on biomass production also be discussed.

# Chapter 20
# Application of the Tolerance to Extreme Environment to Land Plants

**Osami Misumi, Tsuneyoshi Kuroiwa, and Shunsuke Hirooka**

**Abstract** The ability of the primitive red alga *Cyanidioschyzon merolae* to survive in acidic environments at high temperatures was applied to produce acid- and heat-tolerant transgenic plants. The *C. merolae* gene encoding a plasma membrane $H^+$-ATPase was introduced into *Arabidopsis thaliana*, and the acid tolerance of the resulting transgenic plants was investigated. The transgenic seedlings were more acid-tolerant than the wild-type seedlings following a transient acid treatment. We also observed that the roots of transgenic plants grew longer than the wild-type roots. These results suggest that the *C. merolae* $H^+$-ATPase may reinforce the acid tolerance of higher plants by enhancing proton pump activities. Furthermore, we produced transgenic *A. thaliana* plants overexpressing the *C. merolae* gene encoding the stromal ascorbate peroxidase (*CmstAPX*). Soluble APX activities were higher in *CmstAPX*-expressing plants than in the wild-type plants. Compared with the wild-type, the *CmstAPX*-expressing plants were more tolerant to high-temperature stress and oxidative stress induced by methyl viologen. Additionally, the *CmstAPX*-expressing plants retained the highest chlorophyll

O. Misumi (✉)
Graduate School of Science and Technology for Innovation, Faculty of Science, Department of Biological Science and Chemistry, Yamaguchi University, 1677-1 Yoshida, Yamaguchi 753-8512, Japan
e-mail: misumi@yamaguchi-u.ac.jp

T. Kuroiwa
Faculty of Science, Department of Chemical and Biological Sciences, Japan Women's University, 2-8-1 Mejirodai, Bunkyo-ku, Tokyo 112-8681, Japan

S. Hirooka
Department of Cell Genetics, National Institute of Genetics, 1111 Yata, Mishima, Shizuoka 411-8540, Japan

Core Research for Evolutional Science and Technology Program, Japan Science and Technology Agency, Saitama 332-0012, Japan



T. Kuroiwa et al. (eds.), *Cyanidioschyzon merolae*,
https://doi.org/10.1007/978-981-10-6101-1_20

contents after treatments with methyl viologen and high temperatures. Furthermore, the stroma and chloroplasts of the *CmstAPX*-expressing plants remained intact, whereas they disintegrated in the wild-type controls. These findings imply that the increased APX activity in the chloroplasts of *CmstAPX*-expressing plants enhances heat tolerance by increasing reactive oxygen species-scavenging capabilities at high temperatures.

**Keywords** Plasma membrane H + -ATPase • Acid tolerance • Stromal ascorbate peroxidase • Heat tolerance • Reactive oxygen species • *Arabidopsis thaliana* • *Cyanidioschyzon merolae*

## 20.1 Introduction

Environmental issues, including global warming, desertification, and acid rain, are currently affecting various aspects of the global ecosystem. When environmental conditions, such as temperature, drought, salinity, pH, nutrient level, and light, deviate from the optimal range, the growth and productivity of many plant species are significantly inhibited. Developing stress-tolerant transgenic crops may be useful for mitigating the problems caused by environmental fluctuations. Genes responsive to various environmental stresses have been identified mainly in model plants (e.g., *Arabidopsis thaliana*). Oxidative damage is a common consequence of several environmental stresses. Thus, the antioxidant system has been thoroughly investigated to enhance the removal of reactive oxygen species from plant cells (Smirnoff 1998). Several unique genes have also been sought from plants that grow in specialized environments, including those affected by drought, soil acidity, and salinity. To date, many stress-tolerant transgenic plants have been produced by introducing stress-responsive genes into plants using genetic engineering techniques. Additionally, we focused on the characteristics of the unicellular red alga, *Cyanidioschyzon merolae*, which survives at high temperatures (45–55 °C) in highly acidic (pH 1–3) hot springs with high metal ion concentrations. Environmental stress-response genes of Cyanidiales algae have been used as genetic resources for producing stress-tolerant transgenic plants. In this study, we identified several *C. merolae* stress-response genes. The introduction of some of these genes into *A. thaliana* resulted in transgenic plants with increased abiotic stress tolerance under laboratory conditions.

## 20.2 Application of the Tolerance to Extreme Environment to Land Plants

### *20.2.1 Search for Stress-Response Genes by Expressed Sequence Tag Analysis*

The environmental conditions of the hot springs in which *C. merolae* grows are more extreme than the conditions in which the other Cyanidiales algae survive. There are almost no other eukaryotic organisms capable of inhabiting these hot springs. We investigated gene expression levels to determine how *C. merolae* adapts to high temperatures and strongly acidic environments. First, genes that were expressed under normal laboratory culture conditions (i.e., 42 °C and pH 2.3) were examined. Approximately, 33,000 expressed sequence tags were detected, and transcripts for 86.3% of *C. merolae* genes were confirmed. This result revealed that *C. merolae* expresses genes with the smallest variations as much as possible. Table 20.1 lists the 20 most highly expressed genes based on the expressed sequence tag analysis. The encoded proteins are required in relatively large amounts because *C. merolae* lives under hot and acidic conditions. In other words, these are housekeeping genes that are essential for the maintenance and proliferation of *C. merolae* cells. Additionally, many transcripts of the so-called stress-response genes were identified, including those encoding heat shock proteins and antioxidants involved in removing reactive oxygen species. The gene encoding the plasma membrane-type P-type proton ($H^+$) ATPase (PMA) was among the most highly expressed genes. The PMA excretes protons from cells using ATP hydrolysis for energy. Therefore, the *C. merolae* PMA gene may be important for adaptations to the high proton concentrations of acidic environments. The PMA gene is also highly expressed in related species such as *Cyanidium caldarium* (Ohta et al. 1997).

We next focused on the chloroplast stromal-type ascorbic acid peroxidase (APX) gene, which was the fourth most abundantly expressed gene. Chloroplast APX converts the active oxygen species generated in a photochemical reaction into water by incorporating it into the oxidation of ascorbic acid. This enzyme may have an important function related to oxidative stress defense in the chloroplast, which occupies most of the volume of *C. merolae* cells and is required for the autotrophy of these cells. The *C. merolae*-specific stress-tolerance genes may be useful for generating transgenic plants better able to withstand adverse environmental conditions. Therefore, we prepared transgenic plants expressing the *C. merolae* PMA and APX genes and subsequently investigated the resistance of these plants to acid and heat stresses.

In addition, we have selected many gene candidates for stress resistance of *C. merolae* besides PMA and APX genes (Table 20.2). We have confirmed that

**Table 20.1** The top 20 genes with high expression levels by EST analysis

| Rank | Annotation | Number of EST |
|---|---|---|
| 1 | Plasma membrane H+-ATPase | 549 |
| 2 | RuBisCo expression protein cfxQ (cbbX) | 331 |
| 3 | Alpha-tubulin | 320 |
| 4 | Chloroplast ascorbate hydrogen peroxidase, precursor | 295 |
| 5 | Hypothetical protein | 228 |
| 6 | Heat shock protein Hsp70 | 191 |
| 7 | Phycocyanin-associated rod linker protein | 161 |
| 8 | Hypothetical transcript | 158 |
| 9 | DnaK-type molecular chaperone mortalin, mitochondrial precursor | 157 |
| 10 | Phosphoglycerate kinase | 155 |
| 11 | Serine-glyoxylate aminotransferase | 151 |
| 12 | Catalase | 130 |
| 13 | Enolase | 126 |
| 14 | Chroloplast chaperonin CPN60, precursor | 125 |
| 15 | Eukaryotic translation elongation factor 1 alpha (eEF-1a) | 114 |
| 16 | Mitochondrial ribosomal protein S1 precursor | 113 |
| 17 | Heat shock protein of Hsp90 family | 109 |
| 18 | Hypothetical protein | 102 |
| 19 | Fusion protein of AIR carboxylase and purC, cp or mt precursor | 100 |
| 20 | Chloroplast translation elongation factor P (EF-P) | 97 |

Total genes, 4479; total clones, 32,917

some of the candidate genes improve the stress tolerance of *Escherichia coli* or *Physcomitrella patens* (data not shown). Therefore, our results suggest that some kinds of genes of *C. merolae* are beneficial for improving the stress tolerance of land plants.

### 20.2.2 *Acid Tolerance of Plants Expressing the* C. merolae *Gene Encoding the Plasma Membrane $H^+$-ATPase*

The activity of the plasma membrane $H^+$-ATPase is critical for generating membrane potentials and regulating intracellular pH by pumping out protons from the cytoplasm. The resulting electrochemical gradient is then used by secondary transporters to move various ions and metabolites across the plasma membrane. The plasma membrane $H^+$-ATPase is also crucial for the survival of plants exposed to diverse environmental stresses, including salt stress (Vitart et al. 2001), aluminum stress (Ahn et al. 2001), and phosphorus starvation (Shen et al. 2006). The uptake of potassium ions through specific transport proteins is influenced by the plasma membrane $H^+$-ATPase (Hoth et al. 1997). Additionally, phosphorus uptake is postulated to involve proton co-transport (Ullrich-Eberius et al. 1989; Sakano

**Table 20.2** Gene candidates for stress tolerance of *C. merolae*

| Published papers | EST analysis | Microarray | Candidate gene (locus, annotation) | Reference etc. |
|---|---|---|---|---|
| | | | Oxidative stress (ROS, pH, Metal ion) | |
| ✓ | ✓ | | CMQ247C plasma membrane $H^+$-ATPase | Unpublished |
| ✓ | | | CMO102C vacuolar-type $H^+$-translocating pyrophosphatase | Unpublished |
| ✓ | | | CMH036C O-acetylserine (thiol)-lyase 2 | Unpublished |
| ✓ | ✓ | | CMT028C Mn superoxide dismutase | Unpublished |
| ✓ | ✓ | | CMM158C ascorbate hydrogen peroxidase | Hirooka et al. (2009) |
| ✓ | ✓ | ✓ | CMR299C stress-induced phosphoprotein STI1 | Unpublished |
| ✓ | ✓ | | CMB059C aluminum resistance protein | Unpublished |
| ✓ | | | CMS490C malic acid transport protein, malate permease | Unpublished |
| ✓ | | | CMA007C serine acetyltransferase | Unpublished |
| ✓ | | | CMI299C serine O-acetyltransferase | Unpublished |
| ✓ | | | CMI111C probable phytochelatin synthase | Unpublished |
| ✓ | | | CMN023C manganese superoxide dismutase | Unpublished |
| | ✓ | ✓ | CMK024C hypothetical protein | Unpublished |
| | | | Heat stress | |
| ✓ | | | CML274C betaine aldehyde dehydrogenase | Unpublished |
| ✓ | | ✓ | CMV193C ferredoxin | Unpublished |
| | | ✓ | CMJ100C heat shock protein, Hsp20 family | Kobayashi et al. (2014) |
| | | ✓ | CMQ301C alcohol dehydrogenase | Unpublished |
| | | | Drought and salt stress | |
| ✓ | ✓ | | CMQ284C plasma membrane calcium/proton antiporter | Japan Patent Kokai 2011-92109A (2011) |
| | ✓ | ✓ | CMN227C S-adenosylmethionine synthetase | Sakajiri et al. (2008, 2010) |

The check mark (✓) shows the basis for selection of candidate genes. EST and microarray experiments were performed by our research group

et al., 1992). A decrease in cytoplasmic pH and membrane depolarization stimulate the export of cytoplasmic protons by the plasma membrane $H^+$-ATPase to maintain the intracellular pH, thereby providing the proton-driving force for Pi uptake (Ullrich-Eberius et al. 1989). Therefore, activated plasma membrane $H^+$-ATPases may improve plant nutrition by enhancing the electrochemical proton gradient that drives ion transport across the cell membrane via secondary transport systems.

Enami et al. (1986) reported that *Cyanidium* species inhabiting the same environments as *C. merolae* also have a mechanism for excreting intracellular protons to maintain large pH gradients inside and outside cells. Fluctuations to the abundance of cellular ATP produced by photosynthesis and cell respiration due to inhibitors and changes to light conditions affect proton pump activities depending on the cellular ATP concentration. This leads to altered cytoplasmic pH. When the intracellular ATP content is depleted, the proton pump is inactivated, resulting in an acidic cytoplasm.

According to a genome analysis, *C. merolae* contains only one copy of the PMA gene. Therefore, the maintenance of a cytoplasm with a neutral pH under highly acidic conditions is dependent on the PMA encoded by a single gene. In contrast, the genomes of *A. thaliana* and *Oryza sativa* carry several copies of the PMA gene. Different paralogs exhibit tissue- and organ-specific activities. Figure 20.1 presents

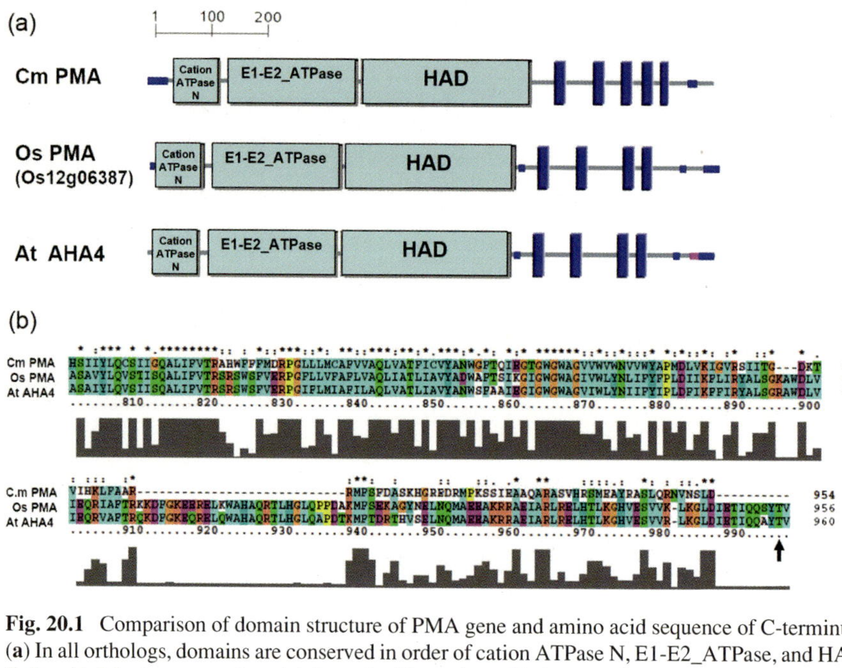


**Fig. 20.1** Comparison of domain structure of PMA gene and amino acid sequence of C-terminus. (**a**) In all orthologs, domains are conserved in order of cation ATPase N, E1-E2_ATPase, and HAD (haloacid dehydrogenase domain) from the N-terminus. Each gene also contains multiple transmembrane domains at the C-terminus. (**b**) *C. merolae* lacks C-terminal ten amino acid residues containing threonine. The *arrow* shows a threonine residue to be phosphorylated. *Cm C. merolae, Os O. sativa, At A. thaliana*

the PMA gene domain structures as well as an alignment of the encoded amino acid sequences of the *C. merolae*, *O. sativa*, and *A. thaliana* PMA C-terminal regions. The *C. merolae* PMA gene exhibits the characteristics of a typical plant-type PMA gene. Additionally, the PMAs of *C. merolae* and land plants are approximately 70% homologous at the gene and protein levels. In the PMA gene of terrestrial plants, the second threonine residue from the C-terminus is highly conserved. The proton pump is normally inactive because it is blocked by the C-terminal region on the cytoplasmic side. The threonine residue must be phosphorylated, and the 14-3-3 protein should be modified to activate the PMA. Because the C-terminal regions of the *C. merolae* and *Cyanidium* species PMAs do not inactivate the pump, these PMAs are believed to be always highly active.

A plant expression construct was prepared in which the *C. merolae* PMA gene was placed under the control of the cauliflower mosaic virus 35S promoter. The construct was then inserted into *Physcomitrella patens* and *A. thaliana* cells, and the acid tolerance of the generated transgenic plants was compared with that of wild-type plants. When *P. patens* was cultured under a constant acidic environment of pH 4.8 (normal pH: 6.7), the wild-type moss died, whereas the PMA-expressing strain remained green and continued to grow (data not shown). The *A. thaliana* transformants were exposed to a 48-h acid treatment (pH 4.0–2.3), after which they were allowed to recover under normal culture conditions (pH 5.8). Most of the wild-type *A. thaliana* plants exposed to the acid treatment died within 3 weeks of the treatment. However, most of the transformants continued to grow (Fig. 20.2). These results imply that the increased tolerance to a transient acid treatment was a result of the overexpression of the *C. merolae* PMA gene.

The relationship between the pH of the growth medium and the elongation of transformant roots was also examined. Immediately after germinating, wild-type and transformant seedlings were placed on agar medium [pH 5.8 (normal) or 3.0 (acidic)]. The seedlings were maintained in an incubator for 3 weeks. The transformant roots grew longer than the wild-type roots at both pHs (Fig. 20.3). Thus, the expression of the *C. merolae* PMA gene appears to promote the elongation of plant roots.

### 20.2.3 Improving Thermotolerance in Plants by Gene Transfer of C. merolae

High-temperature stress caused by global warming is predicted to increase the risk of reduced crop productivity (Bita and Gerats. 2013). As the earth's population increases, the efforts to develop thermotolerant plants are important. One of the ways in which tolerance to high temperature can be achieved is by the transfer of genes involved in thermotolerance from other organisms. Thus far, several studies have successfully produced thermotolerant plants by genetic engineering, for

**Fig. 20.2** Influence of transient acid treatment on the young seedlings of wild-type and two transgenic plant lines. Seedlings grown for 3 weeks were treated with hydroponic for 48 h under each pH condition. Treated seedlings were planted again in the soil and grown for 3 weeks. Three weeks after treatment most of the wild type died, whereas most of the transformants survived and grew. Two seedlings were used for the experiments at each pH condition. Number 2 and number 5 represent *A. thaliana* lines of independent transformants of the CmPMA gene, respectively

example, by overexpressing heat shock transcription factor (Lee et al. 1995) or choline oxidase from *Arthrobacter globiformis* (Alia et al. 1998) in model plant *Arabidopsis thaliana*. However, genes of thermo-acidophilic algae Cyanidiales have scarcely been paid attention, although they have been considered to possess genes involved in adaptation to thermo-acidic environment (pH 0.05–5.0, 35–56 °C). Thus, analysis of their genomes may provide genetic information that could be used for production of thermotolerant plants.

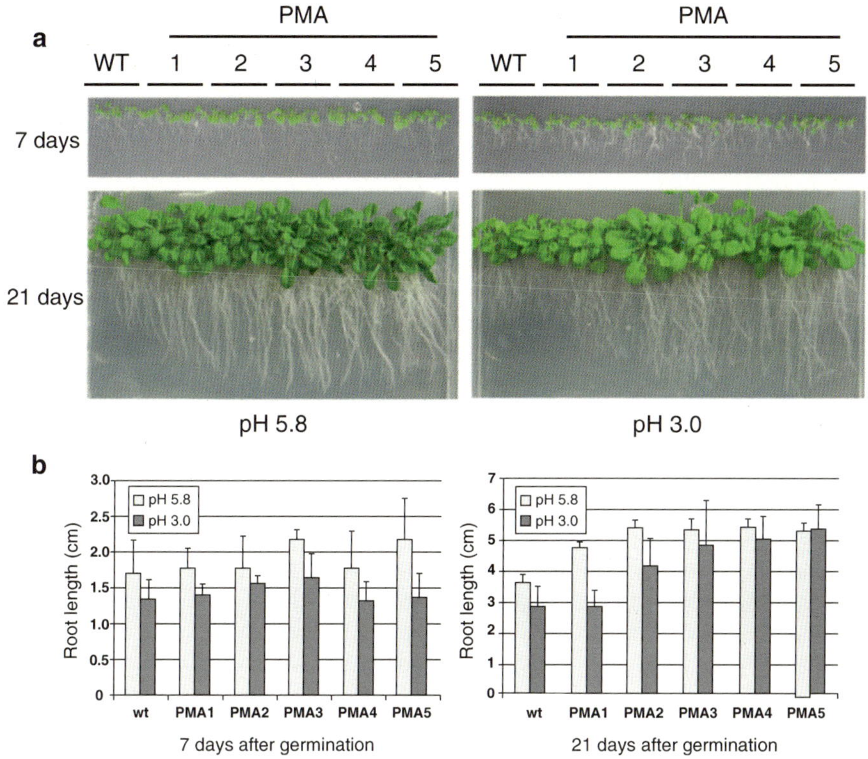


**Fig. 20.3** Effect of pH of medium on root elongation in wild type and transformants of CmPMA gene. (**a**) Root elongation of transformants was promoted in both pH 5.8 and 3.0. The upper part shows seedlings on the seventh day after sowing, and the lower part shows the plants on day 21 after sowing. One to five show *A. thaliana* lines of independent transformants of the CmPMA gene, respectively. (**b**) The results of the measurement of root length under each pH condition. Five individuals per line were used in the experiment

## *20.2.4 High-Level Expression of ROS Scavenging Genes in* C. merolae

Expressed sequence tag (EST) information was used to confirm expression levels of genes involved in stress tolerance in *C. merolae*. The transcripts encoding ROS-scavenging enzymes stromal ascorbate peroxidase (stAPX), catalase, and Mn superoxide dismutase were highly expressed (4th, 12th, and 69th highest among 4479 transcripts, respectively). These results suggested that *C. merolae* might be constantly exposed to higher oxidative stress, and high expression of ROS-scavenging genes might contribute to adaptation to thermo-acidic environment.

Reactive oxygen species (ROS), such as superoxide anion ($O^{2-}$), hydrogen peroxide ($H_2O_2$), hydroxyl radical (·HO), and singlet oxygen ($^1O_2$), are produced as harmful by-products of plant cellular metabolism (Mittler 2002). Thus, plant cell has ROS-scavenging system, and ROS is immediately scavenged through the ROS-scavenging system in chloroplast (Asada 1999). However, the concentrations of all major ROS within the chloroplasts are overproduced simultaneously under abiotic stress such as high temperature, high light, and drought (Apel and Hirt 2004). Overproduced ROS may react with macromolecules such as proteins, lipids, and DNA, causing oxidative damage and, thereby, impairing the normal functions of cells (Mittler 2002). Thus, enhancing ROS-scavenging system will be helpful in producing plants with capacity of tolerance to abiotic stresses such as high temperature.

### *20.2.5 The Characteristic of Ascorbate Peroxidase in Plant Cell*

Among ROS-scavenging genes, stromal APX was a highest expressed in *C. merolae* (Hirooka et al. 2009). APX catalyzes the reduction of $H_2O_2$ to water by using ascorbic acid (AsA) as the specific electron donor and is found mainly in plants, algae, and some protozoa (Ishikawa and Shigeoka 2008). In plants, APX isoenzymes are distributed in distinct cell compartments: the stroma and thylakoid membrane in chloroplasts, mitochondria, cytosol, peroxisomes, and apoplasts (Mittler 2002). In *C. merolae* genome, stromal and cytosolic *APX* genes are encoded (Hirooka et al. 2009). Interestingly, Mano et al. (2001) showed that the chloroplastic APX is the primary target of methyl viologen-induced ROS in *Spinacia oleracea*. Moreover, it was showed that chloroplastic APXs are more labile than cytosolic APXs in the absence of AsA in plants (Amako et al. 1994). On the other hand, algae chloroplastic APX was relatively stable in the absence of AsA compared with the land plants chloroplastic APX (Takeda et al. 2000). Therefore, expressing chloroplastic APX derived from algae is a reasonable strategy to enhance ROS-scavenging system in plants.

### *20.2.6 Introduction of CmstAPX Gene into* **Arabidopsis thaliana**

To overexpress CmstAPX gene in *A. thaliana*, the CmstAPX gene was fused to the cauliflower mosaic virus 35S (CaMV 35S) promoter, and the construct was introduced by *Agrobacterium*-mediated transformation method. The soluble APX activity in 35S:CmstAPX plants was approximately tenfold higher than that in the wild-type plants (Hirooka et al. 2009). Subsequently, we confirmed the localization of the CmstAPX protein in 35S:CmstAPX plants by immunofluorescence using Technovit method (Kuroiwa and Kuroiwa 1992) and showed that CmstAPX protein was co-localized with RuBisCo large subunit (rbcL), which can be used as a compartment marker of chloroplast stroma (Fig. 20.4). Thus, these results showed that CmstAPX protein could be functional in chloroplast stroma of *A. thaliana*.

### 20.2.7 35S:CmstAPX Plants Show Increased Tolerance to High-Temperature Stress

We compared the thermotolerance of the *35S:*Cm*stAPX* plants to that of wild-type plants. Both of wild-type and *35S:*Cm*stAPX* plants were cultured on Murashige-Skoog (MS) agar plates at 23 and 33 °C for 7 days (Fig. 20.5). They germinated and grew normally at 23 °C and had fine chloroplast structure (Fig. 20.5). At 33 °C, however, growth of wild-type plants were inhibited, and membranes of the chloroplasts were disrupted (Fig. 20.5). By contrast, the cotyledons of *35S:*Cm*stAPX*

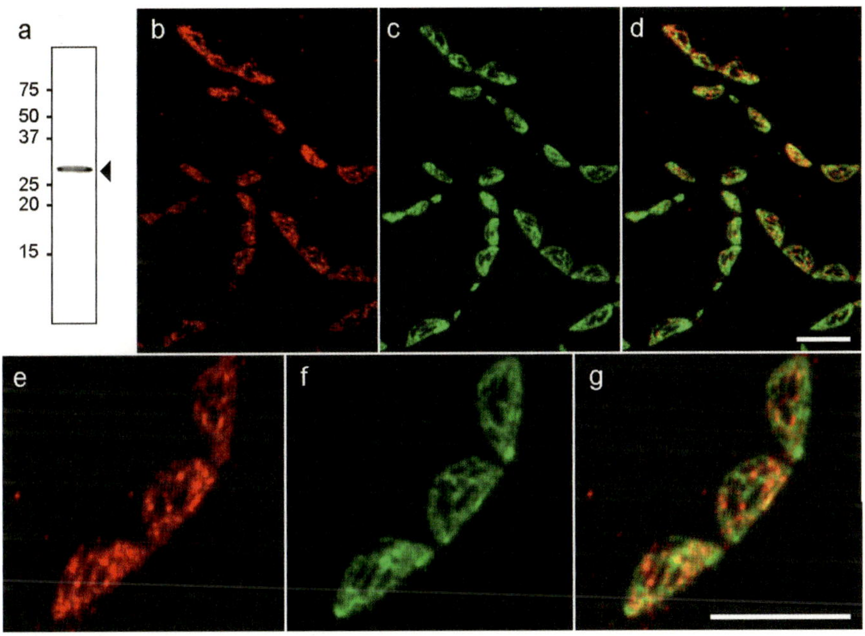


**Fig. 20.4** Subcellular localization of CmstAPX in *35S:*Cm*stAPX* plants. (**a**) Immunoblotting analysis showed that the anti-CmstAPX antibodies specifically detected a 34 kDa band (*arrowhead*) from the soluble proteins of *35S:*Cm*stAPX* plants. (**b–g**) Immunofluorescence microscopy using anti-CmstAPX antibodies in Technovit sections of *35S:*Cm*stAPX* plants. Localization of CmstAPX visualized with an anti-CmstAPX antibody and an Alexa 555-conjugated secondary antibody. Localization of rbcL is visualized with an anti-rbcL antibody and an Alexa 488-conjugated secondary antibody (**b**) and (**e**) which are signals from CmstAPX (*red*), (**c**) (**f**) are signals from rbcL (*green*), and (**d**) and (**g**) are merged images. (**e**), (**f**), and (**g**) are higher magnifications of (**b**), (**c**), and (**d**), respectively (Scale bars represent 5 μm)

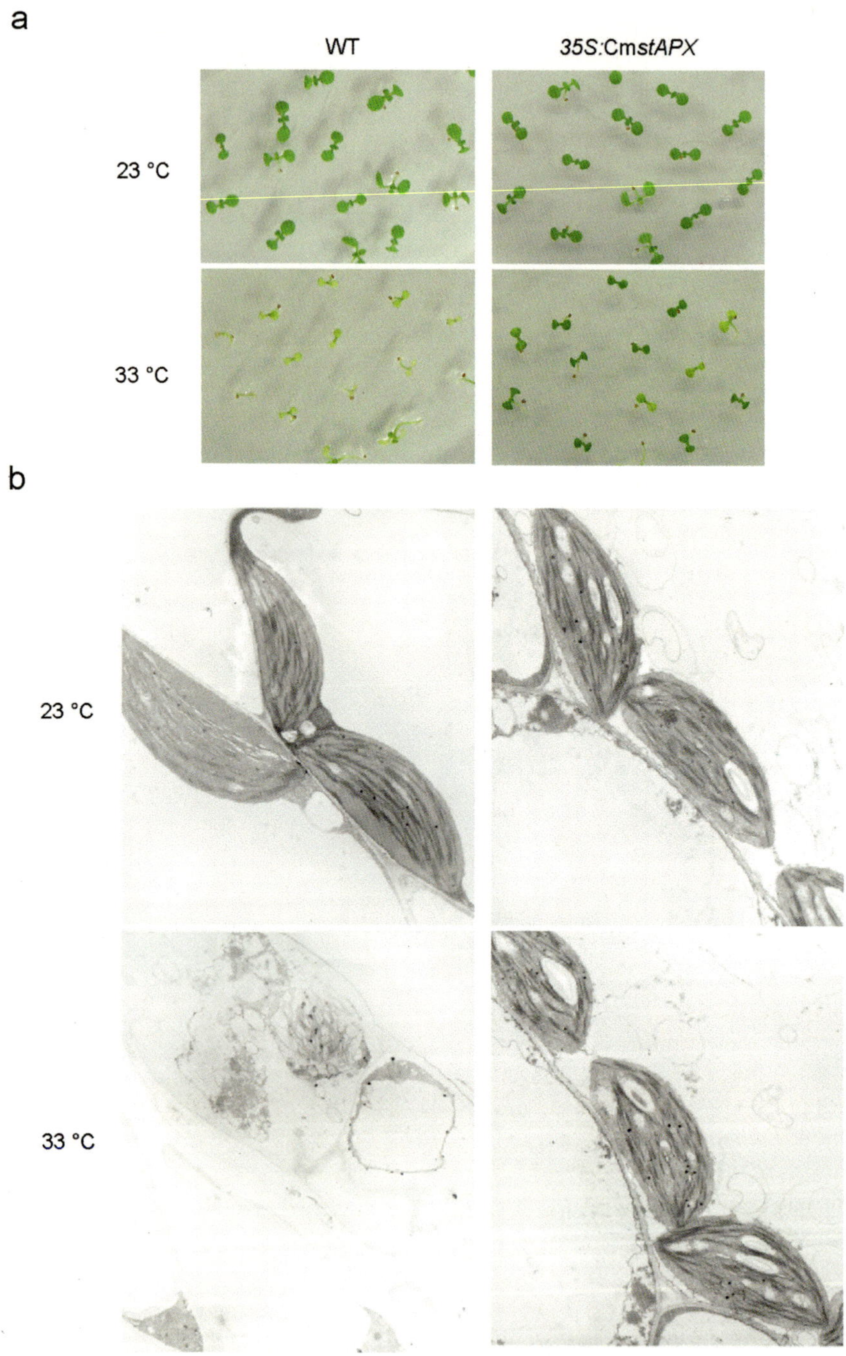


**Fig. 20.5** Thermotolerance of seedlings of wild-type and *35S:*Cm*stAPX* plants. (**a**) Photographs of wild-type (WT) and *35S:*Cm*stAPX* plants were grown on MS agar plates at 23 °C or 33 °C for 7 days. (**b**) Electron microscopy showed that the chloroplasts structure of both WT and *35S:*

plants retained their green color and fine chloroplast structure, although they were light green color compared with the control (Fig. 20.5). Furthermore, both of wild-type and *35S:*Cm*stAPX* plants were grown for 7 days at 33 °C and then were transferred to 23 °C and incubated for a further 7 days. More than 70% of *35S:* Cm*stAPX* plants were survived, while less than 10% of wild-type plants were survived (Fig. 20.6). In addition, we tested effects of high-temperature stress on adult stage of wild-type and *35S:*Cm*stAPX* plants. Wild-type and *35S:*Cm*stAPX* plants were grown at 23 °C for 4 weeks and then were transferred to 33 °C and cultured for 14 days. *35S:*Cm*stAPX* plants survived in comparison with wild-type plants (Fig. 20.6). Thus, it is suggested that expression of the CmstAPX in *A. thaliana* enhances thermotolerance. We aimed to show usefulness of genes derived from thermo-acidophilic algae Cyanidiales to produce stress-tolerant transgenic plants. And we successfully produced a thermotolerant plant by expressing stromal ascorbate peroxidase derived from *C. merolae* (Hirooka et al. 2009). To our knowledge, this is the first report of thermotolerant plants produced by gene transfer of Cyanidiales. Thus, we think that Cyanidiales have been considered to be a possible resource of genes involved in stress tolerance.

## 20.3 Future Perspectives

Our observations indicate that *C. merolae* genes may be useful for enhancing the acid- or heat-stress tolerance of land plants. We also attempted to produce salt-tolerant plants using a *C. merolae* gene. Genes that were specifically expressed at 2 h and 24 h after treatments with a 0.3 M salt solution were examined by microarray analysis. The highly expressed genes included those encoding proteins similar to low molecular weight heat shock proteins, heat shock protein 70, and S-adenosylmethionine (SAM) synthetase (Sakajiri et al. 2008). Four transgenic *A. thaliana* plant lines expressing *CmSAMS* were produced via an *Agrobacterium tumefaciens*-mediated transformation. Compared with the wild-type plants, the *CmSAMS*-expressing transgenic plants were more tolerant to salt stress, implying this gene is associated with salt tolerance (Sakajiri et al. 2010).

We also revealed that the *C. merolae* plasma membrane calcium/proton antiporter (CAX) gene confers stress resistance to transgenic *A. thaliana* plants. The *A. thaliana* seedlings overexpressing the *C. merolae* CAX gene were able to grow in environments with high concentrations of calcium ions (25 mM) or sodium chloride (100 mM) (data not shown; Japan Patent Kokai 2011-92109A, 2011.5.12).

**Fig. 20.5** (continued) Cm*stAPX* plants had a fine structure when grown at 23 °C (*upper panel*). When WT and *35S:*Cm*stAPX* plants were grown at 33 °C for 7 days, thylakoid membranes in chloroplasts were ruptured in WT plants, while fine structures, including thylakoid membranes, remained intact in chloroplasts of *35S:*Cm*stAPX* plants (*lower panel*)

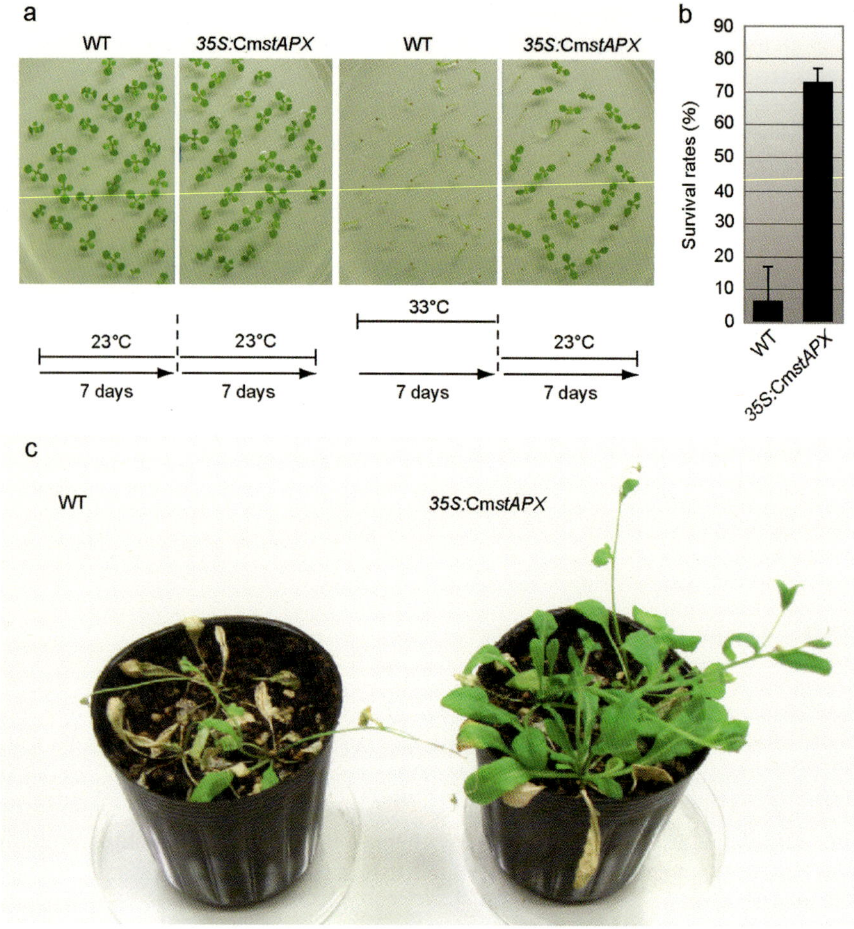


**Fig. 20.6** Recovery of seedlings after high-temperature stress and thermotolerance of adult stage of wild-type and *35S*:Cm*stAPX* plants. (**a**) Photographs of WT and *35S*:Cm*stAPX* plants were grown on MS agar plates at 23 °C for 7 days after cultivation at 23 °C or 33 °C for 7 days. (**b**) Survival rates of the WT and *35S*:Cm*stAPX* plants under high-temperature stress described in (**a**). Bars indicate means ± SD of three individual experiments. (**c**) Photographs of WT and *35S*:Cm*stAPX* plants were grown at 33 °C for 14 days after cultivation at 23 °C for 4 weeks. WT plants withered up and died, while *35S*:Cm*stAPX* plants survived

These results suggest that the genes of Cyanidiales algae living in extreme environments may be useful genetic resources for the development of stress-tolerant land plants.

Thus far, the three genomes of thermo-acidophilic red algae *C. merolae*, *Galdieria sulphuraria*, and *Galdieria phlegrea* have been sequenced (Matsuzaki et al. 2004; Schönknecht et al. 2013; Qiu et al. 2013). Recently, the genome analyses of *G. sulphuraria* showed that horizontal gene transfer (HGT) from various bacteria and archaea has probably played important roles in the adaptation to thermo-acidic environments (Schönknecht et al. 2013). Interestingly, at least 5% of protein-coding genes of *G. sulphuraria* were probably acquired through HGT and are involved in ecologically important processes ranging from heavy metal detoxification to glycerol uptake and metabolism (Schönknecht et al. 2013). These results show that Cyanidiales possess many stress-tolerant genes acquired from other organisms in addition to original genes. Thus, future research on each genomic factor in these algae and further analyses of other extremophile genomes will assist our understanding of the adaptation to extreme environment and help to produce transgenic plants with increased tolerance to various stresses such as high temperature, heavy metals, and low pH.

**Acknowledgments** We thank Dr. Haruko Kuroiwa for kindly teaching microscopy experiments. Our study was partly supported by the Frontier Project "Adaptation and Evolution of Extremophile" (to T.K.) from the Ministry of Education, Culture, Sports, Science, and Technology of Japan and the Program for the Promotion of Basic Research Activities for Innovative Biosciences (PROBRAIN) (to T.K.) and by the Core Research for Evolutional Science and Technology Program of the Japan Science and Technology Agency (T.K and O.M.).

## References

Ahn SJ, Sivaguru M, Osawa H, Chung GC, Matsumoto H (2001) Aluminum inhibits the $H^+$-ATPase activity by permanently altering the plasma membrane surface potentials in squash roots. Plant Physiol 126:1381–1390. https://doi.org/10.1104/pp.126.4.1381

Alia, Hayashi H, Sakamoto A, Murata N (1998) Enhancement of the tolerance of Arabidopsisto high temperatures by genetic engineering of the synthesis of glycinebetaine. Plant J 16(2): 155–161

Amako K, Chen GX, Asada K (1994) Separate assays specific for ascorbate peroxidase and guaiacol peroxidase and for the chloroplastic and cytosolic isozymes of ascorbate peroxidase in plants. Plant Cell Physiol 35:497–504. https://doi.org/10.1093/oxfordjournals.pcp.a078621

Apel K, Hirt H (2004) Reactive oxygen species: metabolism, oxidative stress, and signal transduction. Annu Rev Plant Biol 55:373–399. https://doi.org/10.1146/annurev.arplant.55.031903.141701

Asada K (1999) The water-water cycle in chloroplasts: scavenging of active oxygens and dissipation of excess photons. Annu Rev Plant Biol 50:601–639. https://doi.org/10.1146/annurev.arplant.50.1.601

Bita C, Gerats T (2013) Plant tolerance to high temperature in a changing environment: scientific fundamentals and production of heat stress-tolerant crops. Front Plant Sci 4:273. https://doi.org/10.3389/fpls.2013.00273

Enami I, Akutsu H, Kyogoku Y (1986) Intracellular pH regulation in an acidophilic unicellular alga, *Cyanidium caldarium*: 31P-NMR determination of intracellular pH. Plant Cell Physiol 27:1351–1359. https://doi.org/10.1093/oxfordjournals.pcp.a077233

Hirooka S, Misumi O, Yoshida M, Mori T, Nishida K, Yagisawa F, Yoshida Y, Fujiwara T, Kuroiwa H, Kuroiwa T (2009) Expression of the *Cyanidioschyzon merolae* stromal ascorbate peroxidase in *Arabidopsis thaliana* enhances thermotolerance. Plant Cell Rep 28:1881–1893. https://doi.org/10.1007/s00299-009-0791-2

Hoth S, Dreyer I, Dietrich P, Becker D, Müller-Röber B, Hedrich R (1997) Molecular basis of plant-specific acid activation of $K^+$ uptake channels. Proc Natl Acad Sci U S A 94:4806–4810. https://doi.org/10.1073/pnas.94.9.4806

Ishikawa T, Shigeoka S (2008) Recent advances in ascorbate biosynthesis and the physiological significance of ascorbate peroxidase in photosynthesizing organisms. Biosci Biotechnol Biochem 72:1143–1154. https://doi.org/10.1271/bbb.80062

Kobayashi Y, Harada N, Nishimura Y, Saito T, Nakamura M, Fujiwara T, Kuroiwa T, Misumi O (2014) Algae sense exact temperatures: small heat shock proteins are expressed at the survival threshold temperature in *Cyanidioschyzon merolae* and *Chlamydomonas reinhardtii*. Genome Biol Evol 6:2731–2740. https://doi.org/10.1093/gbe/evu216

Kuroiwa H, Kuroiwa T (1992) Giant mitochondria in the mature egg cell of *Pelargonium zonale*. Protoplasma 168:184–188. https://doi.org/10.1007/BF01666264

Lee JH, Hübel A, Schöffl F (1995) Derepression of the activity of genetically engineered heat shock factor causes constitutive synthesis of heat shock proteins and increased thermotolerance in transgenic Arabidopsis. Plant J 8:603–612. https://doi.org/10.1046/j.1365-313X.1995.8040603.x

Mano JI, Ohno C, Domae Y, Asada K (2001) Chloroplastic ascorbate peroxidase is the primary target of methyl viologen-induced photooxidative stress in spinach leaves: its relevance to monodehydroascorbate radical detected with in vivo ESR. Biochim Biophys Acta 1504:275–287. https://doi.org/10.1016/S0005-2728(00)00256-5

Matsuzaki M, Misumi O, Shin IT, Maruyama S, Takahara M, Miyagishima SY, Mori T, Nishida K, Yagisawa F, Yoshida Y, Nishimura Y, Nakao S, Kobayashi T, Momoyama Y, Higashiyama T, Minoda A, Sano M, Nomoto H, Oishi K, Hayashi H, Ohta F, Nishizaka S, Haga S, Miura S, Morishita T, Kabeya Y, Terasawa K, Suzuki Y, Ishii Y, Asakawa S, Takano H, Ohta N, Kuroiwa H, Tanaka K, Shimizu N, Sugano S, Sato N, Nozaki H, Ogasawara N, Kohara Y, Kuroiwa T (2004) Genome sequence of the ultrasmall unicellular red alga *Cyanidioschyzon merolae* 10D. Nature 428:653–657. https://doi.org/10.1038/nature02398

Mittler R (2002) Oxidative stress, antioxidants and stress tolerance. Trends Plant Sci 7:405–410. https://doi.org/10.1016/S1360-1385(02)02312-9

Ohta H, Shirakawa H, Uchida K, Yoshida M, Matuo Y, Enami I (1997) Cloning and sequencing of the gene encoding the plasma membrane $H^+$-ATPase from an acidophilic red alga, *Cyanidium caldarium*. Biochim Biophys Acta 1319:9–13. https://doi.org/10.1016/S0005-2728(97)00007-8

Qiu H, Price DC, Weber AP, Reeb V, Yang EC, Lee JM, Kim SY, Yoon SH, Bhattacharya D (2013) Adaptation through horizontal gene transfer in the cryptoendolithic red alga Galdieria phlegrea. Curr Biol 23:R865–R866. https://doi.org/10.1016/j.cub.2013.08.046

Sakajiri T, Asano K, Hirooka S, Tashiro K, Misumi O, Fujiwara T, Kuroiwa T (2008) Microarray analysis reveals S-adenosylmethionine (SAM) synthetase involvement in salt tolerance of *Cyanidioschyzon merolae*. Cytologia 73:341–368. https://doi.org/10.1508/cytologia.73.341

Sakajiri T, Asano K, Hirooka S, Ohnuma M, Misumi O, Yoshida M, Kuroiwa T (2010) The overexpression of *Cyanidioschyzon merolae* S-adenosylmethionine synthetase enhances salt tolerance in transgenic *Arabidopsis thaliana*. Cytologia 75:341–352. https://doi.org/10.1508/cytologia.75.341

Sakano K, Yazaki Y, Mimura T (1992) Cytoplasmic acidification induced by inorganic phosphate uptake in suspension cultured *Catharanthus roseus* cells measurement with fluorescent pH indicator and $^{31}$P-nuclear magnetic resonance. Plant Physiol 99:672–680. https://doi.org/10.1104/pp.99.2.672

Schönknecht G, Chen WH, Ternes CM, Barbier GG, Shrestha RP, Stanke M, Bräutigam A, Baker BJ, Banfield JF, Garavito RM, Carr K, Wilkerson C, Rensing SA, Gagneul D, Dickenson NE, Oesterhelt C, Lercher MJ, Weber APM (2013) Gene transfer from bacteria and archaea facilitated evolution of an extremophilic eukaryote. Science 339(6124):1207–1210. https://doi.org/10.1126/science.1231707

Shen H, Chen J, Wang Z, Yang C, Sasaki T, Yamamoto Y, Yan X (2006) Root plasma membrane $H^+$-ATPase is involved in the adaptation of soybean to phosphorus starvation. J Exp Bot 57:1353–1362. https://doi.org/10.1093/jxb/erj111

Smirnoff N (1998) Plant resistance to environmental stress. Curr Opin Biotechnol 9:214–219. https://doi.org/10.1016/S0958-1669(98)80118-3

Takeda T, Yoshimura K, Yoshii M, Kanahoshi H, Miyasaka H, Shigeoka S (2000) Molecular characterization and physiological role of ascorbate peroxidase from halotolerant Chlamydomonas sp. W80 strain. Arch Biochem Biophys 376:82–90. https://doi.org/10.1006/abbi.1999.1564

Ullrich-Eberius CI, Sanz A, Novacky AJ (1989) Evaluation of arsenate-and vanadate-associated changes of electrical membrane potential and phosphate transport in *Lemna gibba* G1. J Exp Bot 40:119–128. https://doi.org/10.1104/pp.65.5.1004

Vitart V, Baxter I, Doerner P, Harper JF (2001) Evidence for a role in growth and salt resistance of a plasma membrane H+−ATPase in the root endodermis. Plant J 27(3):191–201. https://doi.org/10.1046/j.1365-313x.2001.01081.x

# Chapter 21
# Metabolic Engineering of *Cyanidioschyzon merolae*

**Nobuko Sumiya and Shinya Miyagishima**

**Abstract** Algae are expected to be promising alternative sources of biofuels, foods, and cosmetics. The unicellular red alga *Cyanidioschyzon merolae* is potentially useful for producing high concentrations of desirable biomaterials by metabolic engineering. *C. merolae* is genetically traceable and can thrive at low pH (1–5) and high temperatures (25–50 °C), which are harmful to many other organisms. Thus, this alga can be suitable for outdoor cultivation without the risk for contamination from other undesirable organisms. Recent studies regarding *C. merolae* have reported enhanced triacylglycerol (TAG) production, which can be used for biodiesel production, by genetic modification. Introducing cyanobacterial acyl-acyl carrier protein (ACP) reductase in *C. merolae* led to temporary TAG accumulation via an artificial metabolic pathway. The omics analyses showed that acyl-ACP reductase expression resulted in upregulating endogenous aldehyde dehydrogenase and the endogenous fatty acid synthetic pathway in chloroplasts. Another study expressed the 12-kDa FK506-binding protein of *Saccharomyces cerevisiae* in *C. merolae* and succeeded in increasing TAG levels by adding rapamycin. The omics analyses suggested that the target of rapamycin (TOR) regulated the expression of TAG-synthesizing enzymes, glycerol-3-phosphate acyltransferase, and acyl-CoA:diacylglycerol acyltransferase. Therefore, the combination of metabolic engineering and the evaluation of the effects in *C. merolae* by omics analyses will help in understanding the regulatory mechanism of metabolism. In addition, recent studies have started to find culture conditions that increase TAG accumulation while maintaining the cellular growth. Combinations of these cultivation techniques and genetic manipulations will leads to production of desirable biomolecules on a large scale in the future.

N. Sumiya
Department of Biology, Keio University, 4-1-1 Hiyoshi, Kohoku-ku, Yokohama 223-8521, Kanagawa, Japan

S. Miyagishima (✉)
Department of Cell Genetics, National Institute of Genetics, 1111 Yata, Mishima, Shizuoka 411-8540, Japan
e-mail: smiyagis@nig.ac.jp



T. Kuroiwa et al. (eds.), *Cyanidioschyzon merolae*,
https://doi.org/10.1007/978-981-10-6101-1_21

**Keywords** Triacylglycerol • Lipid droplets • Acyl-ACP reductase • Aldehyde dehydrogenase • TOR • Rapamycin • Cyanidiales • *Cyanidioschyzon merolae*

## 21.1 Introduction

Algae are expected to be promising alternative sources of biofuels, foods, and cosmetics (Hlavová et al. 2015). Algae convert 10% of solar energy into biomass, whereas land plants convert only 0.5% (Li et al. 2008). Algal cultivation would complement, rather than compete, with terrestrial plant cultivation because they grow in aquatic environments, even in saline water, heated water, acidic water, and wastewater, where terrestrial crops are difficult to grow (Sheehan et al. 1998; Yan et al. 2016). Thus, many researchers have recently explored and attempted to develop (mostly by genetic manipulation) algae that produce useful materials in high concentrations.

Although studies on the unicellular red alga *Cyanidioschyzon merolae* have just started, the alga is a promising platform organism that can be genetically modified for commercial use. The nuclear (Matsuzaki et al. 2004; Nozaki et al. 2007), mitochondrial (Ohta et al. 1998), and chloroplast (Ohta et al. 2003) genomes of *C. merolae* were completely sequenced. *C. merolae* possesses a simple nuclear genome (16.5 Mbp; 4775 protein-coding genes) with low genetic redundancy (Matsuzaki et al. 2004; Nozaki et al. 2007) (Chap. 5).

Methods for genetic manipulation have been established, such as the transient expression of proteins from plasmids (Ohnuma et al. 2008, 2009; Watanabe et al. 2011) and nuclear or chloroplast gene targeting by homologous recombination (Minoda et al. 2004; Zienkiewicz et al. 2017b). Transgenes are stably overexpressed without any silencing (Fujiwara et al. 2013; Imamura et al. 2013; Miyagishima et al. 2014; Sumiya et al. 2014, 2015, 2016; Watanabe et al. 2014; Zienkiewicz et al. 2017a). Thus, *C. merolae* is a suitable cell factory for metabolic engineering by overexpressing endogenous enzyme genes or introducing artificial metabolic pathways to produce high concentrations of metabolites of interest. In addition, the simple genome architecture of the alga facilitates various omics analyses to evaluate the effects of genetic manipulation on cellular metabolisms.

The hot spring alga *C. merolae* tolerates low pH (>1) and high temperatures (<50 °C) and is resistant to high concentrations of metal ions such as Al, Fe, Cu, Ni, Zn, and Mn (Misumi et al. 2008). Therefore, this alga is potentially suitable for outdoor cultivation without the risk for contamination from other neutrophilic organisms.

Genetically modified strains are usually subjected to regulations by laws; however, *C. merolae* can be genetically modified by self-cloning. In some cases, depending on the country, self-cloning organisms are not under the genetically modified organism (GMO) regulation. Genetic modification produces two categories of organism: GMOs, which have heterologous DNA sequences derived from organisms that are taxonomically different from the host organism, and self-cloning organisms, which do not contain any DNA derived from other organisms. Self-cloning processes are considered the same as naturally occurring gene conversion, such as recombination, deletion, and transposition (Holme et al. 2013). In the case of *C. merolae* transformation using the endogenous *URA* transformation marker, the parental strain is a spontaneous frameshift *URA* mutant or *URA*-deleted mutant (Minoda et al. 2004; Taki et al. 2015). Thus, a gene disruptant or an overexpressor of endogenous genes by endogenous strong promoters does not contain any heterologous DNA sequences and is regarded as a self-cloning organism.

Eukaryotic microalgae accumulate triacylglycerol (TAG) in lipid droplets in cells, which can be used for producing biodiesel (Merchant et al. 2012; Ho et al. 2014). Because TAG comprises a glycerol backbone with three esterified fatty acids, it is a direct precursor of biodiesel via chemical transesterification (Merchant et al. 2012). TAG is synthesized via two pathways. One is the acylation of glycerol-3-phosphate with three acyl-CoAs, which are synthesized via the fatty acid synthesis pathway. In the second pathway, one of the three fatty acids is supplied from membrane lipids. Microalgae usually accumulate TAG when subjected to nutrient stress, such as nitrogen, phosphorus, or sulfur starvation. However, under such nutrient-starved conditions, algal growth ceases. Therefore, for producing TAG without inhibiting cellular growth, algae need to be genetically modified. In this chapter, we introduce two recent studies regarding the metabolic engineering of *C. merolae* as the first step toward producing TAG in the future. In addition to these genetic engineering studies, here we also briefly summarize recent efforts to increase cellular TAG accumulation by changing the culture conditions in *C. merolae*.

## 21.2 Metabolic Engineering of *C. merolae* for Producing Biofuel

### *21.2.1 TAG Accumulation Induced by Expressing Acyl-Acyl Carrier Protein (ACP) Reductase*

In cyanobacteria, the overexpression of acyl-ACP reductase results in free fatty acid accumulation through the subsequent oxidation of fatty aldehydes by an endogenous aldehyde dehydrogenase (Kaiser et al. 2013). Although the *C. merolae* genome does not encode acyl-ACP reductase, it does encode an aldehyde dehydrogenase (Sumiya et al. 2015). Thus, the introduction of a cyanobacteria-like metabolic pathway in *C. merolae* by expressing cyanobacterial acyl-ACP reductase was intended to increase free fatty acid and TAG levels in *C. merolae* (Fig. 21.1).

To test this possibility, the acyl-ACP reductase gene of *Synechocystis* sp. PCC 6803 (sll0209) was inserted in a *C. merolae* chromosomal neutral locus (Fujiwara et al. 2013) to produce a stable transformant (AAR-3HA). The transgene was

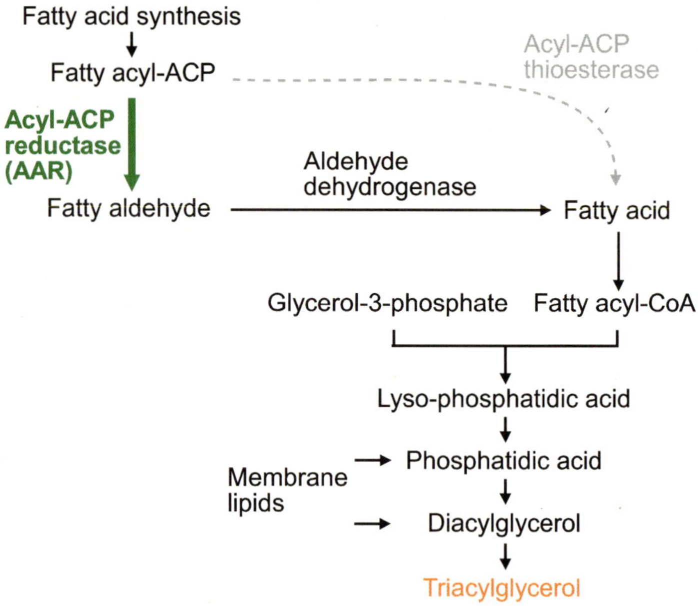


**Fig. 21.1** The fatty acid and TAG synthesis pathway in *C. merolae*. *C. merolae* genome does not encode acyl-ACP reductase (*green*) and acyl-ACP thioesterase (*gray*). The acyl-ACP reductase of *Synechocystis* sp. PCC6803 was expressed in *C. merolae* (AAR-3HA strain)

constitutively expressed under the control of a *C. merolae* catalase promoter (Ohnuma et al. 2009). A cleavable chloroplast transit peptide of CmSecA (Koyama et al. 2011) was fused to the N-terminus of the transgene to transport the protein into the *C. merolae* chloroplast because fatty acid synthesis-related proteins localize in chloroplasts (Mori et al. 2016). To enable detection, a 3 × HA tag was fused to the C-terminus of the enzyme. The S-200 strain, in which the green fluorescent protein (GFP) gene was introduced in the same chromosomal neutral locus, was used as the control strain (Sumiya et al. 2014). Because GFP transcription is driven by a heat shock-inducible promoter, GFP is not expressed at 42 °C, which is the optimal temperature for *C. merolae*.

The staining of neutral lipids with the fluorescent dye BODIPY showed that the number and size of the lipid droplets in AAR-3HA, but not in the control GFP strain, increased 1 day after inoculation (Fig. 21.2a; Sumiya et al. 2015). The TAG level in the AAR-3HA strain also increased to approximately three times that in the control strain 1 day after inoculation (Fig. 21.2b; Sumiya et al. 2015). Moreover, AAR-3HA expression upregulated aldehyde dehydrogenase (CMO345C). In contrast, when AAR-3HA was expressed with a CMO345C knockout background, the TAG level did not increase. These results suggest that AAR-3HA expression in *C. merolae* upregulated aldehyde dehydrogenase, which is also required for increasing the TAG level (Fig. 21.1; Sumiya et al. 2015).

Profiles of the transcriptome (using microarray) and metabolome (by CE-MS analyses) of the AAR-3HA strain showed that AAR-3HA expression upregulated

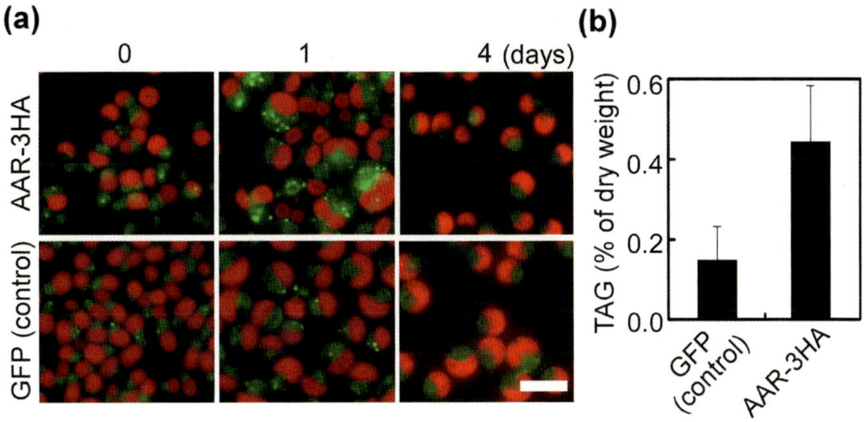


**Fig. 21.2** The effect of cyanobacterial acyl-ACP reductase expression on TAG production in *C. merolae*. (**a**) The lipid droplets stained with BODIPY after inoculation of the steady-state AAR-3HA and the control GFP strain. *Green*, lipid droplets stained with BODIPY; *red*, autofluorescence of the chloroplast. Scale bar = 10 μm. (**b**) TAG level in AAR-3HA and GFP cells 1 day after inoculation. The *error bars* represent the standard deviation ($n = 6$) (Reproduced from Sumiya et al. 2015)

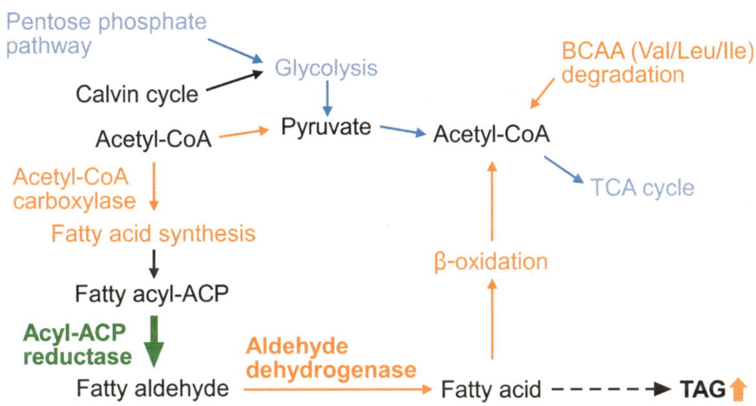


**Fig. 21.3** The effects of cyanobacterial acyl-ACP reductase expression on metabolic pathways in *C. merolae*. *Orange* indicates the upregulated pathways, and *blue* indicates the downregulated pathways (Reproduced from Sumiya et al. 2015)

the fatty acid synthetic pathway in the chloroplast, particularly the rate-limiting enzyme acetyl-CoA carboxylase (Fig. 21.3; Sumiya et al. 2015). The branched-chain amino acid (BCAA; Val/Leu/Ile) degradation pathway, which contributes to TAG accumulation in a diatom (Ge et al. 2014), was upregulated by AAR-3HA expression in *C. merolae*. However, the fatty acid degradation pathway (β-oxidation) was also upregulated in the AAR-3HA strain. The number and size of lipid droplets increased and subsequently decreased in the AAR-3HA strain during culture possibly because of elevated fatty acid degradation (Fig. 21.2a). Thus, further metabolic manipulation is required to maintain elevated TAG levels by expressing acyl-ACP reductase.

### 21.2.2 TAG Accumulation by Target of Rapamycin Inhibition

Although TAG production is accelerated in eukaryotic microalgae by nutrient starvation, the manner in which nutrient starvation is sensed by cells and that in which the signal is transmitted to the TAG production pathway remains unknown. Target of rapamycin (TOR), a serine/threonine protein kinase, is a candidate for sensing nutrient starvation in microalgae because it senses the nutrient status and regulates cellular growth and metabolism in other eukaryotes (Fig. 21.4; Wullschleger et al. 2006; Laplante and Sabatini 2012). In *Saccharomyces cerevisiae*, TOR inhibits the expression of nitrogen-regulated genes, including the global nitrogen transcription regulator Gln3 (Beck and Hall 1999; Cooper 2002).

The function of TORC1 (TOR of *S. cerevisiae*), as well as mTORC1 and mTORC2 (mammalian TORs), is inhibited by a macrolide compound rapamycin (Heitman et al. 1991; De Virgilio and Loewith 2006; Wullschleger et al. 2006). Rapamycin first binds to the 12-kDa FK506-binding protein (FKBP12), and this complex inhibits the TOR serine/threonine kinase by binding to the FRB domain of TOR. The *C. merolae* genome encodes a single TOR protein (CMR018C, CmTOR) with the FBR domain but not any FKBP12 homologs. Thus, the *C. merolae* wild type is not sensitive to rapamycin.

To produce rapamycin-sensitive *C. merolae*, Imamura et al. (2013) produced a *C. merolae* strain (F12 strain) wherein *S. cerevisiae* FKBP12 is constitutively expressed. Lipid droplet formation was observed in the cytoplasm 24 h after adding rapamycin in the F12 culture. The formation was enhanced by further incubating for 24 h, whereas cellular growth was inhibited. The lipid droplet formation was not observed with a drug solvent/DMSO solution treatment (Fig. 21.5; Imamura et al. 2015). The TAG levels in F12 were increased by approximately 8.8-fold in response to rapamycin treatment compared with the control rapamycin-insensitive C12 strain [the parental plasmid used for constructing the F12 strain was introduced into *C. merolae* M4 (Fig. 21.6; Imamura et al. 2015)].

The TAG level in the F12 strain was also increased in response to nitrogen depletion, as in the wild type, although the level was slightly higher than that under rapamycin treatment (Fig. 21.6; Imamura et al. 2015). TAG accumulation following the inhibition of TOR activity is not an indirect effect that results from growth inhibition because growth inhibition by a topoisomerase I inhibitor, camptothecin, does not result in TAG accumulation (Imamura et al. 2015). These results demonstrate that TAG accumulation is controlled by TOR signaling (Imamura et al. 2015).

To understand the association between nitrogen and TOR signaling, Imamura et al. (2015) compared transcriptome between nitrogen-depleted and TOR-

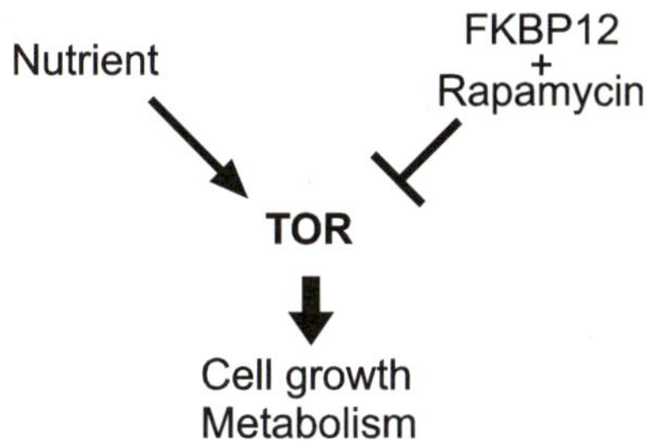


**Fig. 21.4** TOR kinase signaling. TOR senses nutrient concentration and regulates cell growth and metabolism. The rapamycin-FKBP12 complex inactivates TOR

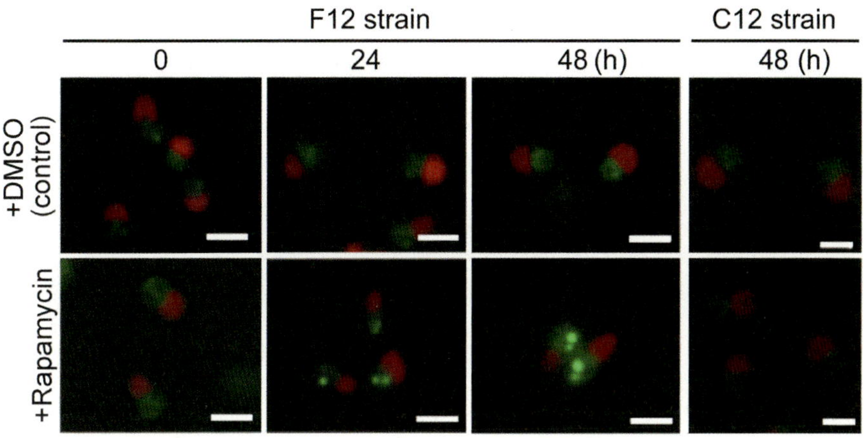


**Fig. 21.5** BODIPY staining of the FKBP12-introduced F12 and control C12 strain after rapamycin or DMSO treatment. *Green*, lipid droplets stained with BODIPY; *red*, autofluorescence of the chloroplast. Scale bars = 2 μm (Reproduced from Imamura et al. 2015)

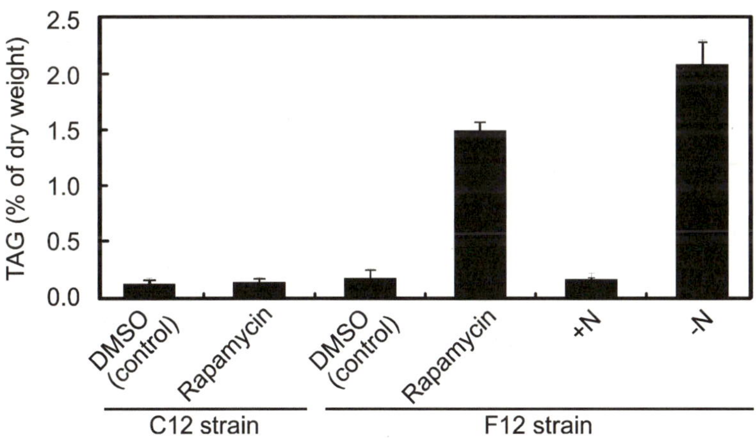


**Fig. 21.6** TAG level 48 h after rapamycin treatment or nitrogen depletion. The *error bars* represent the standard deviation ($n = 3$) (Reproduced from Imamura et al. 2015)

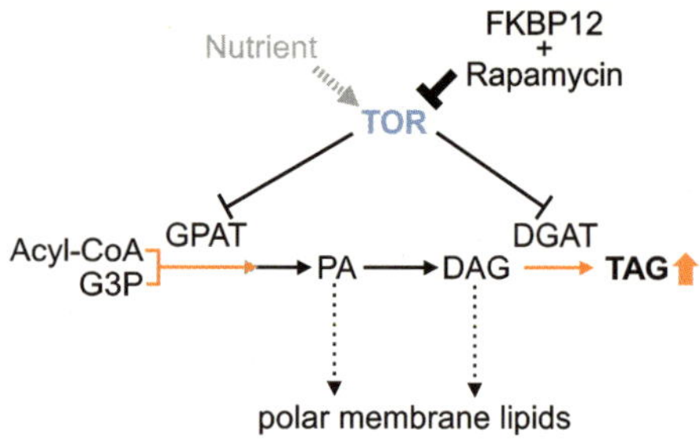


**Fig. 21.7** A proposed model for the relationship between TOR signaling and TAG synthesis. By nutrient depletion or rapamycin treatment, TOR is inactivated, and the TAG synthesis pathway, especially glycerol-3-phosphate acyltransferase (GPAT) and acyl-CoA:diacylglycerol acyltransferase (DGAT), is activated (Reproduced from Imamura et al. 2016)

inactivated conditions using microarray analyses. In the rapamycin-treated F12 strain, 148 and 64 genes were up- and downregulated, respectively. Of the 148 upregulated genes, 71 were also upregulated under nitrogen-depleted conditions. These 71 upregulated genes, which are shared by nitrogen depletion and TOR inactivation under nitrogen-replete conditions, include nitrogen assimilation-related genes, suggesting that TOR is involved in the signaling of nitrogen starvation in *C. merolae* (Imamura et al. 2015). In terms of TAG synthesis, CMA017C and CMK217C, which encode glycerol-3-phosphate acyltransferase (GPAT) and acyl-CoA:diacylglycerol acyltransferase (DGAT), respectively, are upregulated by both nitrogen depletion and TOR inactivation under nitrogen-replete conditions (Imamura et al. 2015). These results suggest that following nitrogen starvation, TOR is inactivated, which in turn upregulates GPAT and DGAT, as well as TAG accumulation (Fig. 21.7; Imamura et al. 2015).

TAG accumulation by TOR inactivation requires the addition of rapamycin and inhibits cellular growth. To enhance TAG synthesis while maintaining cellular growth for commercial use, further studies, particularly in understanding the signaling pathway from TOR to TAG synthesis, are required.

## 21.3 Future Perspectives

Although >10 years have passed following the discovery of a homologous recombination event of introduced DNA in *C. merolae* (Minoda et al. 2004), metabolic engineering trials have only just started. By combining a *URA* transformation marker (Minoda et al. 2004; Fujiwara et al. 2013) and the recently reported new transformation marker chloramphenicol acetyltransferase (Zienkiewicz et al. 2017a), it is feasible to edit multiple chromosomal loci in *C. merolae*. In addition, a chloroplast transformation system was recently developed (Zienkiewicz et al. 2017b), which leads to the expression of larger amounts of transgene products than the nuclear transformation in other organisms. These recently developed techniques should facilitate further metabolic manipulations in *C. merolae*.

In the two examples of metabolic engineering of *C. merolae* described here, the heterologous sequences (i.e., acyl-ACP reductase gene of *Synechocystis* sp. PCC 6803 or *S. cerevisiae* FKBP12 gene) were integrated into a *C. merolae* chromosome. However, metabolically manipulating *C. merolae* by self-cloning should be feasible in the future. In this regard, the group of Drs. N. Sato and T. Moriyama in the University of Tokyo has extensively analyzed the lipid synthetic pathway in *C. merolae* (Chap. 19 ) and the information will be useful for further metabolic engineering.

Besides the genetic manipulations, recent studies have tried to increase cellular TAG accumulation in the wild-type *C. merolae* by adding chemicals to the culture or changing the environment for cultivation. The group of Drs. K. Tanaka and S. Imamura in Tokyo Institute of Technology found that inhibition of histone deacetylase (HDAC) with addition of trichostatin A or other inhibitors increase the number of lipid droplets and TAG contents (Japan patent JP 2014-144756). The group of Dr. Misumi tested the wavelength of illumination that is suitable for TAG accumulation. As a result, it was found that the cultivation under red light (600 nm∼700 nm) increases the cellular TAG accumulation while maintaining the cellular growth. In addition, changing the light for cultivation from blue (400 nm∼500 nm) to red further increased the TAG contents (Japan patent JP 2014-152585). The group of Dr. Kuroiwa tested the formula of media that are suitable for accumulation of lipid droplets. By using media, in which sulfate concentration was reduced and sodium concentration was increased compared to conventional media (e.g. Allen's medium and M4 medium), they succeeded in increasing the cellular storage lipids while maintaining the cellular growth (Japan patent JP 2014-071081). Thus, combinations of these techniques and genetic manipulations will yield high production of TAG in *C. merolae* in near future.

The ability of *C. merolae* to thrive at low pH and in the presence of high metal concentrations is potentially ideal for open pond cultivation without the risk for contamination from undesirable organisms. Further studies on metabolic engineering of *C. merolae* will lead to, for example, the production of biofuels and other biomaterials using acidic drainage derived from hot springs, mines, and industrial plants.

Although here we have focused on issues of *C. merolae*, the group of Drs. Tanaka and Imamura found that TOR inhibition with rapamycin (Japan patent JP 2013-142173) or HDAC inhibition with several inhibitors (Japan patent JP 2014-144756) also leads to TAG accumulation in green algae. Thus, further studies on the mechanisms of TAG accumulation under these conditions will also give important information for biofuel production in other algal lineages.

**Acknowledgments** Our study was partly supported by Japan Society for the Promotion of Science Grant-in-Aid for Scientific Research 25251039 (to S.M.) and by the Core Research for Evolutional Science and Technology Program of the Japan Science and Technology Agency (to S.M.).

# References


Beck T, Hall MN (1999) The TOR signalling pathway controls nuclear localization of nutrient-regulated transcription factors. Nature 402:689–692. https://doi.org/10.1038/45287

Cooper TG (2002) Transmitting the signal of excess nitrogen in *Saccharomyces cerevisiae* from the Tor proteins to the GATA factors: connecting the dots. FEMS Microbiol Rev 26:223–238. https://doi.org/10.1111/j.1574-6976.2002.tb00612.x

De Virgilio C, Loewith R (2006) The TOR signalling network from yeast to man. Int J Biochem Cell Biol 38:1476–1481. https://doi.org/10.1016/j.biocel.2006.02.013

Fujiwara T, Ohnuma M, Yoshida M, Kuroiwa T, Hirano T (2013) Gene targeting in the red alga *Cyanidioschyzon merolae*: single- and multi-copy insertion using authentic and chimeric selection markers. PLoS One 8:e73608. https://doi.org/10.1371/journal.pone.0073608

Ge F, Huang W, Chen Z, Zhang C, Xiong Q, Bowler C, Yang J, Xu J, Hu H (2014) Methylcrotonyl-CoA carboxylase regulates triacylglycerol accumulation in the model diatom *Phaeodactylum tricornutum*. Plant Cell 26:1681–1697. https://doi.org/10.1105/tpc.114.124982

Heitman J, Movva N, Hall M (1991) Targets for cell cycle arrest by the immunosuppressant rapamycin in yeast. Science 253:905–909. https://doi.org/10.1126/science.1715094

Hlavová M, Turóczy Z, Bišová K (2015) Improving microalgae for biotechnology – from genetics to synthetic biology. Biotechnol Adv 33:1194–1203. https://doi.org/10.1016/j.biotechadv.2015.01.009

Ho S-H, Ye X, Hasunuma T, Chang J-S, Kondo A (2014) Perspectives on engineering strategies for improving biofuel production from microalgae – a critical review. Biotechnol Adv 32:1448–1459. https://doi.org/10.1016/j.biotechadv.2014.09.002

Holme IB, Wendt T, Holm PB (2013) Intragenesis and cisgenesis as alternatives to transgenic crop development. Plant Biotechnol J 11:395–407. https://doi.org/10.1111/pbi.12055

Imamura S, Ishiwata A, Watanabe S, Yoshikawa H, Tanaka K (2013) Expression of budding yeast FKBP12 confers rapamycin susceptibility to the unicellular red alga *Cyanidioschyzon merolae*. Biochem Biophys Res Commun 439:264–269. https://doi.org/10.1016/j.bbrc.2013.08.045

Imamura S, Kawase Y, Kobayashi I, Sone T, Era A, Miyagishima SY, Shimojima M, Ohta H, Tanaka K (2015) Target of rapamycin (TOR) plays a critical role in triacylglycerol accumulation in microalgae. Plant Mol Biol 89:309–318. https://doi.org/10.1007/s11103-015-0370-6

Imamura S, Kawase Y, Kobayashi I, Shimojima M, Ohta H, Tanaka K (2016) TOR (target of rapamycin) is a key regulator of triacylglycerol accumulation in microalgae. Plant Signal Behav 11:e1149285. https://doi.org/10.1080/15592324.2016.1149285

Kaiser BK, Carleton M, Hickman JW, Miller C, Lawson D, Budde M, Warrener P, Paredes A, Mullapudi S, Navarro P, Cross F, Roberts JM (2013) Fatty aldehydes in cyanobacteria are a metabolically flexible precursor for a diversity of biofuel products. PLoS One 8:e58307. https://doi.org/10.1371/journal.pone.0058307

Koyama Y, Takimoto K, Kojima A, Asai K, Matsuoka S, Mitsui T, Matsumoto K, Hara H, Ohta N (2011) Characterization of the nuclear- and plastid-encoded secA-homologous genes in the unicellular red alga *Cyanidioschyzon merolae*. Biosci Biotechnol Biochem 75:2073–2078. https://doi.org/10.1271/bbb.110338

Laplante M, Sabatini David M (2012) mTOR signaling in growth control and disease. Cell 149:274–293. https://doi.org/10.1016/j.cell.2012.03.017

Li Y, Horsman M, Wu N, Lan CQ, Dubois-Calero N (2008) Biofuels from microalgae. Biotechnol Prog 24:815–820. https://doi.org/10.1021/bp070371k

Matsuzaki M, Misumi O, Shin IT, Maruyama S, Takahara M, Miyagishima SY, Mori T, Nishida K, Yagisawa F, Yoshida Y, Nishimura Y, Nakao S, Kobayashi T, Momoyama Y, Higashiyama T, Minoda A, Sano M, Nomoto H, Oishi K, Hayashi H, Ohta F, Nishizaka S, Haga S, Miura S, Morishita T, Kabeya Y, Terasawa K, Suzuki Y, Ishii Y, Asakawa S, Takano H, Ohta N, Kuroiwa H, Tanaka K, Shimizu N, Sugano S, Sato N, Nozaki H, Ogasawara N, Kohara Y, Kuroiwa T (2004) Genome sequence of the ultrasmall unicellular red alga *Cyanidioschyzon merolae* 10D. Nature 428:653–657. https://doi.org/10.1038/nature02398

Merchant SS, Kropat J, Liu B, Shaw J, Warakanont J (2012) TAG, You're it! Chlamydomonas as a reference organism for understanding algal triacylglycerol accumulation. Curr Opin Biotechnol 23:352–363. https://doi.org/10.1016/j.copbio.2011.12.001

Minoda A, Sakagami R, Yagisawa F, Kuroiwa T, Tanaka K (2004) Improvement of culture conditions and evidence for nuclear transformation by homologous recombination in a red alga, *Cyanidioschyzon merolae* 10D. Plant Cell Physiol 45:667–671. https://doi.org/10.1093/pcp/pch087

Misumi O, Sakajiri T, Hirooka S, Kuroiwa H, Kuroiwa T (2008) Cytological studies of metal ion tolerance in the red algae *Cyanidioschyzon merolae*. Cytologia (Tokyo) 73:437–443. https://doi.org/10.1508/cytologia.73.437

Miyagishima SY, Fujiwara T, Sumiya N, Hirooka S, Nakano A, Kabeya Y, Nakamura M (2014) Translation-independent circadian control of the cell cycle in a unicellular photosynthetic eukaryote. Nat Commun 5:3807. https://doi.org/10.1038/ncomms4807

Mori N, Moriyama T, Toyoshima M, Sato N (2016) Construction of global Acyl lipid metabolic map by comparative genomics and subcellular localization analysis in the red alga *Cyanidioschyzon merolae*. Front Plant Sci 7:958. https://doi.org/10.3389/fpls.2016.00958

Nozaki H, Takano H, Misumi O, Terasawa K, Matsuzaki M, Maruyama S, Nishida K, Yagisawa F, Yoshida Y, Fujiwara T, Takio S, Tamura K, Chung S, Nakamura S, Kuroiwa H, Tanaka K, Sato N, Kuroiwa T (2007) A 100%-complete sequence reveals unusually simple genomic features in the hot-spring red alga *Cyanidioschyzon merolae*. BMC Biol 5:1–8. https://doi.org/10.1186/1741-7007-5-28

Ohnuma M, Yokoyama T, Inouye T, Sekine Y, Tanaka K (2008) Polyethylene glycol (PEG)-mediated transient gene expression in a red alga, *Cyanidioschyzon merolae* 10D. Plant Cell Physiol 49:117–120. https://doi.org/10.1093/pcp/pcm157

Ohnuma M, Misumi O, Fujiwara T, Watanabe S, Tanaka K, Kuroiwa T (2009) Transient gene suppression in a red alga, *Cyanidioschyzon merolae* 10D. Protoplasma 236:107–112. https://doi.org/10.1007/s00709-009-0056-5

Ohta N, Sato N, Kuroiwa T (1998) Structure and organization of the mitochondrial genome of the unicellular red alga *Cyanidioschyzon merolae* deduced from the complete nucleotide sequence. Nucleic Acids Res 26:5190–5198. https://doi.org/10.1093/nar/26.22.5190

Ohta N, Matsuzaki M, Misumi O, Miyagishima SY, Nozaki H, Tanaka K, Shin IT, Kohara Y, Kuroiwa T (2003) Complete sequence and analysis of the plastid genome of the unicellular red alga *Cyanidioschyzon merolae*. DNA Res 10:67–77. https://doi.org/10.1093/dnares/10.2.67

Sheehan J, Dunahay T, Benemann J, Roessler P (1998) A look back at the US Department of energy's aquatic species program: biodiesel from algae, vol 328. National Renewable Energy Laboratory, Golden

Sumiya N, Fujiwara T, Kobayashi Y, Misumi O, Miyagishima SY (2014) Development of a heat-shock inducible gene expression system in the red alga *Cyanidioschyzon merolae*. PLoS One 9: e111261. https://doi.org/10.1371/journal.pone.0111261

Sumiya N, Kawase Y, Hayakawa J, Matsuda M, Nakamura M, Era A, Tanaka K, Kondo A, Hasunuma T, Imamura S, Miyagishima SY (2015) Expression of cyanobacterial acyl-ACP reductase elevates the triacylglycerol level in the red alga *Cyanidioschyzon merolae*. Plant Cell Physiol 56:1962–1980. https://doi.org/10.1093/pcp/pcv120

Sumiya N, Fujiwara T, Era A, Miyagishima SY (2016) Chloroplast division checkpoint in eukaryotic algae. Proc Natl Acad Sci U S A 113:E7629–E7638. https://doi.org/10.1073/pnas.1612872113

Taki K, Sone T, Kobayashi Y, Watanabe S, Imamura S, Tanaka K (2015) Construction of a *URA5.3* deletion strain of the unicellular red alga *Cyanidioschyzon merolae*: a backgroundless host strain for transformation experiments. J Gen Appl Microbiol 61:211–214. https://doi.org/10.2323/jgam.61.211

Watanabe S, Ohnuma M, Sato J, Yoshikawa H, Tanaka K (2011) Utility of a GFP reporter system in the red alga *Cyanidioschyzon merolae*. J Gen Appl Microbiol 57:69–72. https://doi.org/10.2323/jgam.57.69

Watanabe S, Sato J, Imamura S, Ohnuma M, Ohoba Y, Chibazakura T, Tanaka K, Yoshikawa H (2014) Stable expression of a GFP-reporter gene in the red alga *Cyanidioschyzon merolae*. Biosci Biotechnol Biochem 78:175–177. https://doi.org/10.1080/09168451.2014.877823

Wullschleger S, Loewith R, Hall MN (2006) TOR signaling in growth and metabolism. Cell 124:471–484. https://doi.org/10.1016/j.cell.2006.01.016

Yan N, Fan C, Chen Y, Hu Z (2016) The potential for microalgae as bioreactors to produce pharmaceuticals. Int J Mol Sci 17:962. https://doi.org/10.3390/ijms17060962

Zienkiewicz M, Krupnik T, Drożak A, Golke A, Romanowska E (2017a) Chloramphenicol acetyltransferase—a new selectable marker in stable nuclear transformation of the red alga *Cyanidioschyzon merolae*. Protoplasma 254:587–596. https://doi.org/10.1007/s00709-015-0936-9

Zienkiewicz M, Krupnik T, Drożak A, Golke A, Romanowska E (2017b) Transformation of the *Cyanidioschyzon merolae* chloroplast genome: prospects for understanding chloroplast function in extreme environments. Plant Mol Biol 93:171–183. https://doi.org/10.1007/s11103-016-0554-8

# Handling Techniques for *Cyanidioschyzon merolae*

*Cyanidioschyzon merolae* is a unicellular red alga that lives in sulfate-rich acid hot springs.

1. *Modified Allen's(MA2) medium*

**Stock Solutions**

A: $(NH_4)_2SO_4$ (52.9 g); $MgSO_4.7H_2O$ (10.0 g); conc. $H_2SO_4$ (3 mL). All chemicals are dissolved in 1000 mL distilled water.
B: $KH_2PO_4$ (10.88 g) in 100 mL distilled water.
C: $CaCl_2$ (14.7 g) in 100 mL distilled water.
D: $FeCl_3$ (0.4 g); $Na_2EDTA{\cdot}2H_2O$ (0.7 g); a few drops of conc. $H_2SO_4$ in 100 mL distilled water.
E: $H_3BO_4$ (2.85 g); $MnCl_2.4H_2O$ (1.8 g); $ZnCl_2$ (0.105 g); $Na_2MoO_4.H_2O$ (0.39 g); $CoCl_2.6H_2O$ (0.04 g); $CuCl_2$ (0.043 g). All chemicals are dissolved in 1000 mL distilled water.

**MA2 Medium (1000 mL) A (100 mL) + B (10 mL) + C (1 mL) + D (4 mL) + E (40 mL) + distilled water → 1000 mL**

*MA2 medium is acidic (approximately pH 2.3). For this reason, there is a low risk of other organisms contaminating cell cultures of *C. merolae* strain 10D. In addition, MA2 medium can be stored at room temperature for up to 6 months.

2. *Cell culture and synchronization*

Cells are cultured in 500 mL flasks with shaking under continuous white light (40 $W/m^2$) at 40–42 °C with or without 2–5% $CO_2$. For synchronization of cell and organelle divisions, the cell cultures are subcultured to $<1 \times 10^7$ cells/mL in a flat bottle under a 12-h light/12-h dark photoperiod at 42 °C in an automatic light/dark cycle incubator.


T. Kuroiwa et al. (eds.), *Cyanidioschyzon merolae*,
https://doi.org/10.1007/978-981-10-6101-1

*The division of organelles can be highly synchronized by controlling natural conditions such as the photoperiod without treatment with cell cycle inhibitors.

3. *Cell and organelle division*

*C. merolae* is unicellular and proliferates by binary division. The cells contain a minimum set of eukaryotic organelles, that is, three types of double membrane-bound organelles (one cell nucleus, one mitochondrion, and one plastid (chloroplast)) and four types of single membrane-bound organelles (one simple endoplasmic reticulum (ER), one Golgi apparatus, several lysosomes (vacuoles), and one peroxisome (microbody)). The dynamics of the seven organelles can be studied simultaneously using seven stains such as DAPI and antibodies of each organelle marker. The double membrane-bound organelles arrange in a line in the cell and then divide in order. During cell division, the divisions of some double membrane-bound and some single membrane-bound organelles are linked. Division of the cell nucleus is accompanied by division of the ER and Golgi apparatus, while division of the mitochondrion is accompanied by division of the peroxisome and lysosomes. Chloroplasts divide independently. For highly synchronized peroxisome division, cells are treated with 40 μM oryzalin, a microtubule-disrupting drug, for 12 h during the dark phase of the photoperiod.

*More than three organelles can be traced simultaneously, and many substances produced by *C. merolae* are unique and useful in diverse applications.

4. *Isolation of intact organelles and dividing machineries*

The fact that *C. merolae* lacks a rigid cell wall facilitates the isolation of each organelle from the cells by osmotic pressure and homogenization. Methods to isolate the chloroplast, mitochondrion, vacuole, and peroxisome have been established. Additionally, the division of organelles can be highly synchronized, allowing mass isolation of the chloroplast-dividing machinery, mitochondrion-dividing machinery, and peroxisome-dividing machinery at desired time points. Briefly, synchronized cells are disrupted in a French pressure cell (at 1750 psi for chloroplasts and mitochondria, and at 700 psi for peroxisomes). The lysate is separated by density gradient centrifugation, and intact organelles are collected as a band at the 60%/80% Percoll interface. Further detergent treatment using *n*-octyl-β-D-glucoside or sodium dodecyl sulfate/urea buffer enables the purification of organelle division machineries from isolated dividing organelles. More detailed information is provided in Chap. 4.

*The in vivo structures of isolated organelles and organelle division machineries can be identified using whole-mount immunofluorescence microscopy and electron microscopy. These methods are also applicable in proteomic analyses.

5. *Protein identification by post-genomics (multi-omics analysis)*

*C. merolae* is the first organism for which the genomes of the nucleus, mitochondrion, and plastid were fully sequenced. In total, there are 4775 genes in the nuclear DNA. Most of these genes encode proteins and do not contain introns

(*C. merolae* genome project database: http:merolae.biol.s.u-tokyo.ac.jp, National Center for Biotechnology Information NCBI: https://www.ncbi.nlm.nih.gov/Taxonomy/Browser/wwwtax.cgid=280699). The availability of complete genomic information means that *C. merolae* is an ideal model system for research on the components of organelle division machineries, in combination with post-genomic analyses such as mass spectrometry (MS).

*Low-level redundancy of the genes facilitates highly efficient identification in multi-omics analyses (transcriptomic and proteomic) by using time-of-flight MS (TOF-MS). The *C. merolae* genome lacks gene duplicates, and 95% of genes do not have splicing variants. Consequently, peptide-mass finger printing is more effective for *C. merolae* than for other eukaryotes. Microarray analyses are practical for studying the events that occur during light/dark synchronization because the transcriptome in each cell cycle period can be analyzed. These techniques make it possible to identify candidate proteins in organelle division machineries.

6. Searches for genes in *C. merolae* genomic databases

The *C. merolae* genome project database (http:merolae.biol.s.u-tokyo.ac.jp) and NCBI (https://www.ncbi.nlm.nih.gov/Taxonomy/Browser/wwwtax.cgi?id=280699) store entire genome DNA sequences. By typing any gene name, ID, or keyword into the search box, it is possible to search for genes of interest in the *C. merolae* genome. The search results are shown as a list. Selecting a gene from the list brings up its annotation, relevant comments, nucleotide sequence, deduced amino acid sequence, putative transmembrane region, and several links to external websites.

*Because 90% of *C. merolae* genes lack introns, it is very easy to identify genes from their encoded proteins.

7. *Use of genetic tools to identify gene function*

Several general genetic tools have been established for *C. merolae*. Cells of *C. merolae* can be efficiently transformed using a polyethylene glycol-mediated DNA transfer method. Epitope-tagged fusion proteins can be expressed for protein localization analyses, and antisense RNA for target genes can be introduced via plasmids to analyze phenotypes. Genome editing in *C. merolae* is feasible without engineered nucleases such as CRISPR/Cas9. Homologous recombination-mediated genome-editing technologies are efficient and have been used for gene disruption, for targeted epitope tagging (knock-in), and for modifications of enzymes and gene promoter activities.

*Most genes are present as a single copy in the genome: therefore, genetic modifications are very efficient in *C. merolae*. Since there are a minimum number of organelles in *C. merolae*, genetic modifications lead to very clear phenotypes in organelle biogenesis.

8. Studies on *C. merolae* structural biology

*C. merolae* grows in hot springs and produces large amounts of proteins related to thermotolerance. These proteins have been purified, and their roles in the structural biology of *C. merolae* have been determined. The spherical and rigid structures of each organelle or division apparatus have been shown to be highly effectively in the structural biology of *C. merolae* by in situ cryo-transmission electron microscope tomography. The structure of each protein is correlated with the structures of intracellular machinery, including organelle division machinery, in situ. The ultrastructure and molecular mechanics of the cellular machinery can be analyzed at the atomic resolution scale.

*Because its proteins are stable, *C. merolae* is an important model organism for research at the atomic resolution scale.

# Index




T. Kuroiwa et al. (eds.), *Cyanidioschyzon merolae*,
https://doi.org/10.1007/978-981-10-6101-1

**D**